FUNCTIONALIZED ENGINEERING MATERIALS AND THEIR APPLICATIONS

FUNCTIONALIZED ENGINEERING MATERIALS AND THEIR APPLICATIONS

Edited by
Sabu Thomas, PhD
Nandakumar Kalarikkal, PhD
Pious C. V.
Zakiah Ahmad, PhD
Józef Tadeusz Haponiuk, PhD

Apple Academic Press Inc.	Apple Academic Press Inc.
3333 Mistwell Crescent	9 Spinnaker Way
Oakville, ON L6L 0A2	Waretown, NJ 08758
Canada	USA

Exclusive worldwide distribution by CRC Press, a member of Taylor & Francis Group
No claim to original U.S. Government works

ISBN-13: 978-1-77463-673-2 (pbk)
ISBN-13: 978-1-77188-523-2 (hbk)

Library and Archives Canada Cataloguing in Publication

Functionalized engineering materials and their applications / edited by Sabu Thomas, PhD, Nandakumar Kalarikkal, PhD, Pious C.V. Zakiah Ahmad, PhD, Józef Tadeusz Haponiuk, PhD.

Includes bibliographical references and index.
Issued in print and electronic formats.
ISBN 978-1-77188-523-2 (hardcover).--ISBN 978-1-315-36554-1 (PDF)

1. Polymers. 2. Nanostructured materials. 3. Materials. I. Thomas, Sabu, editor

| TA455.P58F86 2018 | 620.1'92 | C2018-900451-7 | C2018-900452-5 |

CIP data on file with US Library of Congress

Apple Academic Press also publishes its books in a variety of electronic formats. Some content that appears in print may not be available in electronic format. For information about Apple Academic Press products, visit our website at **www.appleacademicpress.com** and the CRC Press website at **www.crcpress.com**

ABOUT THE EDITORS

Sabu Thomas
Prof. Sabu Thomas is a Professor of Polymer Science and Engineering at the School of Chemical Sciences and Director of the International and Inter University Centre for Nanoscience and Nanotechnology at Mahatma Gandhi University, Kottayam, Kerala, India. He received his BSc degree (1980) in Chemistry from the University of Kerala, BTech (1983) in Polymer Science and Rubber Technology from the Cochin University of Science and Technology, and PhD (1987) in Polymer Engineering from the Indian Institute of Technology, Kharagpur. The research activities of Prof.
Thomas include surfaces and interfaces in multiphase polymer blend and composite systems; phase separation in polymer blends; compatibilization of immiscible polymer blends; thermoplastic elastomers, phase transitions in polymers, nanostructured polymer blends; macro-, micro-, and nanocomposites; polymer rheology; recycling; reactive extrusion; processing–morphology–property relationships in multiphase polymer systems; double networking of elastomers; natural fibers and green composites; rubber vulcanization; interpenetrating polymer networks, diffusion, and transport; and polymer scaffolds for tissue engineering. He has supervised 68 PhD theses, 40 MPhil theses, and 45 Master's theses. He has three patents to his credit. He also received the coveted Sukumar Maithy Award for the best polymer researcher in the country for the year 2008. Very recently, Prof. Thomas received the MRSI and CRSI medals for his excellent work. With over 600 publications to his credit and over 23,683 citations, with an h-index of 75, Prof. Thomas has been ranked fifth in India as one of the most productive scientists.

Nandakumar Kalarikkal

Dr. Nandakumar Kalarikkal obtained his master's degree in Physics with a specialization in Industrial Physics and his PhD in Semiconductor Physics from Cochin University of Science and Technology, Kerala, India. He was a postdoctoral fellow at NIIST, Trivandrum, and later joined Mahatma Gandhi University, Kerala, India. Currently, he is the Director of the International and Inter University Centre for Nanoscience and Nanotechnology as well as an Associate Professor in the School of Pure and Applied Physics at Mahatma Gandhi University. His current research interests include synthesis, characterization, and applications of various nanostructured materials, laser plasma, and phase transitions. He has published more than 80 research articles in peer-reviewed journals and has co-edited seven books.

Pious C. V.

Mr. Pious C. V. is a research scholar at the School of Chemical Sciences of Mahatma Gandhi University, Kottayam, Kerala, India. He is an MTech graduate in polymer science and engineering. His current area of research is the development of supertough thermosets from self-assembled epoxy/amphiphilic block copolymer systems.

Zakiah Ahmad

Dr. Zakiah Ahmad is a Professor in the faculty of civil engineering and Head of Division (Structure and Materials) of Universtiti Teknologi MARA, Shah Alam, Selangor. She received her BSc in Civil Engineering from Memphis State University, Tennessee, United States (December 1985), Master's in Statistics from University of Memphis, (December 1994), and PhD from University of Bath, UK. The research activities of Prof. Zakiah Ahmad include wood polymer composites, composites from renewable raw materials, polymer-based nanocomposite, adhesive-timber connections/joints, timber engineering, and bonded-in timber connection. She is currently a member of the

Board of Engineers Malaysia (BEM), International Association of Concrete Technology, and Malaysian Forest Product Association.

Józef Tadeusz Haponiuk
Dr. Józef Tadeusz Haponiuk is a Professor and Head of the Department of Polymer Technology, Gdansk University of Technology. He received his MSc (1974) and PhD (1979) from Technische Hochschule Leuna—Merseburg (Germany). He received Habilitation from Gdansk University of Technology in 2004. The research activities of Professor Haponiuk include chemical technology, polymer engineering, polymer characterization, chemistry and technology of polymers and rubber, thermal analysis of polymers, and materials science.

CONTENTS

LIST OF CONTRIBUTORS

S. Joseph Antony
Institute of Particle Science and Engineering, School of Chemical and Process Engineering, University of Leeds, LS2 9JT, Leeds, UK. E-mail: S.J.Antony@leeds.ac.uk

Sampann Arora
Polymer Nanocomposite Laboratory, Materials Physics Division, School of Advanced Sciences, VIT University, Vellore 632014, Tamil Nadu, India

Jyothi Lakshmi Avusula
Department of Chemistry, Birla Institute of Technology and Science, Pilani-Hyderabad Campus, Jawaharnagar, Shameerpet mandal, Ranga Reddy, Hyderabad 500078, Telangana, India

Babitha K. K.
Department of Physics, Nanoscience Research Centre (NSRC), Nirmala College, Muvattupuzha 686661, Kerala, India

Shruti Bhattacharya
Research & Technology Center, Asian Paints Limited, Plot No. C3-B/1, TTC MIDC Pawane, Thane-Belapur Road, Navi Mumbai 400703, Maharashtra, India

Popatrao N. Bhosale
Department of Chemistry, Materials Research Laboratory, Shivaji University, Kolhapur 416004, India

Paulo Cachim
Department of Civil Engineering and LABEST, University of Aveiro, 3810193 Aveiro, Portugal. E-mail: pcachim@ua.pt

George Cordoyiannis
Jožef Stefan Institute, Jamova 39, 1000 Ljubljana, Slovenia
Department of Physics, University of Athens, 15784 Athens, Greece
Department of Physics and Astronomy, KU Leuven, 3001 Leuven, Belgium

Pedro Costa
Department of Materials Engineering and CICECO, University of Aveiro, 3810193 Aveiro, Portugal

Dhawal Desai
Structural Engineering, University of Illinois at Urbana-Champaign, Champaign, IL, United States

Neha D. Desai
Department of Chemistry, Materials Research Laboratory, Shivaji University, Kolhapur 416004, India. E-mail: nehadesai323@gmail.com

Kalim Deshmukh
Polymer Nanocomposite Laboratory (PNL), Material Physics Division, School of Advanced Sciences (SAS), VIT University, Vellore 632014, Tamil Nadu, India

R. R. Deshmukh
Departmentof Physics, Institute of Chemical Technology (ICT), Matunga, Mumbai 400018, India

Arumaikkannu G.
Department of Manufacturing Engineering, Anna University, Chennai, India.
E-mail: arumai@annauniv.edu

Chandrashekara R. Haramagatti
Research & Technology Center, Asian Paints Limited, Plot No. C3-B/1, TTC MIDC Pawane,
Thane-Belapur Road, Navi Mumbai 400703, Maharashtra, India.
E-mail: chandrashekara.haramagatti@asianpaints.com

Hariharan K.
Department of Manufacturing Engineering, Anna University, Chennai, India.
E-mail: hariharancim28@gmail.com

David Holec
Department of Physical Metallurgy and Materials Testing, Montanuniversität Leoben,
8700 Leoben, Austria

Jaisankar V.
PG and Research Department of Chemistry, Presidency College (Autonomous), Chennai 600005,
Tamil Nadu, India. E-mail: vjaisankar@gmail.com

Subbalakshmi Jayanty
Department of Chemistry, Birla Institute of Technology and Science, Pilani-Hyderabad Campus,
Jawaharnagar, Shameerpet mandal, Ranga Reddy, Hyderabad 500078, Telangana, India.
E-mail: jslakshmi@hyderabad.bits-pilani.ac.in

Dalija Jesenek
Jožef Stefan Institute, Jamova 39, 1000 Ljubljana, Slovenia

Girish M. Joshi
Polymer Nanocomposite Laboratory (PNL), Material Physics Division,
School of Advanced Sciences (SAS), VIT University, Vellore 632014, Tamil Nadu, India.
E-mail: varadgm@gmail.com

S. Kalainathan
Polymer Nanocomposite Laboratory (PNL), Material Physics Division,
School of Advanced Sciences (SAS), Crystal Research Centre, VIT University,
Vellore 632014, Tamil Nadu, India

Shantanu Kallakuri
Department of Chemistry, Birla Institute of Technology and Science, Pilani-Hyderabad Campus,
Jawaharnagar, Shameerpet Mandal, Ranga Reddy, Hyderabad 500078, Telangana, India

K. Karthick
FEAT-Annamalai University, Annamalainagar 608002, Tamil Nadu, India.
E-mail: bkarthi_au@yahoo.com

B. Karthikeyan
Department of Chemistry, Annamalai University, Annamalainagar 608002, Tamil Nadu, India

Rohini R. Kharade
Department of Chemistry, Materials Research Laboratory, Shivaji University, Kolhapur 416004, India

Vijay V. Kondalkar
Department of Chemistry, Materials Research Laboratory, Shivaji University, Kolhapur 416004, India

Samo Kralj
Faculty of Natural Sciences and Mathematics, University of Maribor, Koroška 160, 2000 Maribor,
Slovenia. E-mail: samo.kralj@ijs.si
Jožef Stefan Institute, Jamova 39, 1000 Ljubljana, Slovenia

M. Suresh Chandra Kumar
Department of Chemistry, Polymer Nanocomposite Centre, Scott Christian College, Affiliated to
Manonmaniam Sundaranar University, Tamil Nadu, India

P. Sathish Kumar
School of Building and Mechanical Sciences, Kongu Engineering College,
Erode 638052, Tamil Nadu, India
Department of Mining Engineering, Indian Institute of Technology, Kharagpur 721302,
West Bengal, India

Zdravko Kutnjak
Jožef Stefan Institute, Jamova 39, 1000 Ljubljana, Slovenia
Jožef Stefan International Postgraduate School, Jamova 39, 1000 Ljubljana, Slovenia

J. H. Lee
Department of Polymer and Nano Engineering, Chonbuk National University, Jeonju,
Jeonbuk 561756, Republic of Korea

Chin Hsu Lin
General Motors Global R&D, Vehicle Development Research Lab, Vehicle Systems Group,
Mumbai, India

P. P. Lizymol
Biomedical Technology Wing, Sree Chitra Tirunal Institute for Medical Sciences and Technology,
Poojappura, Thiruvananthapuram 695012, Kerala, India. E-mail: lizymol@rediffmail.com

B. Loganathan
Department of Chemistry, Annamalai University, Annamalainagar 608002, Tamil Nadu, India

S. Mahalakshmi
Department of Mechanical Engineering, Erode Sengunthar Engineering College, Erode,
Tamil Nadu 638052, India

Rahul M. Mane
Department of Chemistry, Materials Research Laboratory, Shivaji University, Kolhapur 416004, India

Sumit Mishra
Department of Applied Chemistry, Birla Institute of Technology, Mesra, Ranchi, Jharkhand, India

C. Moganapriya
School of Building and Mechanical Sciences, Kongu Engineering College, Erode 638052,
Tamil Nadu, India

Mohammed E. M.
Department of Physics, Maharajas College, Ernakulam, Kerala, India

A. Mohankumar
School of Building and Mechanical Sciences, Kongu Engineering College, Erode 638052,
Tamil Nadu, India

André Monteiro
Department of Civil Engineering and CICECO, University of Aveiro, 3810193 Aveiro, Portugal

M. Murugavelu
Department of Chemistry, Annamalai University, Annamalainagar 608002, Tamil Nadu, India

Prakash Nanthagopalan
Department of Civil Engineering, IIT Bombay, Mumbai, India. E-mail: prakashn@civil.iitb.ac.in

George Nounesis
Biomolecular Physics Laboratory, National Centre for Scientific Research "Demokritos",
153 10 Aghia Paraskevi, Greece

George Okeke, PhD
Institute of Particle Science and Engineering, School of Chemical and Process Engineering,
University of Leeds, LS2 9JT, Leeds, UK

Nesibe G. Ozerkan, PhD
Center for Advanced Materials, Qatar University, Doha, Qatar

R. Parameshwaran
School of Building and Mechanical Sciences, Kongu Engineering College, Erode 638052,
Tamil Nadu, India

Bhaskar Patham
General Motors Technical Centre India (GMTCI), Vehicle CAE, Material CAE Methods Group,
Mumbai, India
SABIC Research and Technology Pvt. Ltd., Bangalore, India (Present affiliation).
E-mail: bhaskar.patham@sabic.com

Pallavi B. Patil
Materials Research Laboratory, Department of Chemistry, Shivaji University, Kolhapur 416004, India

M. Mohan Prasath
School of Building and Mechanical Sciences, Kongu Engineering College, Erode 638052,
Tamil Nadu, India

Priyanka K. P.
Department of Physics, Nanoscience Research Centre (NSRC), Nirmala College,
Muvattupuzha 686661, Kerala, India

Jayaraj Radhakrishnan
General Motors Technical Centre India (GMTCI), Vehicle CAE, Material CAE Methods Group,
Mumbai, India

R. Rajasekar
Department of Mechanical Engineering, Kongu Engineering College, Erode 638052, Tamil Nadu,
India. E-mail: rajasekar_cr@yahoo.com

Rajendran T. V.
PG and Research Department of Chemistry, Presidency College (Autonomous), Chennai 600005,
Tamil Nadu, India

Saravanan B. N.
GMTCI, Vehicle CAE, FMVSS-201U Group, Mumbai, India

V. Selvam
Department of Chemistry, K. Ramakrishnan College of Technology, Trichy, India.
E-mail: selvam.che@gmail.com

Gautam Sen
Department of Applied Chemistry, Birla Institute of Technology, Mesra, Ranchi, Jharkhand, India

S. M. Senthil
School of Building and Mechanical Sciences, Kongu Engineering College, Erode 638052,
Tamil Nadu, India

Akshath Sharma
Polymer Nanocomposite Laboratory, Materials Physics Division, School of Advanced Sciences,
VIT University, Vellore 632014, Tamil Nadu, India

Subhadip Sikdar
Research & Technology Center, Asian Paints Limited, Plot No. C3-B/1, TTC MIDC Pawane,
Thane-Belapur Road, Navi Mumbai 400703, Maharashtra, India

Sweta Sinha
Department of Applied Chemistry, Birla Institute of Technology, Mesra, Ranchi, Jharkhand, India

Jan Thoen
Department of Physics and Astronomy, KU Leuven, 3001 Leuven, Belgium

Rohan Tibrawala
Polymer Nanocomposite Laboratory, Materials Physics Division, School of Advanced Sciences,
VIT University, Vellore 632014, Tamil Nadu, India

Maja Trček
Jožef Stefan Institute, Jamova 39, 1000 Ljubljana, Slovenia

Biswajit Tripathy
GMTCI, Vehicle CAE, Vehicle Optimization Group, Mumbai, India

M. Vadivel
Department of Chemistry, Polymer Nanocomposite Centre, Scott Christian College,
Affiliated to Manonmaniam Sundaranar University, Tamil Nadu, India

Thomas Varghese
Department of Physics, Nanoscience Research Centre (NSRC), Nirmala College,
Muvattupuzha 686661, Kerala, India. E-mail: nanoncm@gmail.com

C. Vibha
Biomedical Technology Wing, Sree Chitra Tirunal Institute for Medical Sciences and Technology,
Poojappura, Thiruvananthapuram 695012, Kerala, India

Sheena Xavier
Department of Physics, Maharajas College, Ernakulam, Kerala, India

Amarnath Yerramala
Civil Engineering Department, Madanapalle Institute of Technology and Science, Madanapalle,
Andhra Pradesh, India. E-mail: dramarnath.y@mits.ac.in

LIST OF ABBREVIATIONS

AM	additive manufacturing
BAM	bio-additive manufacturing
BNPs	bimetallic nanoparticles
BPs	blue phases
CCB	conducting carbon black
CMOD	crack mouth opening deformation
CVs	cyclic voltammograms
DCR	defect core replacement
DMA	dynamic mechanical analysis
DSC	differential scanning calorimeter
DTS	diametral tensile strength
EDX	energy dispersive X-ray spectroscopy
EFB	empty fruit bunches
FEA	finite element analysis
FEGSEM	field emission gun scanning electron microscopic
FMVSS	Federal Motor Vehicle Safety Standard
FS	flexural strength
FTIR	Fourier transform infrared spectroscopy
GBs	grain boundaries
GGBS	ground granulated blast furnace slag GGBS
GO	graphene oxide
HA	hydroxyapatite
HIC	head injury criteria
HIPEs	high internal phase emulsions
HRR	Hutchinson, Rice, and Rosengren
HSC	high-strength concrete
LC	liquid crystalline
LD-GMT	low-density random glass-fiber-mat thermoplastics
LEFM	linear elastic fracture mechanics
LOI	loss on ignition
MD	molecular dynamics
MMT	montmorillonite
MS	micro silica
MW	microwave

MWCNT	multiwall carbon nanotube
NPs	nanoparticles
NR	normal Raman
OCV	open-circuit voltage
PA	polyamide
PEG	polyethylene glycol
PR	Poisson's ratio
PVA	polyvinyl alcohol
PVC	polyvinyl chloride
PVDF	polyvinylidene fluoride
SEM	scanning electron microscope
SERS	surface-enhanced Raman spectra
SLS	selective laser sintering
SP	superplasticizer
ST-BA	styrene–butyl acrylate
TDs	topological defects
TEM	transmission electron microscopy
TGA	thermogravimetric analysis
TGB	twist grain boundary
THF	tetrahydrofuran
TMOs	transition metal oxides
TNC	trimetallic nanocomposite
TS	tensile strength
TSC	trisodium citrate
UFA	ultrafine fly ash
UTS	ultimate tensile strength
w/b ratio	water/binder ratio
w/c ratio	water-to-cement ratio
XRD	X-ray diffraction

PREFACE

Nowadays people are looking for new smart materials to replace the old or conventional materials to get better performance in their applications. The use of polymeric materials and nanomaterials is increasing due to their wide property spectrum and tunability. It is easier to formulate materials for some special purposes using these materials than other classes of materials. Many commercial products, such as tires which make use of polymer and nanomaterials, are available on the market.

Nowadays scientists, engineers, technologists, and industrialists are exploring the new areas where nanomaterials and polymeric materials can be used to be substituted for conventional materials. They are also trying to find new applications for these materials and their blends/composites. This book presents various research to develop materials for different applications that include electrical applications, biomedical applications, sensing applications, coating applications, etc. Apart from this, there are a few chapters dedicated for the materials for construction applications. Discussions include the preparation, processing, and characterization of various polymeric materials, nanomaterials, and their composites. Some of the authors present theoretical studies of these systems, which can help readers to develop a better understanding in this area.

This book will be a very valuable reference source for university and college faculties, professionals, postdoctoral research fellows, senior graduate students, civil engineers, and researchers from R&D laboratories working to develop new smart materials.

Finally, the editors would like to express their sincere gratitude to all the contributors of this book, who extended excellent support for the successful completion of this venture. We are grateful to them for the commitment and the sincerity they have shown toward their contributions in the book. Without their enthusiasm and support, the compilation of this volume could not have been possible. We would also like to thank all the reviewers who have taken their valuable time to make critical comments on each chapter. We also thank the publisher, Apple Academic Press, for recognizing the demand for such a book and for realizing the increasing importance of the area of polymeric and nanomaterials to develop advanced materials and for starting such a new project.

Pious C. V.
Prof. Sabu Thomas
Dr. Józef T. Haponiuk
Dr. Zakhia Ahmed
Dr. Nanadakumar K.

CHAPTER 1

BEHAVIOR OF CEMENT COMPOSITES WITH MICRO AND NANO MINERAL ADMIXTURES

AMARNATH YERRAMALA[*]

Civil Engineering Department, Madanapalle Institute of Technology and Science, Madanapalle, India

[*]*E-mail: dramarnath.y@mits.ac.in*

CONTENTS

ABSTRACT

Cement, an energy-intensive material in construction industry, can partially be replaced with mineral admixtures for producing cement composites for conserving cement, utilizing waste materials, and improving performance of the cement composites. For the past few decades, mineral admixtures, like fly ash and silica fume, at microlevel particle size were successfully been incorporated in the cement composites. However, the research on using nano-sized mineral admixtures is gaining importance in the recent years to achieve advantages associated with nanotechnology in construction industry. This chapter provides a review on the properties of cement composites with micro and nano-sized fly ash and silica fume admixtures and presents a difference in behavior of cement composites with microparticles and nanoparticles. Furthermore, it discusses potential of further research in the cement composites with nanomaterials.

1.1 INTRODUCTION

Portland cement has a major role in producing cementitious composites like concrete. However, production of cement is responsible for 7% of worlds CO_2 emission.[20] With the advantages such as availability of raw materials, flexibility in using the materials to produce concrete, and easy moldability into different shapes, durability, and economy, the concrete is one of the most consumed construction materials around the world. Concrete has unlimited applications in construction industry. The applications range from rural roads to heavy duty pavements and residential buildings to seashore structures. According to a study, ~6 billion m^3 of concrete is made each year which is equal to 1 m^3 per capita.[18]

Therefore, in the scenario of environmentally friendly construction, it is important to conserve cement by improving performance of the concrete and thereby making it long lasting. Researchers, for the past few decades, have been using different alternative materials for cement to produce concrete. It has been proved that use of by-products like fly ash, silica fume, ground granulated blast furnace slag (GGBS), rice husk ash, and other mineral admixtures in place of cement not only minimizes cement usage but also improves performance of the concrete.[1,4,6,7,11,14,27] However, among all the mineral admixtures, fly ash and silica fume are most widely used in the concrete production.

Fly ash is a by-product of the combustion of pulverized coal in electric power generating plants. It is most widely used supplementary cementitious material in concrete. Fly ash consists of major chemical constituents such as silicon dioxide (SiO_2), calcium oxide or lime (CaO), iron oxide (FeO_2), and aluminum oxide (Al_2O_3). Depending up on the amount of calcium, silica, alumina, and iron content, fly ash is divided into class C and class F type (ASTM C 618). Typical chemical compositions of class C and class F fly ash are shown in Table 1.1. At microlevel, the fly ash particles are rounded in shape. Figure 1.1 shows the fly ash particles at 10 μm.

FIGURE 1.1   Micrograph showing spherical fly ash particles.

Silica fume is a by-product in the manufacture of silicon metal and ferrosilicon alloys. The process involves the reduction of high-purity quartz (SiO_2) in electric arc furnaces at temperatures in excess of 2000°C. Silica fume is a very fine powder consisting mainly of spherical particles or microspheres of mean diameter about 0.15 μ, with a very high specific surface area (15,000–25,000 m^2/kg). Physical and chemical properties of silica fume are shown in Table 1.1.[15]

As mentioned earlier, fly ash and silica fume, as partial replacement for cement, not only minimize cement usage but also improve performance of the concrete. However, further there are many improvements needed in concrete like increased durability, decreased brittleness, and increased tensile strength.[10] Various nanomaterials like nano cement, nano-admixtures, and nanofibers can improve vital characteristics of cement composites such as strength, durability, and lightness (Sanchez and Sobolev, 2010).[12,16,17] Therefore, in the recent-past, research is gaining importance toward use of nanomaterials and understanding performance of the cement composites to utilize advantages associated with nanotechnology in construction industry, and make cement composites as environmentally friendly and sustainable construction materials. This chapter presents a current state of the knowledge of properties of cement composites with micro and nano fly ash and micro and nano-silica fume; brings out performance difference with micro and nanomaterials; and discusses potential of future research in cement composites with nanomaterials.

TABLE 1.1 Chemical and Physical Characteristics of Fly Ash.[15, 22]

Chemical composition (%)	Class C	Class F	Silica fume
Silica (SiO$_2$)	35.4	46.8	>85
Alumina (Al$_2$O$_3$)	17.5	23.7	<2
Ferric oxide (Fe$_2$O$_3$)	5.3	13.2	<1
Calcium oxide (CaO)	26.1	3.1	<1
Magnesium oxide (MgO)	4.6	1	<1
Sulfuric anhydride (SO$_3$)	2.8	1.2	–
Moisture	0.1	0.1	–
Loss on ignition (LOI)	2.4	0.3	<4
Physical characteristics (per ASTM C 618)			
Fineness (retained on 325 sieves, %)	15.9	25.7	–
Specific gravity	2.58	2.34	2.2
Soundness (autoclave expansion, %)	0.11	0	

1.2 MATERIAL PROPERTIES

Physical and chemical properties of the materials used to produce cement composites largely influence behavior of the end product. Physical properties of micro and nanoparticles of fly ash and silica fume are shown in Table 1.2. Change in particle size, from micro to nano, largely influences the specific surface area. Nearly, the increase in specific surface area is 100 times for both fly ash and silica fume, when the particle size decreased from micro to nano. Furthermore, as the size of the fly ash decreases, the shape of the particle becomes round to rough and surface becomes smooth to angular.[9] (Paul et al., 2007) (Fig. 1.2). With decrease in mean particle size, specific gravity of fly ash increases.[9]

Similarly, loss on ignition (LOI) also increases with fineness. With decrease in fly ash particle size, sum of the main components such as SiO$_2$, Al$_2$O$_3$, and Fe$_2$O$_3$ remain nearly same before and after classification.[9] Table 1.3 shows chemical compositions of micro and nano fly ash and silica fume.

TABLE 1.2 Physical Properties of Materials.

Source	Particle size				Specific surface area (m²/kg)			
	Microparticles (μm)		Nanoparticles (nm)		Microparticles (μm)		Nanoparticles (nm)	
	FA	SF	NFA	NSF	FA	SF	NFA	NSF
Li G, 2004[a]				10 ± 5	521			640,000
Tao Ji, 2005	12.3			15 ± 5	691			
Paul et al., 2007	60		60		249		25,530	
He and Shi, 2008				30				440,000
Zhang and Islam, 2012	27.3	0.2		12		21,300		200,100

FA—fly ash; SF—slica fume; NFA—nano fly ash; NSF—nano-silica fume.
[a]The technique used to measure specific surface area were different for micro and nanoparticles [Blainess for FA and Brunauer–Emmett–Teller (BET) for NFA].

FIGURE 1.2 SEM images of fly ash particles with 40 and 5 μm size.[9]

TABLE 1.3 Chemical Compositions of Fly Ash, Silica Fume, and Nano-silica.

Chemical composition	Fly ash[9]			Silica fume[30]	
	Particle size (μm)			Particle size (nm)	
	40	18	5	150	12
SiO_2	45.94	45.75	47.8	95.9	>99.8
Al_2O_3	25.62	25.27	26.97	0.3	–
Fe_2O_3	8.68	8.8	6.66	0.3	–
CaO	9.39	9.32	7.9	0.2	–
MgO	2.36	2.29	2.18	0.4	–
Na_2O	1.43	1.53	1.33	0.05	–
K_2O	2.71	2.79	3	0.6	–
SO_3	1.23	1.82	0.99	0.2	–
Loss on ignition (LOI)	1.22	1.12	2.2	1.5	–

1.3   MIX PROPORTIONING

Mix proportioning of cement composites depends on the strength and work-ability requirements. In general, either sand or cement can be replaced with fly ash in cement composites. However, when silica fume is considered, in majority of the instances it is replaced with cement only. Replacement of fly ash varies from 10–50% with cement. Relatively, small replacements can be noticed with silica fume. Summary of mix proportions with micro and nanomaterials is shown in Table 1.4. Water to cementitious materials ratio determines workability. For higher strengths, water to cementitious ratio is reduced and superplasticizers are added to increase workability (Table 1.4). When nanomaterials are used in mix, effective dispersion is important for uniform mixing of nanomaterials. Apart from normal mechanical mixing, ultrasonic mixing can be used.[30] However, Zhang and Islam (2012) reported that ultrasonic mixing results in higher strengths than normal mechanical mixing when nanomaterials were used in cement composite mixes.

TABLE 1.4   Summary of Mix Proportions.

Source	w/c	Replacement	Remarks
Zhang and Islam, 2012	0.45	1% (NS)	Mortar
		2% (NS)	Concrete
Li, 2004	0.28	50% (FA)	1.5–2% of SP, mortar, and concrete
		1% (NS)	
Bhanja and Sengupta, 2005	0.26–0.42	5–30% (MS)	3.5% of SP, concrete
Ozyildirim, 2010	0.38–0.45	7% (SF)	Concrete
		3% (NSF)	
Arulraj and Carmichael, 2011	–	10–30% (NFA)	Concrete, coarse aggregate replaced with NFA
Carmichael and Arulraj, 2012	–	10–50% (NFA, NS)	Mortar

FA—fly ash; NFA—nano fly ash; MS—silica fume; NS—nano-silica fume; SP—superplasticizer.

1.4   PROPERTIES

1.4.1   CONSISTENCY

Mora et al. (1993) reported that consistency of cement composites depends upon mean particle size and specific surface area of fly ash. However,

Jaturapitakkal et al. (1999) showed that shape of the fly ash particle had significant role in consistency than size and specific surface area. With nearly same shape, change in four times in particle size had no marked influence on workability of fly ash mixes.[9] Furthermore, when cement was replaced with 50% of nano fly ash, no significant effect on consistency was found in the mix.[5] However, when coarse aggregate was replaced with nano fly ash, the workability increased with increase in nano fly ash replacement.[2] With rounded particle shape, addition of micro- and nano-silica fume improves workability.[5,16,17]

1.4.2   SETTING TIME

With coarse fly ash, setting time decreases, and with smaller fly ash particles, the setting time increases.[9] As coarse fly ash is porous, it absorbs free water quickly, and, hence, the setting time is shorter due to reduced free water in the mix. On the other hand, in smaller size fly ash, broken cavities result in less porous; therefore, no free absorbed water is available to the cavities and the setting time increases.[9] Similar observations were reported by Carmichael and Arulraj[5] in case of nano fly ash and nano-silica fume. They reported that with an increase in replacement of cement with nano fly ash and nano–silica, the initial and final setting times increase as the percentage replacement increases.

1.4.3   COMPRESSIVE STRENGTH

Very fine fly ash increases compressive strength, even at very early ages. In contrast, coarse fly ash slows down strength development. As particle size of fly ash decreases, pozzolanic activity increases and results in higher strength. At very small particle size, packing effect and pozzolanic reaction results in higher strength Jaturapitakkul[9] reported that smaller fly ash can give higher strength both at early age and at later age. However, there exists an optimum nano fly ash replacement for cement. Carmichael and Arulraj[5] found that 30% was the optimum nano fly ash content, when it was replaced for cement in cement mortar. When coarse aggregate was replaced with nano fly ash in concrete, the strength increased with fly ash replacement. However, the strength increase is based on concrete grade. As the concrete grade increases, the rate of strength decreases.[2]

Nano-silica can give higher compressive strength for mortars than those containing micro-silica, at 7 and 28 days.[13,16,17] When comparing the strength of nano-silica and microsilica, the priority of nano-silica is inferred specially in primary ages. The strength difference between microsilica and nano-silica decreases with time.[19] The optimum replacement percentage of nano-silica fume was around 30%.[5] The nanoparticles fill up the pores and reduce $Ca(OH)_2$ among the hydrates, which result in higher strength.[16,17]

1.5 FUTURE POTENTIAL

The previous discussion has demonstrated that nano fly ash and nano-silica fume have adequate performance over micro-sized particles. However, the nanomaterials have significant further potential, which could be achieved by carrying out further research in a number of areas. Further investigations are clearly needed on improving cement composite properties, like tensile strength, shrinkage, and durability, to overcome many construction problems.

1.6 CONCLUSIONS

Review on micro and nano fly ash and silica fume in cement composites suggests that little research has been conducted on nanoparticles in cement composites. Furthermore, the review reveals that there are several research areas such as developing standard mix proportioning for nano fly ash and nano-silica fume, effective mixing techniques, fresh state properties, engineering properties, and durability properties of the nano cement composites for increasing future potential of nano mineral admixtures in cement composites.

KEYWORDS

- **nanoparticles**
- **silica fume**
- **fly ash**
- **cement composites**
- **mineral admixtures**

REFERENCES

1. Amarnath, Y.; Babu, G. K. Transport Properties of High Volume Fly Ash Roller Compacted Concrete. *Cem. Concr. Compos.* **2011,** *33,* 1057–1062.
2. Arulraj, P. G.; Carmichael, J. M. Effect of Nanoflyash on Strength of Concrete. *Int. J. Civ. Struct. Eng.* **2011,** *2*(2), 1924–1935.
3. Bhanjaa, S.; Sengupta, B.; Influence of Silica Fume on the Tensile Strength of Concrete. *Cem. Concr. Res.* **2005,** *35,* 743–747.
4. Bouzoubaa, N.; Zhang, M. H.; Malhotra, V. M. Mechanical Properties and Durability of Concrete Made With High-Volume Fly Ash Blended Cements Using a Coarse Fly Ash. *Cem.Concr. Res.* **2001,** *31,* 1393–1402.
5. Carmichael, J. M.; Arulraj, P. G. Influence of Nano Materials on Consistency, Setting Time and Compressive Strength of Cement Mortar. *IRACST – Eng. Sci. Technol.* **2012,** *2*(1), 2250–3498.
6. Chatveera, B.; Lertwattanaruk, P. Evaluation of Sulfate Resistance of Cement Mortars Containing Black Rice Husk Ash. *J. Environ. Manag.* **2009,** *90,* 1435–1441.
7. Chindaprasirt, P.; Kanchanda, P.; Sathonsaowaphak, A.; Cao, H. T. Sulfate Resistance of Blended Cements Containing Fly Ash and Rice Husk Ash. *Constr. Build. Mater.* **2007,** *21,* 1356–1361.
8. Colleparidi, M.; Olagot O. J.; Troli, R.; Simonelli, F.; Collepardi, S. *Combination of Silica Fume, Fly Ash and Amorphous Nano-Silica in Superplasticized High-Performance Concretes,* Enco, Engineering Concrete, Ponzano Veneto, Italy, 2007.
9. Jaturapitakkul, C.; Kiattikomol, K.; Smith, S. A Study of Strength Activity Index of Ground Coarse Fly Ash with Portland *Cem. Sci. Asia.* **1999,** *25,* 223–229.
10. Garboczi, E. J. Concrete Nanoscience and Nanotechnology: Definitions and Applications. *Nat. Inst. Stand. Technol.* http://ciks.cbt.nist.gov/monograph, (accessed on 15 Dec 2012).
11. Goldman, A.; Bentur, A. Properties of Cementitious Systems Containing Silica Fume or Nonreactive Microfillers. *Adv. Cem. Based Matter. I.* **1994,** 209–215.
12. Ji, T. Preliminary Study on the Water Permeability and Microstructure of Concrete Incorporating Nano-SiO$_2$. *Cem. Concr. Res.* **2005,** *35,* 1943–1947.
13. Jo, B. W.; Kim, C. H.; Tea, G. H.; Park, J. B. Characteristics of Cement Mortar with Nano-Sio2 Particles. *Constr. Buil. Mater.* **2007,** *21,* 1351–1355.
14. Khatib, J. M.; Wild, S. Sulphate Resistance of Metakaolin Mortar. *Cem. Concr. Res.* **1998,** *28*(1), 83–92.
15. King, D. *The effect of silica fume on the properties of concrete as defined in concrete society report 74, cementitious materials.* 37th Conference on Our World in Concrete & Structures, Singapore, August 29–31, 2012.
16. Li, H.; Xiao, H. G.; Yuan, J.; Ou, J. Microstructure of Cement Mortar with Nano-Particles. *Compos. Part B: Eng.* **2004,** *35*(2), 185–189.
17. Li, G. Y. Properties of High-Volume Fly Ash Concrete Incorporating Nano-SiO$_2$. *Cem. Concr. Res.* **2004,** *34,* 1043–1049.
18. Lomborg, B. *The Skeptical Environmentalist: Measuring the Real State of the World;* Cambridge University Press: Cambridge, United Kingdom, 2001; pp 512–540.
19. Koohdaragh, M.; Mohamadi, H. H. Comparison of Mechanical of the Concrete Samples Containing Micro-Silica and Nano-Silica. *Aust. J. Basic Appl. Sci.* **2011,** *5*(10), 560–563.
20. Mehta, P. K. Concrete Technology for Sustainable Development. *Concr. Int.* **1999,** *21*(11), 47–52.

21. Naik, T. R.; Ramme, B. W. Effect of High-Lime Fly Ash Content on Water Demand, Time of Set, and Compressive Strength of Concrete. *ACI Mater. J.* **1990,** *87*, 619–626.

22. Naik, T. R.; Ramme, B. W.; John, H. T. Use of High Volumes of Class C and Class F Fly Ash in Concrete. *Cem. Concr. Aggreg.* **1994,** 16(1), 12–20.

23. Mora, E. P.; Paya, J.; Monzo, J. Influence of Different Sized Fractions of a Fly Ash on workability of Mortars. *Cem. Conc. Res.* **1993,** *23*, 917–924.

24. Ozyildirim, C. *Laboratory Investigation of Nanomaterials to Improve the Permeability and Strength of Concrete*, February 2010, http://www.virginiadot.org/vtrc/main/online_reports/pdf/10-r18.pdf (accessed on Dec 15, 2012).

25. Peris, M. E.; Paya, J.; Monzo, J. Influence of Different Sized Fractions of a Fly Ash on Workability of Mortars. *Cem. Conc. Res.* **1993,** *23*, 917–924.

26. Qing, Y.; Zenan, Z.; Deyu, K.; Rongshen, K. Influence of Nano-Sio2 Addition on Properties of Hardened Cement Paste as Compared with Silica Fume. *Constr. Build. Mater.* **2007,** *21*, 539–545.

27. Thomasa, M. D. A.; Shehataa, M. H. Shashiprakasha, S. G.; Hopkinsb, D. S.; Cailb, K. Use of Ternary Cementitious Systems Containing Silica Fume and Fly Ash in Concrete. *Cem. Concr. Res.* **1999,** 1207–1214.

28. Thomas P. K.; Satpathy, S. K.; Manna, I.; Chakraborty, K. K.; Nando, G. B. Preparation and Characterization of Nano Structured Materials from Fly Ash: A Waste from Thermal Power Stations, by High Energy Ball Milling. *Nanoscale. Res. Lett.* **2007,** *2*, 397–404.

29. Xiaodong H.; Xianming S.2008. Chloride Permeability and Microstructure of Portland Cement Mortars Incorporating Nanomaterials, Transportation Research Record: Journal of the Transportation Research Board, No. 2070, Transportation Research Board of the National Academies, Washington, D.C., 13–21.

30. Zhang, M. H.; Islam, J. Use of Nano-Silica to Reduce Setting Time and Increase Early Strength of Concretes with High Volumes of Fly Ash or Slag. *Constr. Build. Mater.* **2012,** *29*, 573–580.

CHAPTER 2

SMART MATERIALS FOR SENSING, CATALYSIS, AND COATING: NANO BI- AND TRIMETALLIC PARTICLES

B. LOGANATHAN[1], M. MURUGAVELU[1], K. KARTHICK[2], and B. KARTHIKEYAN[1*]

[1]Department of Chemistry, Annamalai University, Annamalainagar 608002, Tamil Nadu, India

[2]FEAT-Annamalai University, Annamalainagar 608002, Tamil Nadu, India

*Corresponding author. E-mail: bkarthi_au@yahoo.com

CONTENTS

ABSTRACT

This chapter describes the results of the investigation on the synthesis and characterization of bi- and trimetallic nanocomposites of noble metals Au, Pt, Ag, and Pd. Green chemical methods for the preparation are explained. Various tools which are helpful to solve the structure and composition of the nanoparticles (NPs) are highlighted. Application of these bi- and tri-nanometallic particles in the field of sensing and catalysis is taken into account. As in these multimetallic NPs, the properties can be tuned by varying the composition; the smarter application of these systems in construction materials, for example, steel is considered and the preliminary results obtained are also been discussed.

2.1 INTRODUCTION

Bi- and trimetallic nanocomposite (TNC) particles have been attracting immense attention due to their novel chemical and physical properties arising from their size and cooperative effect. These nanometallic particles are used in many applications, such as catalysis to electronic devices. The electronic structure of nanometallic particles undergoes major deviation from that of bulk. Nanocomposite based on nanosized inorganic particles and clusters represents an attractive field of research activity because of the possibility to tailor and optimize the properties of the resulting materials for various applications.[5] Heterogeneous bimetallic nanoparticles (BNPs) have recently received much attention as electrocatalysts with enhanced activities[6,7] and electrochemical reversibility for redox reactions.[4] These bimetallic alloys can retain the functional properties of each component and possibly offer synergistic effects via cooperative interactions, resulting in important features such as increased surface area, enhanced electrocatalytic activity, improved biocompatibility, promoted electron transfer, and better invulnerability against intermediate species.

Here, we report the catalytic efficiency of Ag–Pd BNPs tested for the reduction of 4-nitrophenol (4-NPh) in aqueous solution at room temperature. On the other hand, the preparation of Ag–Pt core–shell NPs and the efficiency of the prepared BNPs toward the sensing of adenine are elaborated using opto and electrochemical studies. Also, we report the synthesis and characterization of Au/Pt/Ag TNCs, synthesized through microwave (MW) irradiation method. To identify the surface-enhanced Raman spectral (SERS) activity of this metal nanocomposite, 7-azaindole (7-AI) was proposed as a probe molecule. (Karthikeyan, 2012)[2]. Stainless steels contain sufficient chromium to form a passive film of chromium oxide, which prevents further surface corrosion by blocking oxygen diffusion to the steel surface and blocks corrosion from spreading into the metal's internal structure, and due to the similar size of the steel and oxide ions, they bond very strongly and remain attached to the surface. The corrosion occurs when there is an overall breakdown of the passive film. Halogens penetrate the passive film of stainless steel and allow corrosion to occur at the surface of stainless steel. It is now established that in the environments that result in the exposure of concrete reinforced with carbon steels and the environments that allow transport of chlorides to the reinforcement can result in serious corrosion.

2.2 EXPERIMENTAL SECTION

2.2.1 SYNTHESIS OF Ag–Pt BNPs

In a typical synthesis of Ag–Pt BNPs, 0.024 g of the silver nitrate was added to 100 mL of sterile double-distilled water and the solution was heated for 1 h at 90°C. Thereafter, 2 mL of 1% trisodium citrate was added to the solution. After 30 min of boiling, 280 μL of 0.02 M $H_2PtCl_6.6H_2O$ was added dropwise directly to the round-bottomed flask, and heating was continued for another 1.5 h. The visible change of color to dark brownish solution indicates the formation of colloidal NPs.

2.2.2 SYNTHESIS OF Ag–Pd BNPs

In a typical synthesis of Ag–Pd BNPs, 0.024 g of the silver nitrate was added to 100 mL of sterile double-distilled water and the solution was heated for 1 h at 90°C. Thereafter, 2 mL of 1% trisodium citrate was added to the solution. After 30 min of boiling, 280 μL of 0.02 M $PdCl_2$ was added dropwise directly to the round-bottomed flask, and heating was continued for another 1.5 h. The visible change of color to the dark color solution indicates the formation of colloidal NPs.

2.2.3 SYNTHESIS OF Au/Pt/Ag TNCs

Trisodium citrate-reduced Au NPs were first prepared according to the reported method.[2,3] First, 10 mL of aqueous 0.1% metal salt ($HAuCl_4$ $3H_2O$)

was heated to boil and 2 mL of 1% trisodium citrate was then added with stirring. The reaction mixture was heated for 240 s and cooled down to room temperature. The solution turned to vivid magenta from slight yellow, indicating the formation of Au NPs. Then, 10 mL of 0.1% metal salt (H_2PtCl_6 xH_2O) was added to the Au NPs followed by the addition of 2 mL of 1% trisodium citrate with stirring. Finally, 10 mL of 0.1% metal salt ($AgNO_3$) was added into the Au/Pt NPs. The overall reaction time for the synthesis of Au/Pt/Ag TNCs is 420 s. Then, the synthesized colloidal solution was sonicated for 1 h for the fine dispersion of NPs. The preparation procedure was carried under a MW irradiation method.

2.2.4 SAMPLE PREPARATION FOR THE SURFACE-ENHANCED RAMAN SPECTROSCOPIC (SERS) ANALYSIS

7-Azaindole (7-AI) analyte sample was prepared in a concentration range of 0.001 M. To collect the SERS of 7-AI adsorption on Au/Pt/Ag TNCs, the 7-AI sample and synthesized colloidal solutions were mixed in the ratio of 0.5:1.

2.2.5 PREPARATION OF THE STEEL FOR ALUMINUM (AI) COATING BY CHEMICAL VAPOR DEPOSITION

Oil was removed from the three high-strength steel substrates, and then they were cleaned ultrasonically with absolute alcohol. The steel substrates were placed in a self-built chemical vapor reactor. Nitrogen was fed into the reactor at 35 mL/min, which was heated to 100°C and maintained at this temperature for 60 min. The precursor was fed into the reactor to coat the steel samples.

2.3 RESULTS AND DISCUSSION

2.3.1 CHARACTERIZATION OF BI- AND TRIMETALLIC NPs

TEM was utilized to characterize the size, morphology, and structure of the nano bi- and trimetallic particles. Figure 2.1 depicts the TEM images of bi- and trimetallic NPs to confirm the structure of the synthesized colloidal solutions. Figure 2.1 clearly shows that the NPs are having well-defined spherical-shaped structure. As observed in Figure 2.1, each NP exhibits its homogeneous electron density with a dark core surrounded by a lighter shell (the black in color core and light color shell are clearly seen).[8] In the obtained bimetallic structure (Fig. 2.1a), the darker nucleus corresponds to the initial Ag and the lighter shell corresponds to the Pt NPs.

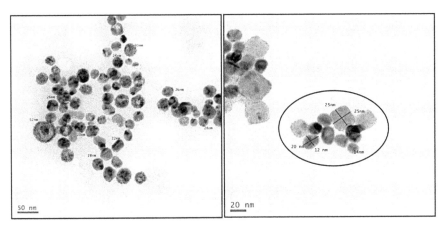

FIGURE 2.1   Transmission electron microscopic images of (a) Ag–Pt BNPs and (b) Au/Pt/Ag TNCs.

2.3.2 CATALYTIC PROPERTIES OF Ag–Pd BNPs

Figure 2.2 shows the catalytic performance of the Ag–Pd tested for the reduction of 4-NPh to 4-amino phenol (4-APh) with an excess amount of $NaBH_4$. 4-NPh is inert to $NaBH_4$ and the reaction did not occur in the absence of the synthesized nanobimetal catalysts, even for a period of 1 day. But, Ag–Pd effectively catalyzes the reduction of 4-NPh by acting as an electron relay system; electron transfer takes place between 4-NPh and BH_4^- through the BNPs. The reaction was readily monitored by using UV–vis spectroscopy, when a small amount (0.1 mL) of Ag–Pd BNPs nanocatalyst was introduced into the reaction solution; the absorption peak at 400 nm completely vanished within 60 s and concomitant appearance of a new peak was observed at 300 nm (Fig. 2.2). This catalytic effect is much higher from the other reported catalysts.[1] The reduction action occurred via relaying electrons from the donor BH_4^- to the acceptor 4-NPh after the absorption of both onto the Ag–Pd BNPs' surface. The hydrogen atom, which was formed from the hydride, after electron transfer to the Ag–Pd BNPs, attacked 4-NPh molecules and reduced it. The UV–vis spectra shows an isosbestic point (313 nm), illustrating that the catalytic reduction of 4-NPh yields only 4-APh without any by-product.

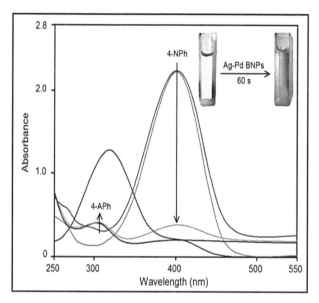

FIGURE 2.2   UV–vis absorbance spectra of the catalytic performance of Ag–Pd for the reduction of 4-Nitrophenol to 4-amino phenol.

2.3.3   *ELECTROCHEMICAL SENSING PERFORMANCE OF Ag–Pt BNPs*

Generally, cyclic voltammograms (CVs) of the reversible redox pair $[Fe(CN)_6]^{3-/4-}$ are valuable and convenient to monitor the properties of the modified electrode because the electron transfer between the solution species and the electrode must occur by tunneling either through the layer or through the defects in the layers. Figure 2.3 shows the CVs of different modified electrodes in 5 mM $[Fe(CN)_6]^{3-/4-}$ containing 0.1 M KCl at 100 mV/s. Curve *a* shows two pairs of well-defined redox peaks for bare GCE, which is due to the reversible one-electron redox behavior of $[Fe(CN)_6]^{3-/4-}$. Figure 2.3 compares the CV response of a bare GCE, adenine/GCE (curve *b*), Ag–Pt/GCE (curve *c*), and Ag–Pt adenine/GCE (curve *d*). As expected, ferricyanide exhibits reversible behavior on a bare Ag–Pt electrode with a peak-to-peak separation (ΔEp) of 70 mV at a scan rate of 100 mV/s. The ΔEp value observed for

Ag–Pt–adenine/GCE (10 mV) reveals the kinetic hindrance exerted on the electron transfer process. *Curve d* shows the sensing of adenine by the Ag–Pt/ GCE electrode as an enhanced anodic current and the (Epa) peak potential of 0.863 V (inset Fig. 2.3). Adsorption of an adenine monolayer by an electrostatic interaction through the amine group in the synthesized Ag–Pt BNPs on GCE and the uniform dispersion of Ag–Pt BNPs enhanced the amount of uniform adsorption on the surface. Whereas in only the adenine-modified GCE electrode (curve *b*), no significant anodic peak was observed, indicating that the Ag–Pt–adenine-modified GCE had larger adsorptions than other electrodes. The reproducibility and stability of the sensing ability were evaluated by using 2 mM adenine at a scan rate of 100 mV/s. The stability of the modified electrode was tested by scanning the electrode continuously in 2 mM adenine. It was found that there was no apparent decrease in the current response for 20 consecutive cycles, indicating that the modified electrode is relatively stable.

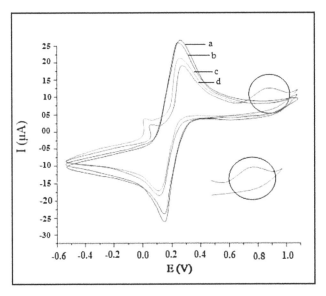

FIGURE 2.3   Cyclic voltammograms of different modified electrodes in 5 mM [Fe(CN)$_6$]$^{3-/4-}$ containing 0.1 M KCl at 100 mV/s.

2.3.4   7-AZAINDOLE AS A PROBE MOLECULE

The synthesized Au/Pt/Ag TNCs was further characterized by SERS using 7-AI as a probe molecule. FT-Raman spectrum of 7-AI, 1 mM of aqueous 7-AI and SERS spectrum of 7-AI-adsorbed monometallic Au, Pt, Ag NPs, and trimetallic Au/Pt/Ag TNCs are displayed in Figure 2.4. The most prominent bands in the normal Raman (NR) spectrum of 7-AI are in the region of 500–1600 cm⁻¹. Some of the bands, which appear very weak in the NR spectrum are found to be enhanced in the observed SERS spectrum. The intense band at 793 cm⁻¹ in the SERS spectrum can be assigned to the 7-AI breathing vibration, which appears at 766 and 795 cm⁻¹ in the NR spectrum of 7-AI and 1 mM aqueous solution, respectively. The results from a NR spectrum of 7-AI, 1 mM 7-AI aqueous solution, and SERS of 7-AI-adsorbed mono and trimetallic NPs are compared with the reported results.[2,3] The similarity in Pt NPs and the SERS spectrum of TNCs suggested that higher amount of Pt NPs are available at the surface for the 7-AI adsorption. This observation reveals that Pt forms the shell.

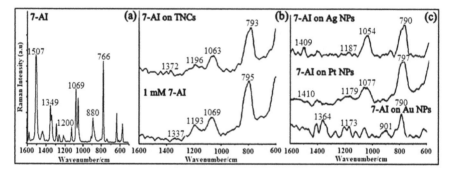

FIGURE 2.4 (a) Normal Raman spectrum of 7-AI and (b) 1 mM 7-AI aqueous solution, (c–f) SERS spectrum of 7-AI adsorbed Au/Pt/AG TNCs, monometallic Au, Pt, and Ag NPs, respectively.

2.3.5 INSPIRED Al COATING ON STEEL

Nano aluminum layer is coated on a passivated-304 stainless steel, the aluminum protects where they touch the stainless steel, because it is much more anodic than stainless steel; this approach can improve the performance, it is an inherently durable solution for the problem of chloride-induced corrosion. SEM image shows the deposition of aluminum layer on the substrate (Fig. 2.5). The pitting corrosion was evaluated with the start potential as −150 mV, reverse potential as 300 mV, and sweep rate as 300 mV/min. The corrosion values obtained are as follows: Icorr (mA/cm²), 0.0256715; corrosion rate (mm/year), 523290.25; corrosion rate (mm/year), 2.06E + 07 (Fig. 2.6).

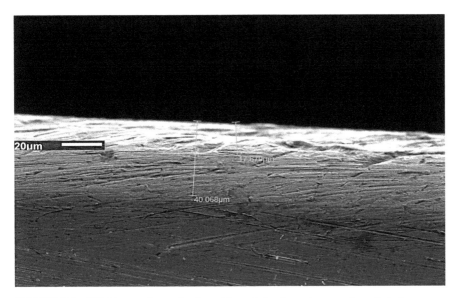

FIGURE 2.5   SEM image shows the deposition of aluminum layer on the substrate.

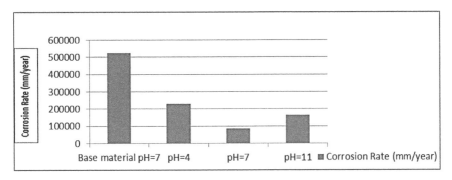

FIGURE 2.6   Comparison of pitting corrosion values for steel and Al-coated specimen in various pH.

2.4 CONCLUSIONS

Ag–Pt and Ag–Pd bimetallic NPs and Au/Pt/Ag TNCs have been synthesized using thermal and MW irradiation method. Bimetallic NPs-based novel electrochemical sensor for adenine and efficient catalytic reduction of 4-nitrophenol by the BNPs are discussed. Au/Pt/Ag TNCs were also synthesized and characterized. Detailed investigations of catalytic activity and their applicability were obtained by understanding the theoretical mechanism. The results inspired to extend their macro applications in steel coating, which will be used as the construction materials.

KEYWORDS

- **nanometallic**
- **electrochemical**
- **catalytic**
- **trisodium**

REFERENCES

1. Jiang, D.; Xie, J.; Chen, M.; Li, D.; Zhu, J.; Qin, H. J. *Alloys Compd.* **2011,** *509,* 1975–1979.
2. Karthikeyan, B.; Loganathan, B. *Mater. Lett.* **2012,** *85,* 53–56.
3. Karthikeyan, B.; Loganathan, B.; *Physica E.* **2013,** *49,* 105–110.
4. Katz, E.; Willner, I.; Wang, J. *Electroanalysis.* **2004,** *16,* 19–44.
5. Li, B.; Liu, T.; Wang, Y.; Wang, Z. J. *Colloid. Interface Sci.* **2012,** *377,* 114–121.
6. Zhang, J.; Sasaki, K.; Sutter, E.; Adzic, R. R. *Science.* **2007,** *315,* 220–222.
7. Zhang, S.; Shao, Y.; Yin, G.; Lin, Y. *Angew. Chem. Int. Ed.* **2010,** *49,* 2211–2214.
8. Zhang, X.; Wang, H.; Su, Z. *Langmuir.* **2012,** *28,* 15705–15712.

CHAPTER 3

MACROSCOPIC AND IMPEDANCE STUDY OF PVC/GRAPHENE OXIDE NANOCOMPOSTIES

GIRISH M. JOSHI[1*], KALIM DESHMUKH[1], and S. KALINATHAN[1]

[1]*Polymer Nanocomposite Laboratory (PNL), Material Physics Division, School of Advanced Sciences (SAS), VIT University, Vellore 632014, Tamil Nadu, India*

Corresponding author. E-mail: varadgm@gmail.com

CONTENTS

ABSTRACT

A new protocol for the preparation of GO-PVC was developed. Morphology, thermal stability, mechanical properties, dielectric properties, and the impedance of the prepared composites were studied. Studies showed that the GO occupied and entangled with polymer PVC. Variation of optical properties as function of GO loading, suit for tailoring desired dielectric medium. Furthermore, the studies of structural, thermal, and electrical, properties were reported for suitable applications as high-performance nanocomposite.

3.1 INTRODUCTION

Graphene derivatives are intensively investigated as high-performance nanofillers in the polymer composites. Graphene oxide (GO) is an oxygen-rich carbon allotrope, which is produced by controlled oxidation of graphite. GO consists of intact graphitic regions interspersed with sp^3-hybridized carbons, containing hydroxyl and epoxide functional groups on the top and bottom surfaces of each sheet, and sp^2-hybridized carbons, containing carboxyl and carbonyl groups almost at the sheet edges.[1] The oxidation of graphite to GO break up the sp^2-hybridized structure of the stacked graphene sheet, which increases the distance between adjacent sheets from 3.35 Å in graphite powder to 6.8 Å for GO powder.[2] This increase in spacing varies significantly, depending on the amount of water intercalated within the stacked sheet structure,[3] and reduces the interaction between sheets; thus, facilitating the delamination of GO into individual GO sheets upon exposure to low-power sonication in water.[4] The combination of the unique properties of GO and good process ability of polymer would open up possibilities for making advanced polymer composites, with achievement of new physicochemical properties. In order to have efficient reinforcement in polymer composites, it is important to have molecular-level dispersion of nanofiller in the polymer matrix.[5] The challenge is to achieve significant improvement in interfacial adhesion between polymer matrix and reinforcing fillers.[6] It has been demonstrated that the composites of polymers and nanofillers can combine the ductile property of polymer matrix and high strength of nanofiller.[7] GO-reinforced polymer composites are well studied and documented by researchers across the globe.[8–12] The unique properties of polymer/GO composites, such as enhanced electrical and mechanical properties and superior electrochemical properties, have been reported for various applications.[8–16]

Polyvinyl chloride (PVC) is highly produced commercial polymer with vast applications in low-voltage and low-frequency electrical insulation, construction, clothing, water supplying pipes, flat sheets, sports, and toys.[17] Several reports show an improvement in the properties of PVC, such as mechanical strength, decomposition temperature, optimization of conductivity, and refractive index, when its composites were prepared with different carbon materials. The uniqueness of PVC is that it is the best substitute for metals for engineering application point of view.[18–22] Recently, the biomedical application of GO was reported with an aim to overcome the antibacterial activities of biological samples.[23] The GO-inducted polymer composites have also shown improved superhydrophobicity, conductivity, and luminescence properties.[24] The hybrid graphene nanocomposite based on solution in situ technique probably opens the door for wide scientific interest. The role of GO is very crucial, which act as chemical hook with PVC moiety in its functional form.[25] GO exhibit surface activity for stabilizing oil droplets, forming pickering emulsions, formation of large-scale, two-dimensional nanostructure possible for various surface science applications.[13] With this motivation, we fabricated the PVC nanocomposites with GO by colloidal blending method. We readily dispersed both GO and PVC in tetrahydrofuran (THF) to achieve molecular level colloidal dispersion. The issue of complete GO-inducted PVC protocol was developed and optical microscopy and electrical properties were elaborated. The prospect of this study is seen in view of tailoring the desired dielectric medium for various optical applications for tuning the surface adhesion properties improved by two-dimensional nanostructure.

3.2 EXPERIMENTAL

3.2.1 MATERIALS

PVC granules were supplied by Techno vinyl polymers, Pvt. Ltd. Mumbai, India. THF, extra-pure AR grade was purchased from Sisco Research Laboratory Chemicals, Pvt. Ltd. Mumbai, India. Natural graphite powder was supplied by Carbotech Engineers, Pvt. Ltd. Jaipur, India. All other chemicals used in this study were of analytical grade and used without further purification.

3.2.2   SYNTHESIS OF GO

GO was produced from natural graphite according to the modified Hummers method.[26] In a typical synthesis process, graphite powder (5 g) and NaNO$_3$ (5 g) were mixed with 125 mL concentrated H$_2$SO$_4$ into a 1000 mL round bottom flask. The mixture was kept in an ice bath for 4 h, with stirring, at temperature 5°C. Afterwards, KMnO$_4$ (9 g) was gradually added with constant stirring, and the rate of addition was controlled carefully to avoid sudden increase in temperature. The ice bath was then removed and the mixture temperature was maintained at 40°C and stirred for 2 h until it became pasty brownish. Subsequently, 250 mL deionized water was added for dilution. Finally, the mixture was stirred at 98°C for about 1 h until the color changed from brown to yellow. Further, this solution was diluted by adding 500 mL deionized water. After 1 h, 30 mL of H$_2$O$_2$ (30 wt%) solution was added to the mixture to reduce the residual KMnO$_4$. Finally, the solution was then filtered and washed with deionized water several times until pH was 7 and dried at 60°C under vacuum for 12 h to obtain GO powder.

3.2.3   FABRICATION OF PVC/GO NANOCOMPOSITES

Colloidal blending method was used to fabricate PVC/GO nanocomposites. A protocol of synthesis is provided in Figure 3.1. PVC solution was prepared by dissolving in THF at 70°C and the solution was further stirred for 2 h

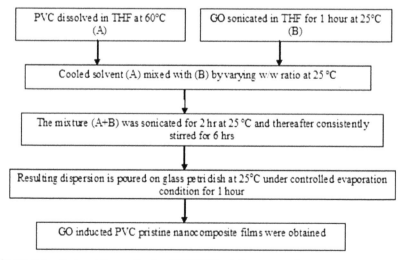

FIGURE 3.1   Protocol for synthesis of PVC/GO pristine composites.

on magnetic stirrer. GO powder was separately dispersed in the THF and sonicated for 90 min at room temperature before introducing to PVC solution. The mixture was further sonicated for 1 h. The resulting homogeneous dispersion was casted on glass petri dish and kept in an oven at 60°C for slow evaporation of the solvent to get pristine PVC/GO composite films. The films were peeled off from the glass plate and used for further study. The thickness of the films can be controlled by the content of the solutions. Figure 3.2 shows the photographs of different compositions of PVC/GO composite solutions in THF, indicating formation of homogenous dispersion. The percentage loading of GO in PVC/GO composite varied from 0.5 wt% to 2.5 wt%.

FIGURE 3.2 Photographs showing dispersion of PVC/GO in THF after sonication. (a) Pure PVC solution, (b) 0.5, (c) 1, (d) 1.5, (e) 2, and (f) 2.5 wt% GO.

3.2.4 CHARACTERIZATIONS

The morphology of GO-reinforced PVC composites was investigated by Carl Zeiss A ×10 vision LE optical microscope. All measurements were carried out at ambient temperature of 25°C.

The impedance measurements were carried out using a Newton's 4th impedance analyzer (N4L, 50 Hz–35 MHz, UK). The samples were tested as a function of frequency and temperature. The samples were heated up to 150°C with an accuracy of about ± 0.1°C by using temperature control system.

3.3 RESULTS AND DISCUSSION

3.3.1 MORPHOLOGY OF PVC/GO COMPOSITES

The purpose of optical microscopy is to understand process control; failure analysis; quantitative and qualitative microstructure analysis; shape, size, and number of reinforcement particles; fiber and resin volume fractions; tears and cracks on fracture surface; layer thickness; void content; and aggregation/clustering of dopant.[27–30] The uniform dispersion of 1% functional GO reveals its presence in the PVC system shown in Figure 3.3(a, b) for 100× and 200× resolution. Furthermore, the increase in reinforced percentage of GO demonstrate the occupied PVC system, which improves the surface morphology shown in Figure 3.3(c, d) at different resolutions. The surface adhesion is basic requirement for various surface science applications. The tuning of surface adhesion is practically possible by functional GO.[13] The GO functional composites showed improvement with respect to morphology and thermal, electrical, mechanical, and optical properties,

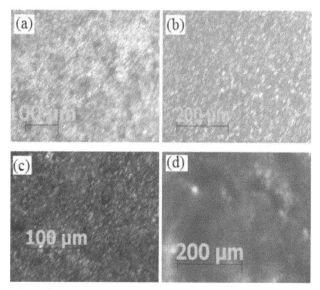

FIGURE 3.3   Optical microscope image of (a) 100× and (b) 200× resolution for PVC + 1% GO nanocomposites and (c) 100× and (d) 200× resolution for PVC + 2% GO nanocomposites.

which are tabulated in Table 3.1. This is because of GO acting as a functional chemical hook with polymer moieties at nanoscale.[27–30] This investigation of varying reinforcement of functional GO with polymer system PVC is crucial for various applications such as optical filters, surface science, packaging industry, mass-level water transportation in civil engineering, and upcoming composite as novel synthetic metal.

TABLE 3.1   Property Improvement of Virgin Polymers Inducting Functional GO.[9, 18–21]

Effect of functional GO composites on properties	Morphology	Thermal	Electrical	Mechanical	Surface adhesion
	↑	↑	↑	↑	↑
	Microporisty increase				

3.3.2   MORPHOLOGY OF PVC/GO NANOCOMPOSITES

The degree of exfoliation of GO was studied by TEM and it was observed that GO was well dispersed in PVC matrix. The morphology of the GO-reinforced PVC composites was studied and the results are already published in our previous article.[14] It was observed that GO was homogeneously dispersed in PVC matrix without any evidence of large agglomeration. It was also observed that increasing GO content in PVC matrix resulted in microporous morphology.

3.3.3   THERMOGRAVIMETRIC ANALYSIS (TGA) OF PVC/GO NANOCOMPOSITES

We studied thermal stability of PVC/GO composites, and the results are already documented.[14] It was observed that the thermal stability of composites has significantly improved when GO was incorporated into PVC matrix, indicating strong interaction between PVC and GO.

3.3.4   MECHANICAL PROPERTIES OF PVC/GO NANOCOMPOSITES

The mechanical properties of PVC/GO composites were also studied and it was observed that the mechanical properties of composites were significantly improved as a function of GO loading. The tensile strength and Young's Modulus of the composite films were increased whereas (%) elongation was decreased by 32.15% as a function of GO loading. The improvement in mechanical properties is may be due to homogeneity of the composites and strong interfacial interaction between GO and PVC matrix.[14]

3.3.5   DIELECTRIC PROPERTIES OF PVC/GO NANOCOMPOSITES

The dielectric properties of PVC/GO composites were studied to evaluate the influence of GO loading on dielectric constant and dielectric loss (tanδ) of composites. The dielectric constant of PVC/GO composite films decreases with increasing frequency up to 1 MHz at all the temperatures investigated. These results are published in our previous report.[14] Such decrease in dielectric constant with increase in frequency can be attributed to the increase in interfacial polarization. The dielectric constant remains nearly a constant in the frequency range from 1 MHz to 35 MHz, which may be due decrease in polarization which abruptly decreases the dielectric constant. This shows an excellent frequency stability of the composites. The dielectric loss (tanδ) of composite films decreases with increase in frequency up to 1 kHz. From 1 kHz to 10 MHz, the values of dielectric loss were almost constant at all temperatures investigated. With further increase in frequency up to 20 MHz, the dielectric loss values increase and further decrease, when frequency increases to 35 MHz. Thus, PVC/GO composites show high dielectric constant and low dielectric loss. Such materials can be useful for fabrication of flexible electronic devices.

3.3.6   IMPEDANCE ANALYSIS OF PVC/GO COMPOSITES

The impedance measurement as function of temperature was done at low frequency of 50 Hz. Figure 3.4 shows the effect of GO loading on impedance of PVC/GO nanocomposites. The GO loading from 0.5% to 2% demonstrates the decrease in impedance. Further increase in loading offers higher impedance of 20 MΩ. For virgin and GO-loaded PVC show decreasing trend after 90°C, it indicates the effect of the glass transition temperature.[31] The optimization of impedance is crucial for utilization of this composite for several electronic and electrical applications.[32] The temperature stability of these composites is crucial for further development of these materials for suitable application.

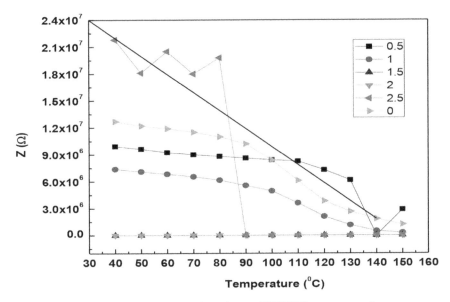

FIGURE 3.4   Effect of GO loading on impedance of PVC/GO nanocomposites.

3.4   CONCLUSION

In the present investigation, protocol for synthesis of GO-reinforced PVC composites was developed and optimized. The physical characterizations of uniform dispersion of GO with PVC was verified by using optical microscopy. The functional GO mixed and entangle with polymer PVC demonstrated. The varying optical properties, as function of GO loading, suit for tailoring desired dielectric medium. Furthermore, the studies of structural, thermal, and electrical, properties were reported for suitable applications as high-performance nanocomposite.

ACKNOWLEDGMENT

The authors are highly thankful to Naval Research Board (NRB), DRDO, New Delhi for financial support to this work through the Project No. 259/Mat./11-12.

KEYWORDS

- graphene oxide
- polyvinyl chloride
- nanocomposites
- tetrahydrofuran
- polymer

REFERENCES

1. Stankovich, S.; Piner, R. D.; Nguyen, S. T.; Ruoff, R. S. *Carbon.* **2006**, *44*, 3342–3347.
2. Bourlinos, A. B.; Gournis, D.; Petridis, D.; Szabo, T.; Szeri, A.; Dekany, I. *Langmuir.* **2003**, *19*, 6050.
3. Scholz, W.; Boehm, H. P. Z. *Anorg. Allg. Chem.* **1969**, *369*, 327.
4. Compton, O. C.; Nguyen, S. T. *Small.* **2010**, *6*, 711.
5. Vadukumpully, S.; Paul, J.; Mahanta, N.; Valiyaveettil, S. *Carbon.* **2011**, *49*, 198.
6. Achaby, M. E.; Arrakhiz, F. Z.; Vaudreuil, S.; Essassi, E. M.; Qaiss A. *Appl. Surf. Sci.* **2012**, *258*, 7668.
7. Bonderer, L. J.; Studart, A. R.; Gauckler, L. J. *Science.* **2008**, *319*, 1069.
8. Kim, H.; Abdala, A. A.; Macosko, C. W. *Macromolecules.* **2010**, *43*, 6515.
9. Sydlik, S. A. *J. Polym. Sci. Part B. Polym. Phys.* **2013**, *51*, 997.
10. Wang, J. Y.; Yang, S. Y.; Huang, Y. L.; Tien, H. W.; Chin, W. K.; Ma, C. C. *J. Mater. Chem.* **2011**, *21*, 13569.
11. Wu, J.; Tang, Q.; Sun, H.; Lin, J.; Ao, H.; Huang, M.; Huang, Y. *Langmuir.* **2008**, *24*, 4800.
12. Wei, H. G.; Zhu, J. H.; Wu, S. J.; Wei, S. Y.; Guo, Z. H. *Polymer.* **2013**, *54*, 1820.
13. Deshmukh, K.; Khatake, S. M.; Joshi, G. M. *J. Polym. Res.* **2013**, *20*, 286.
14. Deshmukh, K.; Joshi, G. M. *Polym. Test.* **2014**, *34*, 211.
15. Joshi, G. M.; Deshmukh, K. *J. Electron. Mater.* **2014**, *43*, 1161.
16. Deshmukh, K.; Joshi, G. M. *RSC Advances.* **2014**, *4*, 37954.
17. Rusen, E.; Marculescue, B.; Butac, L.; Preda, N.; Mihut, L. *Fuller. Nanotub. Carbon Nanostructures.* **2008**, *16*, 178.
18. Jaeton G. A.; Cai, M.; Overdeep, K. R.; Kranbuehl, D. E.; Schniepp, H. C. *Macromolecules.* **2011**, *44*, 9821.
19. Kulthe, M. G.; Goyal, R. K. *Adv. Mat. Lett.* **2012**, *3*, 246.
20. Tikka, H. K.; Suvanto, M.; Pakkanen, T. A. *J. Coll. Inter. Sci.* **2004**, *273*, 388.
21. Jung, K. E.; Suk, C. J. *J. Nanosci. Nanotech.* **2012**, *12*, 5820.
22. Shen, H.; Zhang, L.; Liu, M.; Zhang, Z. *IVYS, Int. Pub.* **2012**, *2*, 283.
23. Huang, S.; Ren, L.; Guo, J.; Zhu H.; Zhang C.; Liu, T. *Carbon.* **2012**, *50*, 21 6.
24. Mukhopadhyay, P.; Gupta, R. K. *Plast. Engg.* **2012**, *1*, 20132.
25. Kim, F.; Cote, L. J.; Huang, J. *Adv. Mater.* **2010**, *22*, 1954.

26. Hummers, W. S. J.; Offman, R. E. *J. Am. Chem. Soc.* **1958,** *80*, 1339.
27. Pham, V. H.; Dang, T. T.; Hur, S. H.; Kim, E. J.;Chung, J. S. *J. Nanosci. Nanotechnol.* **2012,** *12*, 5820.
28. Gao, W.; Huang, R. J. *Phys. D: Appl. Phys.* **2011,** *44*, 452001.
29. Zeiss, C. Germany http://microscopy.zeiss.com/microscopy/en_us/home.html.
30. Kuilla, T.; Bhadra, S.; Yao, D.; Kim, N. H.; Bose, S.; Lee, J. H. *Prog. Polym. Sci.* **2010,** *35*, 1350.
31. Joshi, G. M.; Pawde, S. M. *J. Appl. Polym. Sci.* **2006,** *102*, 1014.
32. Joshi, G. M.; Khatake, S. M.; Kaleemulla, S.; Rao, N. M.; Cuberes, T. *Curr. Appl. Phys.* **2011,** *11*, 1322.

CHAPTER 4

STUDY OF MORPHOLOGY AND ELECTRICAL PROPERTIES OF PURE AND HYBRID POLYMER COMPOSITES

KALIM DESHMUKH[1], GIRISH M. JOSHI[1*], AKSHATH SHARMA[1], SAMPANN ARORA[1], ROHAN TIBRAWALA[1], S. KALAINATHAN[2], and R. R. DESHMUKH[3]

[1]Polymer Nanocomposite Laboratory, Materials Physics Division, School of Advanced Sciences, VIT University, Vellore 632014, Tamil Nadu, India

[2]Crystal Research Centre, VIT University, Vellore 632014, Tamil Nadu, India

[3]Departmentof Physics, Institute of Chemical Technology (ICT), Matunga, Mumbai 400018, India

*Corresponding author. E-mail: gm_joshi@rediffmail.com

CONTENTS

ABSTRACT

Polyvinyl alcohol (PVA) is a water-soluble polymer that has been studied extensively because of several interesting physical properties and vast applications in various domains. In the present study of solution casting, pure and hybrid composites were obtained by PVA/conducting carbon black (CCB) and PVA/CCB + montmorillonite (MMT) clay. The morphology of these composites was studied by optical polarizing microscope (OPM) and scanning electron microscope (SEM). The optical microscopy studies reveal the phase transformation of crystalline to amorphous, which is confirmed by SEM. We demonstrated the impact of fillers on improved electrical properties of hybrid composites compared with pure composites. The effect on increased impedance magnitude is $\geq 57\%$, dielectric constant $\geq 50\%$, and reverse current phase (θ) response of signals. Hence, these composites may serve as a remedy for various industrial applications.

4.1   INTRODUCTION

Polyvinyl alcohol (PVA) has been used for more than four decades because of its unique chemical and physical properties. The important features of PVA are the presence of crystalline and amorphous regions and its physical properties which result from the crystalline–amorphous interfacial effect.[1] Many polymers have two coexistent phases: crystalline and amorphous. When such polymers are blended or mixed with suitable entity to form a composite, they may interact either in the amorphous region or in the crystalline region of the polymers and in both cases, the polymeric properties will be altered. By disclosing the fact of loading filler, the desired performance properties may be achieved to overcome the application objectives.[2,6]

MMT is a clay mineral containing stacked silicate sheets measuring ~10 Å in thickness and ~2200 Å in length.[7] The stacked silicate sheets provide improved thermal stability, mechanical strength, fire retardant, and molecular barrier properties. MMT has high swelling capacity, which is significant for efficient intercalation of polymer.[7] Polymer–clay nanocomposites have attracted great academic and technological interest due to improvement in their mechanical, thermal, gas permeable, electrical, optical, and biological properties as compared to virgin polymers.[8–11] Most of these properties of polymer–clay nanocomposites are widely controlled by adjusting compositions, nanophase size and the H-bond interaction (organic material), and clay (inorganic material) phases in different preparation routes which result in intercalation and exfoliation of the clay.[12]

Polymer composites based on electrically conductive fillers such as carbon black, carbon nanotubes, graphite exhibit many interesting features due to their resistivity change with mechanical, thermal, electrical, or chemical solicitations.[13–16] The properties of conducting carbon black (CCB)-filled polymer composites usually depend on the features of both polymers (such as degree of crystallinity, melting viscosity, and surface tension) and CCB (surface area, chemical groups, etc.).[17] The distribution of CCB in polymer matrix and the interaction between CCB and polymer significantly affect many properties such as percolation threshold of polymer composites to a great extent.[18]

In the present study, we disclosed the morphological and electrical properties of PVA composites and phase transformation due to CCB and mixed CCB + MMT.

4.2   EXPERIMENTAL

4.2.1   MATERIALS

PVA powder of grade RS-2117 (Lot No. 484649) was made by Kurray Exceval, Japan and supplied by Associated Agencies, Mumbai. CCB powder (N991 Thermax) supplied by Sepulcher Brothers Pvt. Ltd. Chennai, India. Clay MMT halloysite was procured from Imerys, New Zealand.

4.2.2   PROTOCOL TO OBTAIN PURE AND HYBRID COMPOSITES

By varying different weighing proportions of PVA as a host polymer and CCB, MMT as a filler, composites were prepared using solution casting. The total weight of PVA + filler was taken as 2 g and weight of PVA and filler was adjusted according to the proportions by its weight. PVA was dissolved in 40 mL water by heating at 80°C. MMT clay powder was initially suspended in deionized water in separate glass beaker and stirred vigorously and (then sonicated for 15 min) to get clay hydrocolloid. Afterwards, PVA solution was mixed with MMT hydrocolloid and stirred for 1 h Furthermore, CCB was added to the mixture of PVA + MMT and stirred for additional 2 h The resulting solution was then cast into galvanized iron plates of 15 cm × 15 cm size. These plates were then kept in a hot air oven at 45°C for a period of 24 h for evaporation of the solvent. The dry composite films were then peeled off for further characterization.

4.2.3 CHARACTERIZATION

OPM was used to record the phase morphology of polymer composites by using Carl Zeiss A ×10 vision LE optical microscope. All the measurements were carried out at 25°C.

The Quanta 200 FEG scanning electron microscope (SEM) of versatile high resolution was used to understand polymer composites morphology.

The AC conductivity of polymer composites was carried out by using N-4L, PSM 1735, Impedance Analyzer, UK, under the frequency of 20–20 MHz at room temperature 25°C.

4.3   RESULTS AND DISCUSSION

4.3.1   OPTICAL MICROSCOPY OF PURE AND HYBRID COMPOSITES

Optical images shown in Figure 4.1(a–f) represent the morphology of PVA + CCB composites, indicating the presence of CCB in the PVA system. From Figure 4.1(g–l) which shows the optical images of PVA + CCB + MMT hybrid composites clearly indicate that the dispersion was achieved in comparison to PVA + CCB composites. The white impressions in the form of hexagon impression are clear evidence of crystalline to amorphous phase transformation. Optical microscopy is crucial in polymer composite due to inexpensive and fast method for batch quality control of particulate products before further manufacturing based on filler transparency and refractive index.[19] Optical microscopy is also useful to identify the effect of reinforcing entities such as epoxy with glass fibers.[20] The brightness observed in the images (Fig. 4.1(i, j)) may be due to presence of MMT, which minimizes the effect of PVA and CCB in the hybrid composites. It is well understood that the optical microscopy completely reveals the phase transition of virgin polymer from crystalline to amorphous phase due to CCB and CCB + MMT fillers.

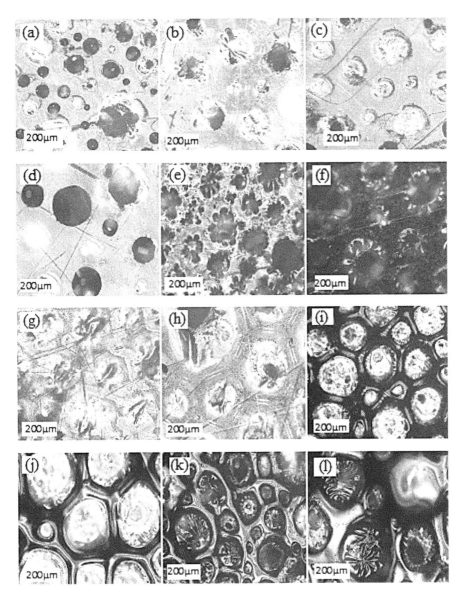

FIGURE 4.1   Optical microscopy images of PVA + CCB + MMT composites. (a) CCB 0.5% (5×), (b) CCB 0.5 wt% (10×), (c) CCB 1 wt% (5×), (d) CCB 1 wt% (10×), (e) CCB 3 wt% (5×), (f) CCB 3 wt% (10×), (g) 0.5 wt% CCB + 2.5% MMT (5×), (h) 0.5 wt% CCB + 2.5% MMT (10×), (i) 1.5 wt% CCB + 1.5wt% MMT (5×), (j) 1.5 wt% CCB + 1.5wt% MMT (10×), (k) 1 wt% CCB + 2 wt% MMT (5×), and (l) wt% CCB + 2 wt% MMT (10×).

4.3.2 MORPHOLOGY OF PURE AND HYBRID COMPOSITES

The morphological investigation was carried out to identify the phase and aggregation of single and mixed nano entity in polymer system. The SEM images of 3% CCB and CCB + MMT mixed (1.5:1.5 weight percent ratio of each) were recorded for 400× and 200 μm resolution and depicted in Figure 4.2(a, b). The phase separation was seen due to the presence of filler and the effect was more dominating for the mixed loading of nano entity. However, the modified morphology of PVA-inducted CCB and CCB clay improved the electrical properties. Nano entity CCB occupied PVA network and the varying resolution image confirmed that filler aggregates may be due to association with PVA. The CCB + MMT-incorporated PVA composites show bright and bulky aggregation and more smooth morphology at 40× resolution. Phase conversion of this pristine composites exhibit transformation of phase from crystalline to amorphous nature by destroying crystalline orientation which is also verified by XRD results.[21]

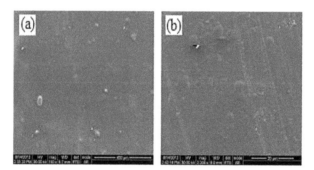

FIGURE 4.2 SEM images of (a) 3% CCB and (b) 1.5 wt% CCB + 1.5 wt% MMT-inducted PVA composites.

4.3.3 ELECTRICAL PROPERTIES OF PURE AND HYBRID COMPOSITES

All the electrical measurements were carried out at 27°C by using an impedance analyzer. Impedance of PVA + CCB and PVA + CCB + MMT hybrid composites as a function of CCB and MMT loading is shown in Figure 4.3(a and b). For PVA + CCB composites, the impedance increases up to 1 wt% of CCB loading and decreases with further increase in CCB wt%. However, for PVA + CCB + MMT hybrid composites, there is a linear increase in the impedance value by 57% with respect to MMT percent loading. Figure 4.3(c and d) shows the values of dielectric constant as a function of CCB and MMT loading. There is an improvement in dielectric constant values in PVA + CCB + MMT hybrid composites by 50% as compared with PVA + CCB composites which may be due to crystalline to amorphous phase transformation of hybrid composites. The plot of phase angle (θ) as a function of CCB and MMT percent loading is given in Figure 4.3(e and f), which shows reverse current phase (θ) response of signals in PVA + CCB and PVA + CCB + MMT composites. The optimized properties of virgin versus hybrid composites are more important to biomedical devices as well as for various electronic gazettes in the form of lightweight potentiometer.

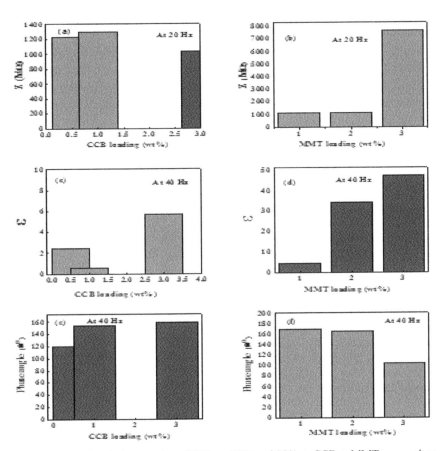

FIGURE 4.3 Electrical properties of PVA + CCB and PVA + CCB + MMT composites: (a) impedance of PVA + CCB composite, (b) impedance of PVA + CCB + MMT composite, (c) dielectric constant of PVA + CCB composite, (d) dielectric constant of PVA + CCB + MMT, (e) phase angle of PVA + CCB composite, and (f) phase angle of PVA + CCB + MMT composites.

4.4 CONCLUSIONS

In the present investigation, synthetic composite was prepared by solution casting. The role of filler in view of phase transformation from crystalline to amorphous nature morphology was identified by both optical and SEM microscopes. The magnitude of electrical properties was optimized in the range of increased impedance (Z) by $\geq 57\%$, dielectric constant (ε) by $\geq 50\%$, and reverse current phase (θ) response of signals due to phase transition

ACKNOWLEDGMENTS

Author greatly thank to NRB, DRDO, for providing the electrical measurement facility under the project No.259/Mat/11-12. The authors are highly thankful to Sophisticated Analytical Instrumental Facility (SAIF) at Indian Institute of Technology (IIT), Madras, India for providing scanning electron microscope facility. Authors also wish to thank Dean of School of Advanced Science at VIT University, Vellore, India, for providing in house facilities under summer research scheme.

KEYWORDS

- **hybrid composites**
- **conducting carbon black**
- **polyvinyl alcohol**
- **montmorillonite**
- **optical microscopy**

REFERENCES

1. Pawde, S. M.; Deshmukh, K.; Parab, S. *J. App. Poly. Sci.* **2008,** *109,* 1328.
2. Zidan, H. M. *J. Appl. Polym. Sci.* **2003,** *88,* 104–111.
3. Lobo, B.; Ranganath, M. R.; Ravi Chandran, T. S. G.; Venugopal Rao, G.; Ravindrachary, V.; Gopal, S. *Phys. Rev. B* **1999,** *59,* 13693.
4. Joshi, G. M.; Khatake, S. M.; Kaleemulla, S.; Rao, N. M.; Teresa Cuberes, *Curr. App. Phy.* **2011,** *11*(6),1322.
5. Joshi, G. M.; Cuberes, M. T. *Ionics.* **2013,** *19*(6), 947.

6. Joshi, G. M.; Deshmukh, K. *Ionics*. **2014,** *20*(4), 529.

7. Yano, K.; Usuki, A.; Okada, A. *J. Polym. Sci: Poly. Chem. Ed.* **1997,** *35*, 2289.

8. Yuan-Hsiang, Y.; Ching-Yi, L.; Jui-Ming, Y.; Wei-Hsiang, L. *Polymer*. **2003,** *44*, 3553.

9. Pinnavaia, T. J.; Beall, G. W. *Polymer-Clay Nanocomposites;* Wiley: Chichester, 2000.

10. Tanaka, T.; Montanari, G. C.; Mulhaupt, R. *IEEE Trans. Dielectr. Elec. Insul.* **2004,** *11*, 763.

11. Yeh, J. M.; Huang, H. Y.; Chen, C. L.; Su, W. F.; Yu, Y. H. *Surf. Coat. Tech.* **2006,** *200*, 2753.

12. Sengwa, R. J.; Sankhla, S.; Choudhary, S. *Ind. J. Pure Appl. Phys.* **2010,** *48*, 196.

13. Thongruang, W.; Spontak, R. J.; Balik, C. M. *Polymer*. **2002,** *43*, 2279.

14. Di, W.; Zhang, G. J. *J. Mater. Sci.* **2004,** *39*, 695.

15. Schueler, R.; Petermann, J.; Schulte, K.; Wentzel, H. P. *J. Appl. Polym. Sci.* **1997,** *63*, 1741.

16. Segal, E.; Tchoudakov, R.; Narkis, M.; Siegmann, A. *J. Mater. Sci.* **2004,** *39*, 5673.

17. Breuer, O.; Tchoudakov, R.; Narkis, M.; Siegmann, A. *J. Appl. Polym. Sci.* **1997,** *64*, 1097.

18. Dai, K.; Xu, X. B.; Li, Z. M. *Polymer*. **2007,** *48*, 849.

19. George, S.; Thomas, S. M. *Comp. Sci. Tech.* **2009,** *69*, 1298.

20. Dunkers, J. P.; Parnas, R. S.; Zimba, C. G.; Peterson, R. C.; Flynn, K. M.; James, G. F.; Bouma, B. E. *Composites A.* **1999,** *30*, 139.

21. Joshi, G. M.; Sharma, A.; Tibrawala, R.; Arora, S.; Deshmukh, K.; Kalainathan, S.; Deshmukh, R. R. *Poly. Plastic Tech. Engg.* **2014,** *53*(6), 588.

CHAPTER 5

NUMERICAL MODELING OF BUCKLING PHENOMENON IN CARBON NANOTUBES FILLED WITH ZnS

PAULO CACHIM[1*], ANDRÉ MONTEIRO[2], PEDRO COSTA[3], and DAVID HOLEC[4]

[1]*Department of Civil Engineering and LABEST, University of Aveiro, 3810193 Aveiro, Portugal*

[2]*Department of Civil Engineering and CICECO, University of Aveiro, 3810193 Aveiro, Portugal*

[3]*Department of Materials Engineering and CICECO, University of Aveiro, 3810193 Aveiro, Portugal*

[4]*Department of Physical Metallurgy and Materials Testing, Montanuniversität Leoben, 8700 Leoben, Austria*

*Corresponding author. E-mail: pcachim@ua.pt

CONTENTS

ABSTRACT

Computational simulation study of ZnS filled CNT has been performed to evaluate, understand, and compare the ability of both numerical tools to simulate the mechanical behavior of filled carbon nanotubes under uniaxial compression. Two different modeling tools are being used. MD simulations demonstrated that ZnS filling has shown to affect, rather weakly, the axial rigidity of the nanotubes. ZnS increases the stability of the system, leading to higher buckling forces and strain. MD simulations have also demonstrated the tubes' Euler's behavior. FEA simulations were also performed based on real Zn 0.92 Ga 0.08 S@SWCNT experiments. The real geometry and loading conditions were reproduced by CAD and FEA tools, constituting the first study that reproduces such mechanical experiments with computational approaches. Results confirm the applicability of continuum mechanics to such nanostructures thus precluding the need of expensive computational tools, particularly for large MWCNTs systems.

5.1   INTRODUCTION

Filling the cavities of carbon nanotube scan generate new hybrid materials, leading to an improvement on their already outstanding physical and chemical properties. The first record dates back to 1993,[1] when the encapsulation of multiwall carbon nanotubes (MWCNT) with lead oxide using capillary methods was reported. The prospects for applications of these core-shell nanostructures are very ambitious, e.g., drug delivery,[2,3] storage of sensitive materials,[4] and nanopipetting.[5,6] Therefore, the importance of the knowledge regarding their mechanical properties and response to applied forces, as the buckling phenomenon, is understandable.[7] Due to the considerable heterogeneity in structural and geometric parameters of the produced samples, bulk analysis of CNTs leads to averaged mechanical responses. Therefore, it is essential to evaluate their individual mechanical behavior.

Two main computational approaches have been used: molecular dynamics (MD) and finite-element analysis (FEA). Each one of the approaches is based on different concepts. Nevertheless, both approaches have been used in materials science, presenting advantages and disadvantages in terms of system size and simulation time. On one hand, MD was developed with the aim to study up to nanoscale systems, considering atoms as the base particles, being possible to consider over 100 billion of them. This, however, involves long calculation times and high computational capacity. The interaction between atoms is defined by interatomic potentials, which simulate the force fields between atoms.[8] On the other hand, FEA is, in simple terms, a method that turns complicated continuous engineering problems into small elements that can be solved in relation to each other by solving a set of algebraic equations.[8] FEA models have the advantage to be faster than MD models. Nevertheless, its applicability to such nanostructures is under development since this method was initially developed for continuum-medium problems.

There exist numerous computational works performed on empty CNTs. In contrast, computational modeling of filled CNTs is less explored. While simulation of filled CNTs through MD has been already reported in litera-ture, the use of FEA for filled CNTs has not yet been mentioned. In this communication, modeling results of the mechanics of hollow and hybrid CNTs will be presented. Both numerical tools were used (MD and FEA); MD analysis was based on theoretical systems, while FEA was based on existing Transmission Electron Microscope (TEM) experiments.

5.2 CLASSICAL MOLECULAR DYNAMICS

MD calculation proceeds by the integration of Newton's equation of motion over time to obtain the time evolution of an atomistic system, using provided interatomic potentials,[9] which approximate the energetic interaction between atoms as a function of relative distances. The advantage of MD vs. FEA is the possibility of studying materials behavior with atomic resolution, espe-cially when the simulations concern filled CNTs, where the interactions between both components have to be studied. Additionally, MD does not require previous knowledge and definition of mechanical properties of the materials such as Young's modulus, Poisson ratio, or yield/crack stress or strain. Nevertheless, simulations of filled carbon nanotubes with diameters larger than 5 nm leads to a large number of atoms, which makes the simula-tions extremely time consuming.

Classical MD simulations were carried out with the aim of reproducing the mechanical behavior of ZnS nanowires filled in Single Wall Carbon Nano Tubes (SWNT) (ZnS@SWCNT), under axial compressive loading.[10] We used LAMMPS MD package, an open source MD code. In order to study the system response in function of several L/d ratios, filled and hollow zigzag SWCNTs were modelled with diameters, d, ranging from 1 nm (13, 0) to 5 nm (64, 0) and lengths, L, varying between 10 and 50 nm, yielding in total 50 models.

The atomic interaction between C–C atoms was modelled by the Tersoff potential.[9] The forces acting between the atoms of the ZnS nanowire were also described by the Tersoff potential, adopting the parameterization defined in[11.] This pair potential allowed modeling the inserted nanowire as a homogeneous crystal, replacing the Zn and S by single-type atom and adopting a diamond structure with a lattice parameter of 5.42 Å. The C–ZnS interaction was described using the Lennard-Jones potential (L-J)[12] which provides an approximate description of the van der Waals interactions. Since no previous studies about L-J potential parameterizations for C–ZnS interactions were found, the Lorentz–Berthelot mixing rules[13,14] were used to determine the vdW interaction parameters ($\varepsilon_{C\text{-}ZnS}$ and $\sigma_{C\text{-}ZnS}$). These rules are widely used and can be applied if the L-J parameters ε and σ[12] are known for C–C and ZnS–ZnS interactions.[15, 16] All the adopted Tersoff and L-J parameters can be found in the supporting work developed by Monteiro et al.[10]

The uniaxial compression of the CNTs was simulated creating a group of fixed atoms in both tips, and moving one of them toward the other along the tube axis. The loading was applied at a rate of 0.001 nm/ps, maintaining a constant temperature of 0.1 K. The compression force is determined through the sum of the axial reaction force of each fixed atom and calculating the average value between both tips. Before the application of the compression, the CNTs were relaxed by running a 2 ps simulation. Figure 5.1a shows an example of the typical behavior found for axial force in function of the tube strain concerning both empty and filled (26, 0) CNT with 20 nm long. Figures 5.1b and 5.1c show the results of buckling forces for all the models tested. A predictable convergence for large diameters was found since the buckling force ceases to be dependent of the tube length, as described by Euler theory.[17]

An initial elastic response followed by an abrupt collapse was noticed in both empty and filled CNT, which means that the axial force increases linearly with the strain until it reaches the buckling force and the respective critical strain. Such behavior can be observed in Figure 5.2, where four different simulations steps are shown: the initial configuration, just before the buckling, immediately after, and at the end of the simulation ($\varepsilon = 0.1$). The slopes of both curves (Fig. 5.1) are similar; however, the critical strain is higher in ZnS (26, 0). Such fact allows concluding that the ZnS filling does not change, significantly, the axial stiffness of CNTs; nevertheless, an increase in the critical buckling force and strain is noticeable for filled tubes, which is directly related to the stability increase provided by the filling, visible when comparing the deformations in steps 3 and 4 of Figure 5.2.

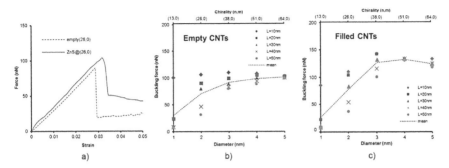

FIGURE 5.1 (a) Axial force vs. strain of empty and ZnS filled (26, 0) SWCNT under axial compression at T = 0.1 K., (b) and (c) represent the buckling force vs. diameter for all SWCNT lengths of empty and filled nanotubes, respectively.[10] (Reprinted with permission from Monteiro, A. O.; Costa, P. M. F. J.; Cachim, P. B.; Holec, D. Buckling of Zns-Filled Single-Walled Carbon Nanotubes-The Influence of Aspect Ratio. *Carbon* **2014**, *79*(1), 529–537. © 2014 Elsevier.)

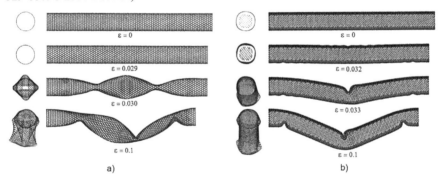

FIGURE 5.2 Snapshots of the deformation process at the critical instants of (a) hollow SWCNT and (b) ZnS@SWCNT. From the top to the bottom: initial state, immediately before buckling, thereupon the buckling, and the deformation at ε = 0.1.[10] (Reprinted with permission from Monteiro, A. O.; Costa, P. M. F. J.; Cachim, P. B.; Holec, D. Buckling of Zns-Filled Single-Walled Carbon Nanotubes-The Influence of Aspect Ratio. *Carbon* **2014**, *79*(1), 529–537. © 2014 Elsevier.)

5.3 FINITE-ELEMENT ANALYSIS

The FEA study was performed with ABAQUS and consisted of the repro-duction of an uniaxial compressive experiment on a discrete SWCNT filled with wurtzite-type gallium doped zinc sulfide ($Zn_{0.92}Ga_{0.08}S@SWCNT$), performed in a TEM for two different situations: one with a continuous ZnS core (Fig. 5.3a) and another with a discontinuous core created by the appli-cation of an electrical pulse (Fig. 5.4a).[18–20] Modeling consisted of a combi-nation of shell elements, for the carbon layers, and solid elements, for the inner filling. Both fully and partially filled models were built conforming to the real structures, using TEM two-dimensional images and CAD software (Figs. 5.3b and 5.4b). This tool allowed to measure an average diameter of 90 nm, a length of 2242 nm and a multiwall carbon layer thickness of 6.5 nm, leading to average cross-sectional areas of 21,904 and 3543 nm², for the doped ZnS filling and the turbostratic carbon coating, respectively. Combining these values with the experimental $F\text{--}d$ data, the estimation of the Young's modulus of both materials was possible. Hooke's law was applied to determine the Young modulus of carbon layers using the data of partially filled CNT experiments. With the determined carbon Young modulus and the $F\text{--}d$ data from filled CNT experiments, mechanics laws from hybrid bars theory were applied to estimate the elastic modulus of the semiconductor filling.

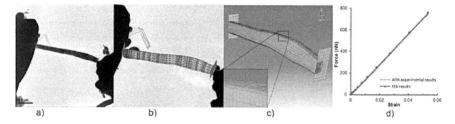

FIGURE 5.3   Filled CNT modeling: (a) TEM snapshot at the initial state, (b) CAD model, (c) longitudinal view of ABAQUS model, and (d) $F\text{--}d$ curves obtained.

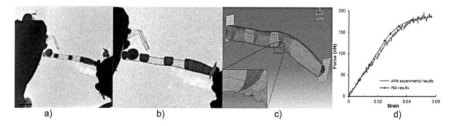

FIGURE 5.4   Partially filled CNT modeling: (a) TEM snapshot at the initial state, (b) CAD model, (c) longitudinal view of ABAQUS model, and (d) $F\text{--}d$ curves obtained.

Figures 5.3c and 5.4c show longitudinal views of the models obtained in ABAQUS by Computer Aided Design (CAD) sketches. Rigid plates were included simulating the clamped tip and the force sensor. In order to reproduce the turbostratic texture of the carbon shell,[11] randomly distributed imperfections were introduced in the models (inset of Figs. 5.3c and 5.4c).

The applied loads in the FEA simulation corresponded to the displacements measured with the in situ TEM experiments. As it is well known that the two-dimensional TEM images do not allow checking perspective errors, which can influence the mechanical analysis. Assuming a carbon thickness of 6.5 nm, the optimized modulus of elasticity of the materials were for E_{ZnS} the value of 0.68 GPa and for the carbon the value $E_{Carbon} = 1.28$ GPa. Figures 5.3d and 5.4d indicate that both filled and partially filled structures fit well the experimental curves.

5.4   CONCLUSIONS

In summary, computational simulations of ZnS filled CNT have been performed, using two different modeling tools, in order to evaluate, understand and compare the ability of both numerical tools to simulate the mechanical behavior of filled carbon nanotubes under uniaxial compression. On one hand, MD simulations of ZnS@SWCNT demonstrated that ZnS filling shown to affect, rather weakly, the axial rigidity of the nanotubes; however, it increases the stability of the system, leading to higher buckling forces and strain. MD simulations have also demonstrated the tubes Euler's behavior and revealed some effects that must be taken into account in future FEA simulations, such as the interaction distances between several materials, and the contact conditions between both components. On the other hand, FEA simulations were performed based on real Zn 0.92 Ga 0.08 S@SWCNT experiments. The real geometry and loading conditions were reproduced by CAD and FEA tools, constituting the first study that reproduces such mechanical experiments with computational approaches. The results obtained confirm the applicability of continuum mechanics to such nanostructures thus precluding the need of expensive computational tools, particularly for large MWCNTs systems.

KEYWORDS

- molecular dynamics
- multiwall carbon nanotubes
- zinc sulfide
- numerical modeling
- buckling

REFERENCES

1. Ajayan, P. M.; Lijima, S. Capillarity-Induced Filling of Carbon Nanotubes. *Nature.* **1993**, *361*(6410), 333–334.
2. Liu, Z.; Chen, K.; Davis, C.; Sherlock, S.; Cao, Q.; Chen, X., et al. Drug Delivery with Carbon Nanotubes for in Vivo Cancer Treatment. *Cancer Res.* **2008**, *68*(16), 6652–6660.
3. Cho, K.; Wang, X.; Nie, S.; Chen, Z.; Shin, D. M. Therapeutic Nanoparticles for Drug Delivery in Cancer. *Clin. Cancer Res.* **2008**, *14*(5), 1310–1316.
4. Dillon, A. C.; Jones, K. M.; Bekkedahl, T. A.; Kiang, C. H.; Bethune, D. S.; Heben, M. J. Storage of Hydrogen in Single-Walled Carbon Nanotubes. *Nature.* **1997**, *386*(6623), 377–379.
5. Mani, R. C.; Li, X.; Sunkara, M. K.; Rajan, K. Carbon Nanopipettes. *Nano. Lett.* **2003**, *3*(5), 671–673.
6. Svensson, K.; Olin, H.; Olsson, E. Nanopipettes for Metal Transport. *Phys. Rev. Lett.* **2004**, *93*(14), 145901.
7. Monteiro, A. O.; Cachim, P. B.; Costa, P. M. F. J. Mechanics of Filled Carbon Nanotubes. *Diam. Relat. Mater.* **2014**, *44*(0), 11–25.
8. Lee, J., *Computational Materials Science;* USA: CRC Press, 2011.
9. Tersoff, J. Modeling Solid-State Chemistry: Interatomic Potentials for Multicomponent Systems. *Phys. Rev. B.* **1989**, *39*(8), 5566–5568.
10. Monteiro, A. O.; Costa, P. M. F. J.; Cachim, P. B.; Holec, D. Buckling of Zns-Filled Single-Walled Carbon Nanotubes-The Influence of Aspect Ratio. *Carbon.* **2014**, *79*(1), 529–537.
11. Benkabou, F.; Aourag, H.; Certier, M. Atomistic Study of Zinc-Blende Cds, Cdse, Zns, and Znse from Molecular Dynamics. *Mater. Chem. Phys.* **2000**, *66*(1), 10–16.
12. Lennard Jones, J. E. *On the Determination of Molecular Fields. II. From the Equation of State of a Gas,* Proceedings of the Royal Society of London, 1924.
13. Lorentz, H. A. Ueber die Anwendung des Satzes vom Virial in der kinetischen Theorie der Gase. *Ann. Der. Physik.* **1881**, *248*(1), 127–136.
14. Berthelot, D. Sur le Mélange des Gaz. *Compt. Rendus.* **1898**, *126*, 1703–1706.
15. Grünwald, M.; Zayak, A.; Neaton, J. B.; Geissler, P. L.; Rabani, E. Transferable Pair Potentials for Cds and Zns Crystals. *J. Chem. Phys.* **2012**, *136*(23), 234111.
16. Rappe, A. K.; Casewit, C. J.; Colwell, K. S.; Goddard, W. A.; Skiff, W. M. UFF, a Full Periodic Table Force Field for Molecular Mechanics and Molecular Dynamics Simulations. *J. Am. Chem. Soc.* **1992**, *114*(25), 10024–10035.
17. Timoshenko S.; Gere, J.*Theory of Elastic Stability;* McGraw-Hill: USA, 1961.
18. Costa, P. M. F. J.; Gautam, U. K.; Wang, M.; Bando, Y.; Golberg, D. Effect of Crystalline Filling on the Mechanical Response of Carbon Nanotubes. *Carbon.* **2008**, *47*(2), 541–544.
19. Costa, P. M. F. J.; Cachim, P. B.; Gautam, U. K.; Bando, Y.; Golberg, D. The Mechanical Response of Turbostratic Carbon Nanotubes Filled with Ga-Doped Zns: II. Slenderness Ratio and Crystalline Filling Effects. *Nanotechnology.* **2009**, *20*(405707), 1–9.
20. Monteiro, A. O.; Costa, P. M. F. J.; Cachim, P. B. Finite Element Modelling of the Mechanics of Discrete Carbon Nanotubes Filled with Zns and Comparison with Experimental Observations. *J. Mater. Sci.* **2014**, *49*(2), 648–653.

CHAPTER 6

EXPERIMENTAL INVESTIGATIONS OF THE INFLUENCE OF INDUSTRIAL BY-PRODUCTS ON HIGHLY FLOWABLE HIGH-STRENGTH CONCRETE

DHAWAL DESAI[1] and PRAKASH NANTHAGOPALAN[2*]

[1]*Structural Engineering, University of Illinois at Urbana-Champaign, Champaign, IL, United States*

[2]*Department of Civil Engineering, IIT Bombay, Mumbai, India*

[*]*Corresponding author. E-mail: prakashn@civil.iitb.ac.in*

CONTENTS

ABSTRACT

In recent years, due to the increased demand for housing and office spaces in urban areas, the number of high-rise buildings increased, in turn increased the usage of high-strength concrete (HSC) by many folds. To increase strength of concrete, there is a need to minimize void content and reduce the water/binder ratio as low as possible. Despite the progress in the research and application of HSC, it is observed that, there are possibilities to improve the properties of HSC by using conceptual design incorporating industrial by-products. The present study, focused on the influence of industrial by-products on highly flowable HSC using the concept of particle packing. For this, Andreassen particle packing model was used. Further, to make HSC more sustainable, two types of ultrafine materials (by-products from the industries) were used, namely, Micro Silica (MS) and Ultrafine Fly Ash (UFA) which are finer than cement particles and fills the voids between the cement particles thereby increasing the packing density of the mixtures. A very low water/binder ratio of 0.22–0.23 was used. The properties of fresh and hardened concrete were investigated. From the results of three different combinations of MS:UFA, the slump flow was achieved in the range from 650 mm to 700 mm and average 28-days compressive strength (cubes) was in the range from 80 MPa to 91 MPa. Among the three mixtures investigated, the MS:UFA (2:2) ratio resulted in maximum compressive strength of 90 MPa with 650–700 mm of slump flow. The use of industrial by-products contributed the overall strength and early age strength as well due to high rate of pozzolanic reaction.

6.1   INTRODUCTION

High-strength concrete (HSC) is currently used in high-rise buildings, pylons of super bridges, long tunnel and also helps in reduction of column area to increase living space. It has been produced in labs with a compressive strength of about 180 MPa with high cementitious content.[1] HSC, by definition according to American Concrete Institute (ACI), is the concrete with compressive strengths more than 60 MPa. To achieve this, there is a need to use high cement content, very low water to cement (w/c) ratio (less than 0.25), use of high range water reducer (or superplasticizer (SP))and fine particles (micro- or nano-sized particles) such as MS and UFA.[2,3] This results in reduction of voids leading to good packing density. Apart from improving the packing density of the particulate system, the ultrafine materials react chemically and influence fresh and hardened properties of concrete as well.[2] From the literature,[4-6] it was observed that ultrafine materials improved the fresh and hardened properties of concrete by increasing the particle packing of the system. Earlier works by Long[7] shows that HSC has cement content higher than 400 kg/m[3].

To achieve excellent strength and slump flow the following factors shall be considered[1-3]:

a) Usage of high cement content with low w/c ratio.
b) Usage of ultrafine materials (micro- or nano-sized particles) to increase the packing density.
c) Usage of high range water reducers (polycarboxylic ether-based SP) is essential to achieve low w/c ratio.
d) Limit the maximum size of coarse aggregate to 10 mm.
e) Time of mixing has to be prolonged or use of high performance concrete mixer.
f) Steam curing may be applied in order to gain more early strength.

6.2   RESEARCH SIGNIFICANCE

Considering the applications of HSC in large number of buildings, bridges, and highways all over the world, there is a need to study its performance and behavior. On the other hand, there is shortage of available raw materials and serious concerns on the environmental impact created by the millions of tonnes industrial by-products (fly ash from thermal power station, MS from ferrosilicon industry, slag from steel industry) produced every year. In this regard, a sustainable solution to overcome this scenario would be to use industrial by-products for the production of HSC. The objective of the present study is to achieve HSC (more than 80 MPa) using industrial by-products like MS and UFA. Although previous studies[1,8] have been carried out in this field, the current research focus on the usage of conceptual approach, that is, particle packing for the development of HSC using industrial by-products leading to better fresh, hardened properties and economy as well. Therefore, in the present study MS and UFA was optimized based on particle packing models and the optimized combinations were used for investigating the fresh (slump flow) and hardened properties (compressive strength) of concrete.

6.3 MATERIALS USED

Ordinary Portland cement of 53 grade (Ambuja Cements Ltd., Mumbai) conforming to IS 12269-2004[10] was used. The chemical composition (X-Ray Fluorescence (XRF)/thermal analysis—obtained from Ambuja Cement Ltd., Mumbai) results of cement are shown in Table 6.1. Two industrial by-products viz. densified MS conforming to American Society for Testing and Materials ASTM (American Society for Testing and Materials) C1240-12[11] and UFA (class F) conforming to IS 3812-1999[12] were also used. Coarse aggregate with maximum particle size of 10 mm and fine aggregate (natural river sand) conforming to IS 383-1997[13] were used for the investigation. Polycarboxylic ether-based SP (SIKA India pvt. Ltd., Mumbai) with solid content of 31.75% conforming to IS 9103-2004[14] was used. The particle size distribution of cement, MS and UFA was determined using laser diffractometer and the results are shown in Figure 6.1. The specific gravity of fine and coarse aggregates was determined according to IS 2386 (Part 1 and Part 3).[15] The physical properties of materials are shown in Table 6.2.

TABLE 6.1	Chemical Test Results of the Cement.

Property	Lime saturation factor	Ratio of % alumina to iron oxide	Insoluble residue (% by mass)	Magnesia (% by mass)	Sulfuric anhydride (% by mass)	Total loss on ignition (%)	Chloride content (% by mass)
Value	0.94	1.47	0.96	1.45	2.40	1.40	0.049

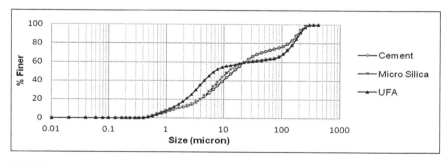

FIGURE 6.1	Particle size based distribution of cement, MS, and fly ash.

TABLE 6.2   Physical Properties of Materials.

Properties	Cement	MS	UFA	4.75 mm	10 mm
Specific gravity	3.15	2.25	2.30	2.35	2.97
Mean diameter	14.43 μm	11.76 μm	6.72 μm	652 μm	7689 μm

The actual average size of MS particles is 0.1 μ (as per the manufac-turers data sheet), however, as these particles are densified for convenient transportation, dispersion of these particles are difficult during measurement (despite ultrasonication) which leads to relatively higher measured values. These difficulties in dispersion of MS (or silica fume) particles were also reported in reference [16].

6.4   EXPERIMENTAL METHODOLOGY

6.4.1   OPTIMIZATION OF MATERIAL COMPOSITIONS FOR ACHIEVING MAXIMUM PARTICLE PACKING

From the literature, it was evident that high binder content of nearly 800 kg/m³ with low water/binder (w/b) ratio required for achieving HSC. In the present investigation, the cement content for the mixtures D1, D2, and D3 was deter-mined (for three different combinations of MS:UFA) based on maximum packing density according to the particle packing software EMMA Mix Analyzer. The cement content was replaced by 20% (by mass) of industrial by-products (MS and UFA) collectively. The quantity of fine aggregates and coarse aggregates were then calculated from Andreassen model of particle packing using the software "EMMA Mix Analyzer" to achieve maximum particle packing. Many combinations were investigated for packing density, out of which three ratios by volume of MS and UFA were selected based on the higher packing density. The combinations are given in Table 6.3.

TABLE 6.3   Ratio of MS with UFA in Different Samples.

Sample	D1	D2	D3
MS:UFA (by volume)	1:3	2:2	3:1

The w/b was selected in the range from 0.22 to 0.23 and SP amount was decided based on the slump flow and strength requirement. Table 6.4 summarizes the mixture proportions developed in this study.

TABLE 6.4 Proportions for Concrete Mixes (kg/m³).

Sample	Cement	MS	UFA	4.75 mm	10 mm	Water	SP
D1	640.3	39.4	120.7	560.3	843.0	184.0	18.4
D2	632.0	78.0	79.9	632.0	795.2	174.0	16.4
D3	646.5	120.6	41.1	646.5	748.5	178.0	16.2

6.4.2 MIXING AND CASTING OF CONCRETE

Initially, some difficulties were encountered while mixing the concrete mixture with such low w/b ratio in addition to the ultrafine materials (despite the usage of SP). So, a systematic procedure was developed and adopted for proper blending of the mixture in the Pan mixer. For homogenization, all the dry ingredients were mixed thoroughly in a Pan mixer (at 25 rpm) for 3 min. Then 80% of the total water content was added and mixed for 2 min. Required SP was uniformly mixed with the remaining amount of water, then added into the mixture and subsequently mixed for 6 min. The cubes of size 100 × 100 × 100 mm were cast and demolded after 24 h and cured in water for 28 days.%%

6.4.3 TESTING FRESH AND HARDENED PROPERTIES OF CONCRETE

The slump flow test was performed for all the mixtures by using the slump cone (IS 1199-1999).[17] The compressive strength test (IS 516-2002)[18] was performed at 7 and 28 days. Three cubes were tested for each combinations and average strength values reported in Table 6.5.

6.5 RESULTS AND DISCUSSIONS

6.5.1 SLUMP FLOW TEST RESULTS

The slump flow obtained was in the range from 650 mm to 700 mm for all these mixtures (D1, D2, and D3) without bleeding and segregation. Figure 6.2 illustrates the slump flow for D1. To achieve the desired range of slump flow for all three mixtures, w/b ratio and SP% had to be slightly modified in D1 due to higher UFA content and its fineness. This is in compliance to the fact that for same amount of water content, the water film thickness is smaller when more amount of finer material (with higher surface area) is present and for smaller water film thickness, flowability is lower.[9]

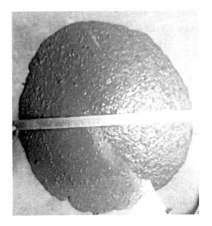

FIGURE 6.2   Slump flow with no segregation.

6.5.2   EFFECT OF MS AND UFA CONTENT ON COMPRESSIVE STRENGTH

The results of the compressive strength test were shown in Table 6.5. From the results, it could be observed that the average 28 days compressive strength increases from sample D1 to D2, however, decreases from sample D2 to D3. The strength of D1 was least among the three combinations due to the usage of relatively higher w/b ratio (0.23) and higher UFA for obtaining the desired slump flow (650–700 mm). These observations are in line with findings of researchers,[19] that the compressive strength decreases with increase in w/b ratio. The strength increment observed in D2 could be due to increased silica content in concrete coming from MS contributing to better particle packing and pozzolanic reaction as well. This is in line with observations reported by Babu and Babu.[20] From the D3 results, it was evident that the compressive strength was higher than D1 however, slightly less than D2. This could be attributed to the increase in MS content and, also, decrease in UFA content, which caused loosening effect and lowered the particle packing of concrete. Similar findings were given by authors,[21] reporting that strength of concrete increases as the silica fume content increases, but only up to a certain limit, and then decreases for further increment in silica fume content. Generally, in concrete at the age of 7 days, 65–70% strength gain of the 28 days strength is noticed. However, the presence of MS and UFA contributed significantly to the early age strength gain which was evident from the 7 days strength results (Table 6.5). Approximately 75–90% of 28-days strength was reached in 7 days for the three different mixtures. This could be attributed to the high rate of pozzolanic reaction of MS and UFA.

TABLE 6.5 Compressive Strength Test Results.

Sample	Avg. 7 day (MPa)	Standard deviation (MPa)	Avg. 28 day (MPa)	Standard deviation (MPa)
D1	72.57	5.79	80.90	8.60
D2	79.97	9.90	91.02	2.56
D3	67.61	6.78	89.73	6.43

6.6 CONCLUSION

The present investigation focused on understanding the influence of industrial by-products on highly flowable HSC using the concept of particle packing. The binder (Cement + MS:UFA) combinations were optimized using the Andreassen particle packing model. Based on the packing density, the combinations of MS:UFA (1:3, 2:2, 3:1) were used for the development of HSC along with aggregates. Collectively, MA + UFA was used to replace the cement by 20%. w/b ratio (0.22–0.23) and SP dosage was used based on the requirement of strength (>80 MPa) and slump flow (650–700 mm), respectively. A systematic procedure was developed for proper blending of materials for HSC mixtures in Pan mixers. Based on the proportions of materials optimized using concept of particle packing, the results of the three mixtures (D1, D2, and D3) indicated that the slump flow was in the range from 650 mm to 700 mm and the compressive strength was in the range from 80 MPa to 91 MPa. It is evident that, the D2 combinations of materials (with optimum combination of MS and UFA leading to better particle packing coupled with early strength due to pozzolanic action) resulted in the maximum compressive strength of 91 MPa with the slump flow of 650–700 mm. Further, it was also observed that the combination of MS and UFA resulted in early age strength gain which was clear from the 7-days strength. In future, studies will be carried out to understand durability and shrinkage properties for all combination of materials.

ACKNOWLEDGMENT

The authors would like to express their gratitude to the following companies for providing materials for this research: Ambuja Cements Ltd., ELKEM, and SIKA India Pvt. Ltd.

KEYWORDS

- **high-strength concrete**
- **fly ash**
- **superplasticizer**
- **slump flow**
- **compressive strength**

REFERENCES

1. Wang, C.; Yang, C.; Liu, F.; Wan, C.; Pu, X. Preparation of Ultra-High Performance Concrete with Common Technology and Materials. *Cem. Concr. Compos.* **2012,** *34,* 538–544.
2. de Larrard, F. Ultrafine Particles for the Making of Very High Strength Concretes. *Cem. Concr. Res.* **1989,** *19,* 161–172.
3. Longa, G.; Wanga, X.; Xie, Y. Very-high Performance Concrete with Ultrafine Powders. *Cem. Concr. Res.* **2002,** *32,* 601–605.
4. de Larrard, F.; Sedran, T. Optimization of Ultra-high Performance Concrete by the Use of a Packing Model. *Cem. Concr. Res.* **1994,** *24*(6), 997–1009.
5. Mehta, P. K. Advancements in Concrete Technology. *ACI Concr. Int.* **1999,** *21*(6), 69–76.
6. Aitcin, P. C.; Neville, A. High Performance Concrete Demystified. *Concr. Int.* **1993,** *15,* 21–26.
7. Long, T. P. High Strength Concrete at High Temperature—An Overview (2008). http://fire.nist.gov/bfrlpubs/build02/pdf/b02171.pdf (accessed Jan 20, 2012).
8. Deeb, R.; Ghanbari, A.; Karihaloo, B. L. Development of Self-compacting High and Ultra-high Performance Concretes with and Without Steel Fibres. *Cem. Concr. Compos.* **2012,** *34,* 185–190.
9. Kwan, A. K. H.; Ng, P. L.; Fung, W. W. S. *Research Directions for High-Performance Concrete,* HKIE Civil Division Conference, 2010.
10. IS 12269-2004. Specification for 53 Grade Ordinary Portland Cement, Bureau of Indian Standards, New Delhi.
11. ASTM C1240-12. Standard Specification for Silica Fume Used in Cementitious Mixtures, American Society for Testing and Materials.

12. IS 3812-1999. Specification for Fly Ash for Use as Pozzolana and Admixture, Bureau of Indian Standards, New Delhi.
13. IS 383-2002. Specification for Coarse and Fine Aggregate from Natural Sources for Concrete, Bureau of Indian Standards, New Delhi.
14. IS 9103-2004. Concrete Admixtures—Specification, Bureau of Indian Standards, New Delhi.
15. IS 2386-2002 (Part 1 & Part 3). Methods of Test for Aggregates for Concrete, Bureau of Indian Standards, New Delhi.
16. Diamond, S.; Sahu, S. Densified Silica Fume: Particle Sizes and Dispersion in Concrete. *Mater. Struct.* **2006,** *39,* 849–859.
17. IS 1199-1999. Methods of Sampling and Analysis of Concrete, Bureau of Indian Standards, New Delhi.
18. IS 516-2002. Method of Tests for Concrete, Bureau of Indian Standards, New Delhi.
19. Wille, K.; Naaman, A. E.; Parra-Montesinos, G. J. Ultra-high Performance Concrete with Compressive Strength Exceeding 150 MPa (22 ksi): A Simpler Way. *ACI Mater. J.* **2011,** *108,* 46–54 (Technical Paper, Title No. 108-M06).
20. Babu, K. G.; Babu, D. S. Behaviour of Lightweight Expanded Polystyrene Concrete Containing Silica Fume. *Cem. Concr. Res.* **2003,** *33*(5), 755–762.
21. Amudhavalli, N. K.; Mathew, J. Effect of Silica Fume on Strength and Durability Parameters of Concrete. *Int. J. Eng. Sci. Emerg. Technol.* **2012,** *3*(1), 28–35.

CHAPTER 7

SYNTHESIS, CHARACTERIZATION, AND ELECTRICAL PROPERTIES OF CERIUM OXIDE NANOPARTICLES

PRIYANKA K. P.[1], BABITHA K. K.[1], SHEENA XAVIER[2], MOHAMMED E. M.[2], and THOMAS VARGHESE[1*]

[1]Department of Physics, Nanoscience Research Centre (NSRC), Nirmala College, Muvattupuzha 686661, Kerala, India

[2]Department of Physics, Maharajas College, Ernakulam, Kerala, India

[*]Corresponding author. E-mail: nanoncm@gmail.com

CONTENTS

ABSTRACT

Cerium oxide nanoparticles or nanoceria were synthesized by using cerium nitrate hexahydrate and ammonium carbonate as precursors. Structural characterizations were performed by using X-ray diffraction and transmission electron microscopy. The dielectric properties of cerium oxide have been studied as a function of frequency and temperature. It is found that dielectric constant and dielectric loss for all temperatures have high values at low frequencies, which decrease rapidly as frequency increases and attain a constant value at higher frequencies. The same results were observed for AC conductivity also. These properties make cerium oxide useful for applications in microelectronics and optics.

7.1 INTRODUCTION

CeO_2 nanoparticles are of great interest due to their novel properties and variety of potential applications such as fuel cells, gas sensors, NO removal, counter electrodes in smart window devices, and humidity sensors.[2–4,14] CeO_2 has a high refractive index, strong adhesion, mechanical abrasion, and stability toward high temperature. CeO_2 has been developed as a fuel additive to improve the efficiency of combustion.[1] It was reported that cerium oxide-containing coatings significantly improve the oxidation and spallation resistance of high-alloy steels in oxidizing environments.[16] It was suggested that the adsorption of Ce^{4+} ions on the oxide grain boundaries hinders grain growth under oxidizing circumstances, resulting in a better adherence to the substrate due to enhanced plasticity of the oxide.

In the present work, we report a successful synthesis of cerium oxide nanoparticles at room temperature, adopting careful control of the reaction kinetics of aqueous precipitation; structural characterizations were done by using X-ray powder diffraction (XRD) as well as transmission electron microscopy (TEM), and dielectric properties of the material have been investigated as a function of frequency and temperature. The dielectric properties and AC conductivity of the nanomaterials differ from those of bulk materials due to increased interfacial atoms or ions and sinking of large amount of defects at or near the grain boundaries. Each interface acts as a capacitor, thereby changing the dielectric values of the material.

7.2 EXPERIMENTAL

Cerium nitrate, ammonium carbonate, and EDTA (99.9% purity) were purchased from Sigma Aldrich Chemicals and were used without further purification. Nanosized powders of cerium oxide were prepared by reacting aqueous solutions of cerium nitrate and ammonium carbonate (0.01 M each) at room temperature. The precipitate formed was centrifuged, filtered, and washed with distilled water a number of times, and dried in an oven to get fine powders of cerium oxide. The precursor is annealed in air at 400°C for 3 h. The scheme of preparation of nanocrystalline CeO_2 used in this work has been presented schematically as shown in Figure 7.1.

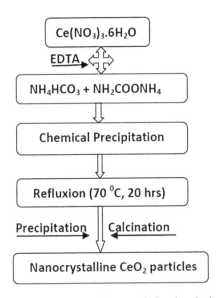

FIGURE 7.1   The scheme of preparation of nanoceria by chemical precipitation method.

The structural characteristics of the synthesized CeO_2 nanoparticles have been studied by XRD using Bruker D8 advance X-ray diffractometer (λ = 1.5406 Å) and TEM using TECNAI F30 FEG model instrument operated at an accelerating voltage of 300 kV. XRD analyses were carried out for the identification of the crystal phase and the estimation of the average crystallite size. The crystallite size was estimated from the Debye–Scherrer equation: $t = 0.9\ \lambda/\beta \cos \theta$; where, λ is the X-ray wavelength, β is the full width at half maximum of the peak, and θ is the Bragg's angle.[15]

For TEM studies, the CeO_2 powder was dispersed in ethanol using an ultrasonic bath. A drop of suspension was placed on a copper grid coated with carbon film. After drying, the copper grid containing nanoparticles was placed on the holder for the imaging process. The powder was then consolidated in the form of cylindrical pellets of diameter 11 mm and thickness $d =$ 1.2 mm at a pressure of ~7 GPa using a hydraulic press for dielectric studies. Both the faces of the pellets were coated with air-drying silver paste. Dielectric measurements as a function of frequency in the range from 100 Hz to 10 MHz are measured at various selected temperatures from 303 to 423 K using a four-probe LCR meter (Wayne Kerr H-6500B model) in conjunction with a portable furnace and temperature controller (±1 K). The dielectric constant was calculated by using the formula $\acute{\varepsilon} = Cd/\varepsilon_0 A$, where A is the surface area and C, the measured capacitance of the pellet. AC conductivity (σ_{ac}) is obtained from the data of dielectric constant ($\acute{\varepsilon}$) and loss (tanδ) using the relation $\sigma_{ac} = \acute{\varepsilon}\varepsilon_0 \omega \tan\delta$, where ε_0 is the permittivity of vacuum and ω the angular frequency.[15]

7.3   RESULTS AND DISCUSSION

Figure 7.2(a) displays the XRD pattern of the cubic CeO_2 samples annealed at 400°C for 3 h. From XRD patterns, the average size of the particles corresponding to annealing temperature estimated using Scherrer equation was 9 nm. The principal d values taken from the JCPDS file No. 75-0076 for CeO_2 are in close agreement with the observed d values. Figure 7.2(b) shows the TEM image of the synthesized CeO_2 particles. The particle size obtained from TEM images is about 10 nm, which is in agreement with the XRD analyzed value. The TEM image reveals that particles are not exactly spherical in shape. Selected area electron diffraction is shown in the inset of Figure 7.2(b), which clearly indicates that the CeO_2 nanoparticles are highly crystalline in nature.

Figure 7.3(a) shows the variation of dielectric constant with frequency for temperatures from 303 K to 423 K. It can be seen that the real part of dielectric constant $\acute{\varepsilon}$ for all temperatures has high values at low frequencies which decrease rapidly as frequency increases and attains a constant value at higher frequencies. In dielectric nanostructured materials, interfaces with large volume fractions contain a large number of defects, such as dangling bonds, vacancies, vacancy clusters, and microporosities, which can cause a change of positive and negative space charge distribution in interfaces. When subjected to an electric field, these space charges move. When they are trapped by defects, a lot of dipole moments are formed. At low-frequency region, these dipole moments easily follow the variation of electric field[11] (Rao et al., 1997). So the dielectric loss and hence the dielectric constant show large values at low frequency. As temperature is increased, more and more dipoles are oriented, resulting in an increase in the value of dielectric constant for a given value of frequency[9]. At very high-frequency (MHz) region, the charge carriers would have started to move before the field reversal occurs and $\acute{\varepsilon}$ falls to a small value at higher frequencies. Space charge polarization and reversal of the direction of polarization contribute much to the $\acute{\varepsilon}$.[6]

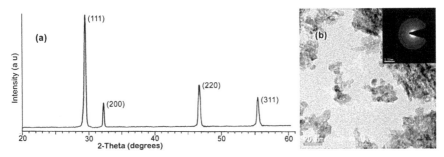

FIGURE 7.2   (a) XRD pattern and (b) TEM image of CeO_2 nanoparticles.

The variation of $\tan\delta$ with frequency of sample is shown in Figure 7.3(b). The variation of $\tan\delta$ is similar to that of $\acute{\varepsilon}$. In nanomaterials, the inhomogeneities present in the interface layers produce an absorption current resulting in dielectric loss. This absorption current decreases with increase in frequency of the applied field. The hopping probability per unit time increases with increase in temperature. Correspondingly, the loss tangent also increases with increase in temperature.[8,10,13] The loss in CeO_2 can be explained by electronic hopping model, which considers the frequency dependence of the localized charge carriers hopping in a random array of centers. This model is applicable for materials in which the polarization responds sufficiently rapidly to the appearance of an electron on any one site so that the transaction may be said to occur effectively into the final state.[7,9] In the high-frequency region, $\tan\delta$ becomes almost constant because the electron exchange interaction (hopping) between Ce^{3+} and Ce^{4+} cannot follow the alternatives of the applied AC electric field beyond a critical frequency.[5] Nanoceria with low values of dielectric constant and loss tangent at higher frequencies is important for the fabrication of materials for ferroelectric, photonic, and electro-optical devices.

The variation of AC electrical conductivity as a function of frequency of sample is shown in Figure 7.3(c). At low frequencies, σ_{ac} has a small value which increases at higher frequencies. The values are shifted upwards as the temperature is raised. It is clear from the figure that the conductivity increases as frequency is increased conforming small polar on hopping. When the temperature is increased, there will be easy transition of charge carriers from valence band to conduction band due to small size of particles in the sample, and hence conductivity increases.[7,12] The high values of σ_{ac} for particles with small grain size are a direct confirmation of the theory.

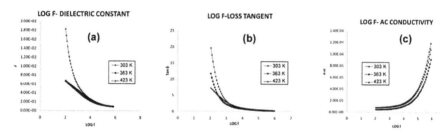

FIGURE 7.3 Variation of (a) dielectric constant with frequency, (b) tanδ with frequency, and (c) σ_{ac} with frequency of CeO$_2$ for temperatures from 303 to 423 K.

7.4 CONCLUSIONS

CeO$_2$ nanoparticles with average size 9 nm were synthesized using chemical precipitation method. The dielectric properties of CeO$_2$ were determined as a function of frequency from 100 Hz to 10 MHz for temperatures ranging from 303 K to 423 K. At lower frequency, $\acute{e}$ and tanδ have higher values while at higher frequency, the values reach steady lower values. Similar variation is observed when the temperature is raised but the values are shifted upwards. Conductivity increases as frequency is increased conforming small polaron hopping. It is concluded that the material synthesized by room temperature chemical precipitation in the nano range shows enhanced dielectric properties.

ACKNOWLEDGMENT

The authors are grateful to KSCSTE, Thiruvananthapuram for providing financial support.

KEYWORDS

- **nanoparticles**
- **cerium nitrate**
- **dielectric constant**
- **frequency**
- **conductivity**

REFERENCES

1. Ashok, K. G.; Tokeer, A.; Sonalika, V.; Jahangeer, A. *Pure Appl. Chem.* **2008,** *80*(11), 2451–2477.
2. Dimonte, R.; Fornasiero, P.; Graziani, M.; Kašpar, J. J. *Alloys Compd.* **1998,** *877,* 275–277.
3. Elidrissi, B.; Addou, M.; Regragui, M.; Monty, C.; Bougrine, A.; Kachouane, A. *Thin Solid Films* **2000,** *379,* 23.
4. Garzon, F. H.; Mukundan, R.; Brosha, E. L. *Solid State Ion.* **2000,** *633,* 136–137.
5. Indulal, C. R.; Vaidyan, A. V.; Kumar, G. S.; Raveendran, R. *Ind. J. Eng. Mater. Sci.* **2010,** *17,* 299–304.
6. Jiang, B.; Peng, J. L.; Busil, L. A., Zhong, W. L. *J. Appl. Phys.* 2000, *87*(7), 3462.
7. Kumar, E.; Selvarajan, P.; Balasubramanian, K. *Rec. Res. Sci. Tech.* **2010,** *2*(4), 37–41.
8. Mathew, J.; Kurien, S.; Sebastian, S.; George, K. C. *Ind. J. Phys.* **2004,** *78*(9), 947.
9. Mechant, P.; Elbarum, C. *Solid State Commun.* **1978,** *26,* 73.
10. Murthy, V.; Sobhanadri, J. *Phys. Status Solidi (a)* 1976, *36,* K133.
11. Parvatheeswara, R. B.; Rao, K. H. *J. Mater. Sci.* 1997, *32,* 6049-6054.
12. Priyanka, K. P.; Joseph, S.; Smitha, T.; Mohammed, E. M.; Varghese, T. *J. Bas. Appl. Phys.* 2013, *2*(1), 105–108.
13. Ravinder, D.; Vijayakumar, K. *Bull. Mater. Sci.* 2000, *24*(5), 505.
14. Steele, B. C. H.; Heinzel, A. *Nature* **2001,** *414,* 345.
15. Varghese, T.; Balakrishna, K. M. *Nanotechnology: An Introduction to Synthesis, Properties and Applications*; Atlantic Publishers: New Delhi, 2011.
16. Haanappel, V. A. C.; Franzen, T.; Geerdink, B.; Gellings, P. J. *Oxid. Met.* **1988,** *30*(3/4), 201–208.

CHAPTER 8

SYNTHESIS AND CHARACTERIZATION OF SILANE-MODIFIED CHITOSAN/ EPOXY COMPOSITES

V. SELVAM[1*], M. SURESH CHANDRA KUMAR[2], and M. VADIVEL[2]

[1]*Department of Chemistry, K. Ramakrishnan College of Technology, Trichy, India*

[2]*Department of Chemistry, Polymer Nanocomposite Centre, Scott Christian College, Affiliated to Manonmaniam Sundaranar University, Tamil Nadu, India*

*Corresponding author. E-mail: selvam.che@gmail.com

CONTENTS

ABSTRACT

Due to nonbiodegradable nature of plastics, there is accumulation of these materials in the environment. Therefore, material developments without an accompanying disruption of the earth's environment are getting much more important even in the polymer industry. Among them, the natural polymers have undergone a reevaluation regarding their ability to biodegradable. Chitosan is a biodegradable natural polymer. Use of chitosan in epoxy reduces the cost and makes the composite material biodegradable. Poor interaction between chitosan (hydrophilic) and epoxy (hydrophobic), results in lower mechanical strength in chitosan/epoxy composites; therefore, for effective utilization of these materials, coupling agent is added. Prepared (3-glycidyloxypropyl) trimethoxysilane grafted chitosan (GPTS-g-CS) is characterized by Fourier transform infrared spectroscopy (FT-IR). The composites prepared by GPTS-g-CS were blended with bisphenol A diglycidyl ether at different loadings. Mechanical and thermal properties of the composites were studied. The results showed that incorporation of GPTS-g-CS increased thermal stability of the composites. The fracture surfaces of composites after tensile testing are investigated by scanning electron microscope (SEM).

8.1 INTRODUCTION

The commercial production of green products from natural sources, interest in composites preparation shifted toward the use of biopolymer for both matrices and reinforcement with ecofriendly. In order to develop ecofriendly composites with good mechanical strength, it is necessary to impart hydrophobicity to green products by chemical reaction with suitable coupling agents or by coating with appropriate resins. Such surface modification not only decrease moisture adsorption but also concomitantly increases wettability of green products with resin and improve the interfacial bond strength, which are the critical factors for obtaining better mechanical strength of composites. Epoxy resins are widely used in many industrial applications[1] because of their low manufacturing cost, ease of properties, low shrinkage, good chemical and electrical resistance, and good mechanical properties.[2,3,4] Chitosan exhibits multiple functionality including antimicrobial activity and metal binding.[7] Many attempts have been conducted to develop functional materials from chitosan such as films,[8–10] sutures, beads, and hydrogels, and to apply them in the wound healing, drug delivery, metal removal, and antimicrobial food packaging.[9]

In this work, chitosan modified with (3-glycidyloxypropyl)trimethoxysilane (GPTMS) modified chitosan blends with epoxy composites were prepared. Subsequently, the effects of chitosan content on the thermal and mechanical properties of the epoxy composites were investigated.

8.2   MATERIALS

Commercially available diglycidyl ether of bisphenol-A (DGEBA)-based epoxy resin LY556 (Ciba-Geigy Ltd., India) having epoxy equivalent of about 180–190, with viscosity of about 10,000 cP. 4,4'diaminodiphenylmethane (DDM), epoxy curing agent is obtained from Ciba-Geigy Ltd., India. Chitosan (Indian Sea Foods, Cochin, India, powder, molecular mass (Mv) = 180,000 Da, moisture content = 9.46%, viscosity = 229 cps, DD = 86.39) using without further purification. GPTMS was procured from Dow Corning Corporation, USA.

8.2.1   PREPARATION OF (3-GLYCIDYLOXYPROPYL) TRIMETHOXYSILANE GRAFTED CHITOSAN (GPTMS–g–CS)

Four grams of dried chitosan was dissolved in 100 mL of 2% acetic acid and stirred for 6 h. This solution was filtered through a sintered glass filter before use. The solution was stirred in a 500 mL round bottomed 3-necked flask and barged with N_2 for 30 min. 0.05 mole of dibutyltin dilaurate in the minimum volume of acetone was added, followed by the addition of predetermined quantity of GPTMS for 10 min. The reaction was continued for 24 h under nitrogen at 60°C. The contents of the flask were stirred for 15 min at ambient temperature, then filtered followed by the addition of 10% NaOH solution with vigorous stirring until the grafted polymer was completely precipitated. The precipitate was washed with ethanol and deionized water to remove the excess unreacted silane and acetic acid. Finally, the precipitate was dried in a vacuum at 100°C for 24 h, and then kept in a desiccator.

8.2.2   PREPARATION OF GPTMS-G-CS MODIFIED EPOXY COMPOSITES

Epoxy resin was sonicated for 2 h with calculated amount of (2 wt%) GPTMS-g-CS. The mixture was stirred at 12,000 rpm for 10 min. A stoichiometric amount of curative 4,4'-diaminodiphenylmethane corresponding to epoxy equivalents was also added. The product was subjected to vacuum to remove the trapped air and then cast and cured then post-cured.

8.3   ANALYTICAL METHODS

The infrared (IR) spectra were recorded on a SHIMADZU FT-IR spectrometer with KBr pellets. Thermogravimetric Analysis (TGA) was performed on a Perkin-Elmer Pyrisl TGA thermogravimetric analyzer between 0°C and 800°C with a 10°C min^{-1} heating rate under nitrogen with flow rate of 20.0 mL min^{-1}. Surface morphology of fractured surface of the composites was recorded using scanning electron microscope (SEM; JEOL JSM Model 6360).

8.4   RESULT AND DISCUSSION

8.4.1   FT-IR CHARACTERIZATION

Figure 8.1a shows FT-IR spectra for pure chitosan the broad peaks at 3430 cm^{-1} due to the stretching vibration –OH superimposed on –NH stretching

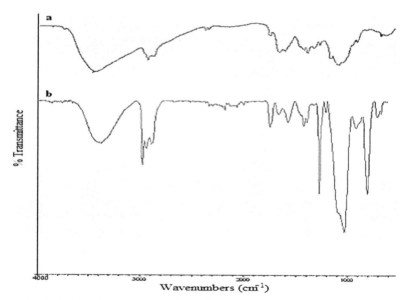

FIGURE 8.1   FT-IR of: (a) chitosan and (b) GPTMS-g-chitosan.

band and broaden due to inter hydrogen bonds of polysaccharides. The 1640 cm^{-1} attributed to the presence of acetyl unit with –C=O stretching; 1586 cm^{-1} attributed to N–H bending, the 1383 cm^{-1} attributed to –CH$_3$ symmetrical angular deformation; 1151 cm^{-1} attributed to β (1, 4) glycosidic bonds; and 1041 cm^{-1} attributed to C–O–C stretching vibration.

The modification of chitosan primary amine groups through their reaction with GPTS epoxy rings, the absorption band at 1596 cm^{-1} is moved to 1560 cm^{-1} (Fig. 8.1b). The C–O–C, C–OH, and C–OH bands appearing at 1157, 1087, and 1026 cm^{-1} in chitosan are superposed on Si–O–Si strong band in membranes. IR spectrum also evidenced at 1258 and 804 cm^{-1} the Si–CH$_3$ characteristic bands of the siloxane chain.[5]

8.4.2   THERMOGARVIMETRIC ANALYSIS

Table 8.1 shows the thermograms of silane-modified chitosan epoxy composites. In the case of biocomposites with GPTMS-g-CS (>1.0 wt%), three-stage degradation are observed. For all these GPTMS-g-CS/EP bionanocomposites, the first stage degradation occurs around 93–95°C is due to the elimination of water molecules. The second step degradation starts at about 280°C and is completed at 340°C. The third stage degradation occurs in the region from 395°C to 440°C. The former stage of degradation is due to the cleavage of GPTMS-g-CS and the later stage degradation is due to scission of epoxy. The peak around 406–408°C is due to scission of epoxy and peaks around 322–324°C correspond to the degradation of GPTMS-g-CS. Furthermore, the decrease in degradation temperature indicates the decrease in interaction between hydroxyl groups of GPTMS-g-CS and DDM due to agglomeration of GPTMS-g-CS.

TABLE 8.1   Thermogram of GPTMS-g-CS/Epoxy Composites.

EP/GPTMS-g-CS	Degradation temperature		
	T_{max1}	T_{max2}	T_{max3}
100/0.1	93	420	–
100/0.5	94	422	–
100/1.0	95	425	–
100/2.0	94	324	406

8.4.3   MECHANICAL PROPERTIES

The tensile strength values of epoxy and GPTMS-g-CS-modified epoxy matrix are presented in the Table 8.2. It is observed that the mechanical properties increase with increase in GPTMS-g-CS concentration and attain the maximum and then decrease. In the case of GPTMS-g-CS/EP bionano-composite systems, introduction of 0.1 wt% of GPTMS-g-CS increases the tensile strength to 70.9 MPa. The highest tensile strength of 77.1 MPa is obtained when the reaction is carried out using 1.0 wt% of GPTMS-g-CS. But, inclusion of 1.5 wt% GPTMS-g-CS decreases the tensile strength to 69.8 MPa. The dramatic decline of mechanical properties at higher concentration of GPTMS-g-CS after attaining the optimum value is due to the agglomeration of the GPTMS-g-CS in the epoxy resin.

TABLE 8.2   Mechanical Properties of GPTMS-g-CS Epoxy Composites.

GPTMS-g-CS/ epoxy composition	Tensile strength (MPa)	Tensile modulus (MPa)
00/100	69.3	3434
0.1/100	70.9	3482
0.5/100	73.2	3590
1.0/100	77.1	3650
1.5/100	69.8	3520
2.0/100	60.7	3310

8.4.4   MORPHOLOGY

Figure 8.2a shows the morphology of chitosan modified epoxy reveals roughness of the structure with some bigger particles. This further supports that there is phase separation between the two components. The morphology of GPTMS-g-CS-modified epoxy matrix (Fig. 8.2b) reveals the encapsulated cross-linked and structured GPTMS-g-CS particle in epoxy matrix.[6]

8.5   CONCLUSION

The (3-glycidyloxypropyl) trimethoxysilane grafted chitosan (GPTS-g-CS) and confirmed by FT-IR. The composites prepared by (GPTS-g-CS) were

(a)

(b)

FIGURE 8.2   (a) SEM image of CS/epoxy and (b) SEM image of GPTMS-g-CS/epoxy.

blend with epoxy. The results showed that the incorporation of GPTS-g-CS increased the thermal stability of the composites. The SEM shows dispersion of modified chitosan in epoxy matrix.

KEYWORDS

- **(3-glycidyloxypropyl) trimethoxysilane**
- **mechanical properties**
- **thermal properties**
- **scanning electron microscope**
- **composites**

REFERENCES

1. Alagar, M.; Thanikaivelan, T. V.; Ashok Kumar, A.; Mohan, V. *Mater. Manuf. Process.* **1999**, *14*, 67–83.
2. Alagar, M.; Ashok Kumar, A.; Mahesh, K. P. O.; Dinakaran, K. *Eur. Polym. J.* **2000**, *36*, 2449–2454.
3. May, C. A.; Tanka, G. Y. *Epoxy Resin Chemistry and Technology*; Marcel Dekker: New York, 1973.
4. Lee, H.; Neville, K. *Handbook of Epoxy Resins*; McGraw-Hill: New York, 1967.
5. Enescu, D.; Hamciuc, V.; Ardeleanu, R.; Cristea, M.; Ioanid, A.; Harabagiu, V.; Simionescu, B. C. *Carbohydr. Polym.* **2009**, *76*, 268–278.
6. Rosati, D.; Perrin, M.; Navard, P.; Harabagiu, V.; Pinteala, M.; Simionescu, B. C. *Macromolecules* **1998**, *31*, 4301–4308.
7. Simionescu, C. I.; Harabagiu, V.; Giurgiu, D.; Hamciuc, V. *Bull. Soc. Chim. Belg.* **1990**, *99*, 11–12.
8. Singh, D. K.; Roy, A. R.; *J. Appl. Polym. Sci.* **1997**, *66*, 869–877.
9. Singh, D. K.; Roy, A. R.; *J. Appl. Polym. Sci.* **1994**, *53*, 1115–1121.
10. Park, B.; You, J. O.; Park, H. Y.; Haam, J. S.; Kim, W. S. *Biomaterials* **2001**, *22*, 323.

CHAPTER 9

HOMOGENIZED CAE REPRESENTATION OF ALD-GMT AUTOMOTIVE HEADLINER SUBSTRATE USING RATE-DEPENDENT PLASTICITY MODELS IN LS-DYNA

JAYARAJ RADHAKRISHNAN[1], BHASKAR PATHAM[1†*],
SARAVANAN B. N.[2], BISWAJIT TRIPATHY[3], and CHIN HSU LIN[4]

[1]*General Motors Technical Centre India (GMTCI), Vehicle CAE,
Material CAE Methods Group, Mumbai, India*
[2]*GMTCI, Vehicle CAE, FMVSS-201U Group, Mumbai, India*
[3]*GMTCI, Vehicle CAE, Vehicle Optimization Group, Mumbai, India*
[4]*General Motors Global R&D, Vehicle Development Research Lab,
Vehicle Systems Group, Mumbai, India*
Corresponding author. E-mail: bhaskar.patham@sabic.com

CONTENTS

†Presently affiliated with SABIC Research and Technology Pvt. Ltd., Bangalore, India

ABSTRACT

Low-density random glass-fiber-mat thermoplastics (LD-GMT) are widely employed in automotive interior trim applications, where they may be required to absorb interior impact energy; optimal design of such composites is, therefore, critical for satisfying the performance requirements for such applications. CAE (Computer Aided Engineering) aided design optimization of automotive interior trim components involving LD-GMT, such as headliner substrates, therefore, requires accuracy in impact simulations in terms of the estimated peak stresses and impact performance criteria where applicable. Since energy absorption during impact primarily is driven by the headliner substrate, it is therefore critical to develop faithful and versatile material models for the headliner substrate in a timely, efficient, and cost-effective manner. This project is aimed at the characterization and LS-DYNA material model development for a commercially available headliner substrate.

Even though LD-GMT headliner substrates are "layered" or laminar thermoplastic composites, a homogenized "smeared" approach was employed in this study to represent the substrate; this was with the aim of rendering the modeling methodology robust and simple enough so that it need not be modified in the event of any change of the supplier or the individual layer composition. In this approach, the headliner substrate was tested as a whole and a single material model input file was developed for the homogenized representation of the substrate (all substrate layers combined). Two LS-DYNA material model candidates were evaluated for such homogenized representation: the first was *MAT_024 (*MAT_PIECEWISE_LINEAR_PLASTICITY) which has been traditionally employed for representing headliner substrates. The second candidate, *MAT_081–82 (*MAT_PLASTICITY_WITH_DAMAGE), in which the plasticity model of *MAT_024 is overlaid with a plastic strain-dependent damage evolution model, is expected to provide better control compared to MAT_024 in predicting post-peak softening behavior.

In this chapter, we will describe the steps followed for developing the *MAT_024 and *MAT_081–082 material models for the homogenized "smeared" representation of the headliner substrate. The focus will be on development of the plasticity model formulation (*MAT_024) using tensile test information obtained at two test velocities. The plasticity model was overlaid with a dynamic localized-out-of-plane load-case-tailored damage model in *MAT_081–082. The resulting *MAT_024 and *MAT_081–082 models are compared and contrasted in terms of representation of subsystem-level headliner head form impact behavior.

9.1   INTRODUCTION

Low-density random glass-fiber-mat thermoplastics (LD-GMT) are widely employed in automotive interior trim applications, where they may be required to absorb impact energy. A combination of light-weight, stiffness, and damping properties makes LD-GMT attractive candidates for automotive interior applications.[1,3] Optimal design of such composites is therefore critical for satisfying the performance requirements, particularly in dynamic impact scenarios, for interior trim components. CAE-aided design optimization of LD-GMT interior trim components, such as headliner substrates, therefore requires accuracy in impact simulations in terms of the estimated peak stresses and accelerations, and impact performance criteria where applicable.

A schematic representation of a typical automotive headliner with an LD-GMT substrate is provided in Figure 9.1. Three distinct layers can be identified in the headliner assembly: an outer foam/fabric layer (3–4 mm), a scrim layer (0.65 mm), and a foamy core (3–5 mm); the scrim and core together constitute the LD-GMT component. All the layers are tightly bonded together post manufacturing and then molded into the required shape. LD-GMT are typically manufactured using a modified paper making process (see, e.g., Ref. 1). For manufacturing a LD-GMT substrate, glass fibers are chopped and mixed with thermoplastic resin along with additives, and the mixture is dispersed in water to make aqueous slurry which is then deposited over a belt substrate moving into a vacuum chamber. The mix is dehydrated in the chamber and cured in an oven. Very light fiber-reinforced thermoplastic foam with high stiffness to weight ratio is thus manufactured. The next step is the lamination which requires the material to be hot and then compressed for creating a good bond between the glass fibers and the resin; this process results in the formation of a skin or scrim, that is different in properties from the foamy core. Reducing the thickness in this way also makes the substrate stiffer. It is then cooled to lock the mesh of closely bonded glass fiber and resin. After the substrate is formed, any additional scrims, fabrics, and other layers required for the specific application are added to the substrate. This laminated assembly is then thermoformed into final component, such as a headliner.[1] In the remaining portion of this document, the LD-GMT core-scrim sandwich laminate will be referred to as "headliner substrate" or simply "substrate." In case of the commercially available LD-GMT headliner substrate that is studied in this report, the substrate core provides the functionalities of stiffness and energy absorption, while the thinner scrim layer provides toughness. This distribution of functionalities is in contrast with, for example, a few other headliners studied by Savic et al.[4] in which the scrim was the stiffer layer which increases the resistance to impact, while the core was the softer composite material. In all automotive headliners, the substrate should ideally absorb enough energy during impact so as to prevent contact with the steel roof or rails. The outer foam/fabric layer that is laminated on the LD-GMT headliner substrate typically consists of two intermediate layers—foam and outer fabric. The foam layer may further absorb a small portion of the impact energy and the outer fabric is used specifically for aesthetic purpose depending on the vehicle type.

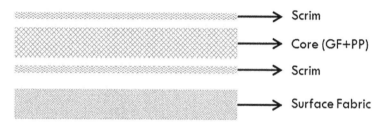

FIGURE 9.1   A schematic showing the typical layer composition of a headliner, composed of a LD-GMT substrate (core + scrim) laminated with outer foam and fabric layers.

9.2   OBJECTIVE AND MODELING APPROACH

The Federal Motor Vehicle Safety Standard (FMVSS) 201U (Occupant Protection in Interior Impact—Upper Interior Head Impact Protection)[5–7] prescribes the minimum performance requirements for automotive components in the upper interior compartment of the automobile during interior impact scenarios. The evaluation of performance of automotive components for FMVSS-201U involves subsystem-level impact testing, in which a Free Motion Head form is propelled at selected velocity to impact several points within the vehicle upper interior area. Deceleration transient of the head form upon impact is recorded using a triaxial accelerometer that is placed inside the center of gravity of the head form; this information is then analyzed for arriving at Head Injury Criteria (HIC). The HIC(d) criterion[5] to assess head injury potential in automobile crash test dummies is derived from the HIC number which is a standardized maximum integral value of the head form acceleration-time response. The performance metrics of the upper interior components are then governed by the requirement that HIC(d) should not exceed 1000.[5] Optimal design of headliner is critical for satisfying the FMVSS (201U) HIC(d) criterion; CAE aided design optimization of headliner components therefore would require accuracy in head form impact simulations in terms of the HIC number and the peak acceleration correlation. Since energy absorption during head form impact primarily is driven by the headliner substrate, it is therefore critical to develop faithful and versatile material models for the headliner substrate in a timely, efficient, and cost-effective manner. This project is aimed at the characterization of a commercially available LD-GMT headliner substrate and the assessment of accuracy of impact simulations carried out in a commercially available Finite Element Method (FEM) code LS-DYNA[8] in which these substrates are represented using selected available material models.

As discussed in the foregoing, the headliner substrate is essentially a layered composite (composed of a core layer sandwiched between two scrim layers). It may be modeled in CAE using either a "layered" or a "smeared" approach.

9.2.1   LAYERED (DETAILED) APPROACH

In the layered approach the scrim and the core of the substrate material are assigned separate material models, which require characterization data of each layer separately. The layered approach is of course detailed and allows for accounting of several possible mechanisms of deformation and damage evolution within the laminated composite. CAE simulation approaches for sandwich composites involving detailed rendition of each layer have been widely employed (see, e.g., Refs 9–17) for understanding their behavior in localized bending,[9,15,16] compression,[14,16,17] as well as impact[10,12,13,16,17] scenarios. The focus of these previous studies has been on development of the theoretical framework,[9–11,13] development of experimental characterization techniques[11,15–17] and optimization of the cellular core layer.[11,14,15] The cellular core in the majority of these studies is composed of metallic honeycomb[16] or polymeric honeycomb structures,[11,13–15] foams,[12] or more complex fold core geometries,[14,17] while the skin is composed of fiber-reinforced polymers[11–13,15–17] or metallic sheets.[9] Of relevance to the context of the present work are studies[12,14–17] which have focused on more applied aspects of material model parameter identification for employing in commercial FEM codes. For example, Feraboli et al.[16] carried out simulations of impact on sandwich composites involving aluminum honeycomb cores and carbon–epoxy face sheets using LS-DYNA;[8] in their study the skin was represented using DYNA *MAT_054[8] and core using a *MAT_HONEYCOMB (*MAT_126),[8] and a tie-break contact[8] was employed to represent the adhesive. While the final correlations achieved by Feraboli et al.[16] using the detailed "building block" approach were good, Feraboli et al.[16] state that the material model and contact formulations required "extensive calibration" at the coupon, element and subcomponent levels of the "building block pyramid" before optimal material model parameters could be identified.

9.2.2   SMEARED (HOMOGENIZED) APPROACH, AND WHY IT WAS EMPLOYED FOR PRESENT STUDY

While it is clear from the foregoing that the detailed layered approach has been widely used for simulations of sandwich composites, the layered approach was not employed to represent the LD-GMT composite in the present study. Instead the LD-GMT was represented in a homogenized or smeared fashion in which the substrate was tested as a whole (core and scrim together) and represented in LS-DYNA using a single material model input file. The homogenized approach was adopted due to the following challenges presented by the use of layered approach in the context of LD-GMT:

1. *Nature of Interface between Layers*: It needs to be noted that in case of the sandwich composites presented in Refs 9–17, the sandwich structures were assembled through application of an additional adhesive layer which could be separately characterized, thus providing parameters for the cohesive zone/tie-break models. By contrast, in case of LD-GMT composite which is formed by a calendaring process, the interface between the scrim and the foam core is physically bonded without adhesives, through inter diffusion of polymer chains. Therefore, characterization and assignment of parameters for the interface is significantly more challenging. Also, the scrim of the LD-GMT material was very thin and difficult to peel out for separate characterization.

2. *Cost*: It is clearly stated by both Feraboli et al.[16] and Heimbs et al.[17] that a "high degree of complexity" and "extensive calibration" is required to assign all the material model parameters associated with the skin and core materials and the interface for developing an accurate layered model. Further, it is obvious from the building-block approach of Feraboli et al.[16] that the detailed experiments that probe the various failure modes of the individual layers require significant investment of experimental effort and cost at the coupon, element and subcomponent levels before an accurate component level representation may be

obtained in FEM. While investing such effort and cost into the development of the layered approach may be required and even reasonable in the context of civil or aerospace structures (which are in general cost-intensive and also entail design longevity), the same may not be practicable in case of automotive structures where cost is the overriding factor and design variations are frequent.

3. *Layer Composition and Supplier Base*: Further, in developing the modeling strategy in the present study, it was recognized at the outset that the methodology needs to be robust and simple enough so that it need not be modified in the event of any change of the supplier or the individual core/scrim layer composition, both of which can occur frequently with respect to the automotive supplier base for headliner substrates. In such scenarios, development of a detailed multilayer model for each new headliner substrate may not be practical or cost-effective.

For homogenized rendering of the LD-GMT headliner substrate in LS-DYNA, *MAT_024[8] was selected as the primary candidate material model. However, it should be noted that *MAT_024 model is developed using tensile stress–strain information, and it may not successfully capture the subtle details of the post-peak softening behavior associated with out-of-plane deformation of the headliner. In general, the limitations of plasticity model formulations in capturing non-monotonic stress–strain characteristics has been well documented (see, e.g., Ref. 18). Therefore, with the aim of exploring the feasibility to overcome some of the limitations of *MAT_024 in estimating softening and post-peak characteristics, an additional model candidate was evaluated; LS-DYNA *MAT_081-82 (*MAT_PLASTICITY_WITH_DAMAGE)[8] is essentially an extension of *MAT_024, in which the plasticity model is the same as that employed in *MAT_024, but an additional plastic strain-dependent damage evolution model is overlaid.

In the subsequent, the steps followed for developing the LS-DYNA *MAT_024 and *MAT_081–082 material models for the homogenized "smeared" representation of a commercially available LD-GMT, for evaluating its use as headliner substrate, will be described. In this report, the focus will be on development of the plasticity model formulation (*MAT_024) using tensile test information obtained at two test velocities. The plasticity model is overlaid with a dynamic localized-out-of-plane impact load-case-tailored damage function in *MAT_081–082. The resulting *MAT_024 and *MAT_081–082 models are compared and contrasted in terms of representation of subsystem-level headliner head form impact behavior.

9.3 EXPERIMENTAL

Mechanical characterization of the LD-GMT substrate was carried out on flat, as-received substrate panels provided by the supplier, without any further processing or compaction. As mentioned earlier, the core and scrim layers were not separated, and the substrate was characterized as a whole with the aim of developing a homogenized CAE material model. Tensile tests were carried on the material with the aim of developing the basic plasticity model applicable to both *MAT_024 and *MAT_081–082. Dynamic localized out-of-plane impact loading test (interchangeably labeled punch test or dart impact test in the subsequent) was carried out with the aim of iterating upon the load-case-specific damage function which can be overlaid on the plasticity model to arrive at the *MAT_081–082 model. The choice of impact loading to probe the damage model was an attempt to achieve kinematic analogy to the headliner head form-impact scenario.

9.3.1 TENSILE TESTING

Tensile tests were carried out as per ASTM D638-03 Standard.[19] The test specimens conforming to ASTM D638-03 Type-I geometry were cut from the as-received substrate plaques. Tests were carried out on five specimens in the 1-1 direction (aka fore-aft or FLOW direction, with the specimen gage length parallel to the direction of the primary orientation of the chopped glass fibers in the plaque) and on six specimens in the 2-2 direction (aka cross-car or CROSS-FLOW direction, with the specimen gage length perpendicular to the direction of the primary orientation of the chopped glass fibers in the plaque). Tensile tests were carried out at two velocities. First, a quasi-static test was devised with a constant velocity of 0.0005 m/s. Another set of tests was carried out at a higher velocity of 2 m/s. The stress–strain responses obtained from the tests are shown in Figures 9.2a and 9.2b for specimens tested in the 1-1 and 2-2 directions, respectively.

As shown in Figures 9.2a and 9.2b, significant spread is observed in the test data. Several factors may have contributed to this spread: the fluctuations in the curves may be typically because of the uneven movement of the cross-head as the specimen is pulled; further, scatter may be due to inherent spatial in homogeneities in the LD-GMT substrate characterized by localized microstructural features such as resin rich regions, regions with fiber clusters, and regions with preferential fiber orientation as well as spatial variation of the "loft." Even after accounting for this data-scatter, the specimens characterized in the 1-1 direction (Fig. 9.2a) are significantly stiffer than those in the 2-2 direction (Fig. 9.2b); this is not unexpected since the 1-1 direction is the flow direction of the LD-GMT manufacturing process and the fibers are expected to preferentially orient in this direction. There is not a significant difference, however, in terms of the strains to failure in the 1-1 and 2-2 specimens.

Figures 9.2a and 9.2b also show median curves that provide a visual estimate or a rough representation of the average of the experimental spread for each testing velocity. The median curves are best polynomial fits for the entire experimental data set corresponding to each test velocity for each direction. From an inspection of the median curves, it is clear that after accounting for the experimental spread, rate dependence is more pronounced in the flow (1-1) direction compared to the cross-flow (2-2) direction. From Figure 9.2a for the 1-1 direction specimens, it may be seen that the median curves for the two test velocities are clearly separated right from the elastic regime, showing a clear stiffening of response at the higher test velocity. On the other hand, there is significant overlap between the median curves for the two test velocities in the cross-flow direction (Fig. 9.2b); the overlap is complete in the elastic regime. No conclusion may therefore be made about rate dependence of the material in the 2-2 direction because of the near complete overlap of experimental spread in the two velocities tested.

The clear difference in flow versus cross-flow characteristics presented a few challenges in identifying the representative curves for development of the rate-dependent piecewise-linear plasticity formulation, as shown in Figure 9.3. In Figure 9.3a, the quasi static data for both 1-1 and 2-2 directions are plotted together, while the data at higher velocity are plotted together in Figure 9.3b. One way of arriving at the representative curves for development of the piecewise linear plasticity model would have been an averaging of the flow and cross-flow responses of the material at each test velocity to arrive at a directionally averaged response (since the piecewise-linear plasticity formulations available in LS-DYNA are inherently isotropic). However, as seen from a visual comparison of the representative median curves plotted in Figures 9.3a (green curve) and 9.3b (brown curve), such averaging would obscure any rate dependence effects in the material.

Therefore, as shown in Figure 9.4, only the 1-1 (flow) direction data was employed for averaging to arrive at the representative curve at the higher test velocity. Test data for all the specimens in both 1-1 and 2-2 directions were averaged to arrive at the quasi-static representative curve.

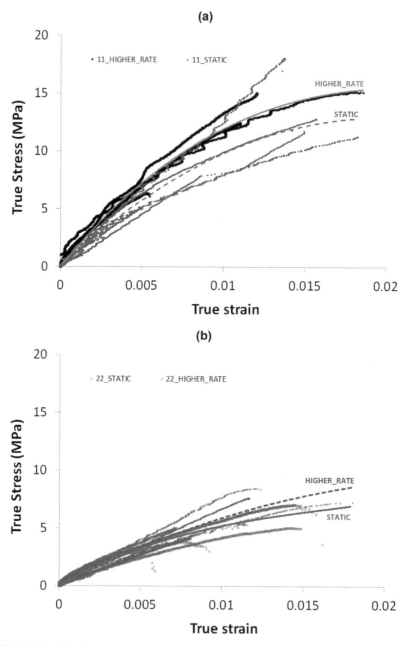

FIGURE 9.2   Tensile stress–strain raw data curves in the (a) 1-1 (flow or fore-aft) direction and (b) 2-2 (cross-flow or cross-car) direction. The graphs show experimental spread in the tensile tests carried out under quasi-static condition and at a higher velocity, along with the representative median curves for each test condition.

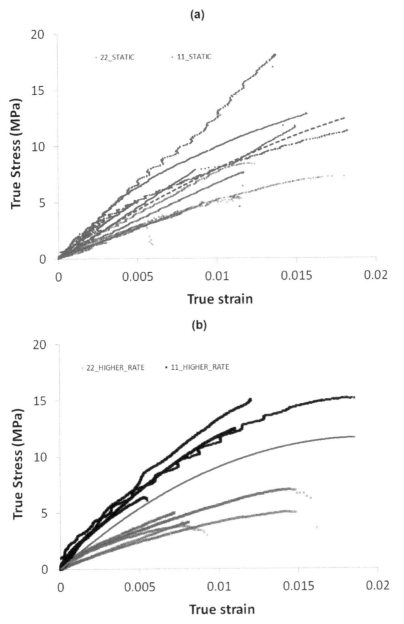

FIGURE 9.3 Directionality in tensile stress–strain characteristics: Spread of experimental data in 1-1 (fore-aft or flow) direction and 2-2 (cross-car or cross-flow) direction in (a) quasi-static tests and (b) tests carried out at a higher velocity. The green curve in 3(a) and the brown curve in 3(b) show the median curves representing all the test data combined from both 1-1 and 2-2 directions for each testing velocity.

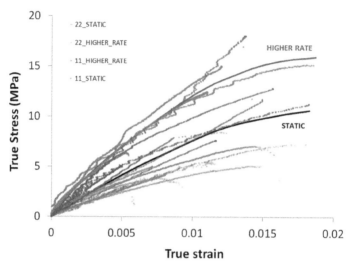

FIGURE 9.4   Considerations underlying choice of representative curves for the development of rate-dependent piecewise linear plasticity model input file (DYNA *MAT_024). This figure shows the experimental spread in quasi-static as well as higher rate test data in both 1-1 and 2-2 directions. For quasi-static data, the representative curve is selected to be the average of experimental spread accounting for information in both 1-1 and 2-2-directions. However, for data at higher rate, only the average of 1-1 direction is selected as the representative curve; this is because in the 2-2 direction, rate dependence is not evident as the experimental spreads at both rates overlap.

9.3.2   PUNCH TESTING

Localized out-of-plane dynamic impact characteristics of the LD-GMT substrate were measured by employing a driven-dart impact test as per the ASTM D3763 standard.[20] Square $4'' \times 4''$ specimens were cut from the as-received LD-GMT substrate sheets for testing. The square specimens were placed above a supporting plate with a circular opening of 90 mm diameter; a metal ring of width 0.5" and inside diameter of 90 mm was then placed on top of the specimen (with the inner circumference of the ring matching with the circumference of the circular opening on the supporting plate) and bolts were used to clamp the specimen between the ring and the bottom support. This resulted in circular specimen test geometry of diameter 90 mm constrained along its circumference. Impact load perpendicular to the specimen surface was applied using a cylindrical steel impactor with a hemispherical end of 25 mm diameter that was connected to a plunger assembly that was driven at a velocity of 200 m/min.

The load–displacement curves obtained for the punch test are plotted in Figure 9.5. Compared to the tensile tests, there is significantly lower scatter observed in punch data. All curves show a peak load of about 500 N; the peak is preceded by portions of reducing slope, and followed by gradual reduction of loads, indicative of softening or damage evolution before rupture.

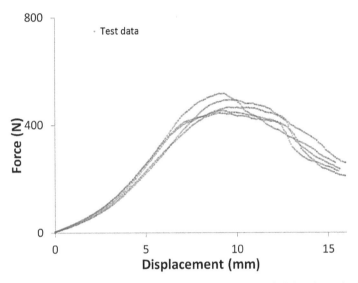

FIGURE 9.5 Experimental load–displacement curves recorded in dynamic localized out-of-plane driven-dart impact (punch) tests.

9.4 MATERIAL MODEL DEVELOPMENT

In the subsequent sections, the development of the CAE model for the LD-GMT substrate—employing the test information described in the foregoing—will be discussed.

*MAT_024 and *MAT_081–082

Brief descriptions of the LS-DYNA Keyword input files[8] for the two material model candidates, *MAT_024 and *MAT_081–082, are provided in Figure 9.6a and 9.6b, respectively. The basic formulation of both models is based on radial-return plasticity governed by von Mises flow rule. In this isotropic elastic-plastic formulation, deviatoric stresses satisfy the von Mises yield function[21]:

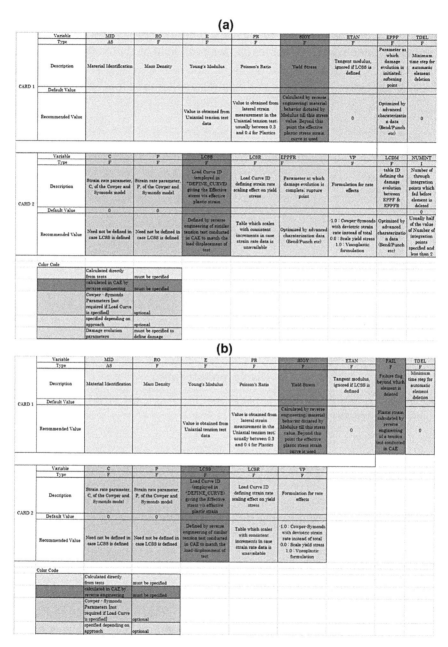

FIGURE 9.6 Brief LS-DYNA keyword descriptions for (a) *MAT_024 (*MAT_PIECEWISE_LINEAR_PLASTICITY) and (b) *MAT_081–082 (*MAT_PLASTICITY_WITH_DAMAGE).

$$\phi = \frac{1}{2} s_{ij} s_{ij} - \frac{\sigma_y^2}{3}$$

In the above equation,

$$\sigma_y = \beta \left[\sigma_0 + f_h \left(\varepsilon_{eff}{}^p \right) \right]$$

The hardening function $f_h \left(\varepsilon_{eff}{}^p \right)$ is taken to be linear by default[21] or can be specified in the form of a table. The parameter accounts for rate dependent effects. In FEM implementation of this model, the deviatoric stresses S_{ij} are updated elastically,[21] and for each update, if the yield function is not satisfied, the plastic strain is updated as shown below:

$$\Delta \varepsilon_{eff}{}^p = \frac{\left(\frac{3}{2} s_{ij}^* s_{ij}^* \right)^{\frac{1}{2}} - \sigma_y}{3G + E_p}$$

where G is the shear modulus and E_p is the current plastic hardening modulus. The trial deviatoric stress is scaled back using the radial return function.[21]

Thus both model candidates, *MAT_024 or *MAT_PIECEWISE_LINEAR_PLASTICITY as well as *MAT_081–082 or *MAT_PLASTICITY_WITH_DAMAGE,[8,21] allow the definition of an elastoviscoplastic material. The elastic regime is characterized by Young's modulus E, and the Poisson's ratio (PR). Elastic-plastic behavior may be represented using a bilinear form employing a tangent modulus ETAN, and yield stress SIGY; alternatively a piecewise linear representation may be employed by defining an appropriate true-stress versus effective plastic strain curve in addition to the definition of E, and PR. *MAT_024 also accounts for strain rate dependent behavior, by allowing the specification of a set of true-stress versus effective plastic strain curves at different effective strain rates. It also allows for definition of material failure based on plastic strain (FAIL). *MAT_024 has been conventionally employed for representation of thermoplastic materials in LS-DYNA. However, it should be noted that *MAT_024 model (which is developed using tensile stress–strain information) may not be able to successfully capture the subtle details of the softening and post-peak behaviors, for example, associated with bending of the headliner.

The second model candidate, *MAT_081–082 or *MAT_PLASTICITY_WITH_DAMAGE, uses the same framework as *MAT_024 for the definition of the piecewise linear plasticity formulation. But in addition, it allows for the consideration of damage evolution before occurrence of rupture. The need for modeling damage evolution in order to faithfully represent non-monotonic stress–strain characteristics (involving softening or post-peak behavior) in plastics has been emphasized in several studies (see, e.g., Refs. 18, 22, 23). In *MAT_081–082, damage may be defined in an arbitrary fashion in terms of plastic strain. Once the magnitude of plastic strains exceed the magnitude ascribed to the plastic strain-at-failure (EPPF) parameter, the evolution of damage and the resulting material softening until attainment of plastic strain-at-rupture (EPPFR) may be defined. This framework provides the possibility of defining the elastoplastic behavior based on uniaxial stress–strain information (same as that in *MAT_024), but independently defining the damage evolution to suit a different load case (with the prescription of EPPF, EPPFR, and the shape of damage evolution function—identified by LCDM—between those two plastic strains). In this manner, *MAT_081–082 is expected to provide more controlled definition of the softening or post-peak characteristics of the material compared to *MAT_024. For further details on the implementation of these models, please refer to the LS-DYNA Keyword Manual[8] and LS-DYNA Theory Manual.[21]

In the current study, the elastoplastic material model definition was common to both *MAT_024 and *MAT_081–082, and was derived using tensile stress–strain information available both at quasi-static condition and at a higher test velocity (refer to Figs. 9.2–9.4). Two versions of *MAT_024 were developed, the first employing only the quasi-static test curve, and the second employing test information at both test velocities to result in a rate-dependent formulation. The two versions of *MAT_024 were then employed without any further modifications to simulate the punch test (Fig. 9.5) to verify their suitability to simulate dynamic loading scenarios analogous to head form impact on a headliner. By contrast, in the case of *MAT_081–082, the simulation of punch was employed to further tune the definition of plastic strain-dependent damage evolution to identify the best combination of the plasticity curve and damage evolution function that would provide a good representation of the headliner material; again two versions of *MAT_081–082 were developed corresponding to the quasi-static and rate-dependent formulations of *MAT_024, each version containing a tailored damage curve. The trends obtained from these four models in the punch simulations as well as subsystem-level headliner head form impact simulations were then compared and contrasted.

9.4.1   PIECEWISE LINEAR PLASTICITY MODEL PARAMETERS

The piecewise linear plasticity model parameters were deduced from tensile test data. The considerations underlying the choice of test curves for averaging to arrive at the representative curve at each test velocity have been discussed earlier (refer to Fig. 9.4). For the averaging process, to ensure perfect match of the abscissa (strain) for averaging of the ordinate (stress), each stress–strain curve was first smoothened with a polynomial fit; each experimental stress–strain curve was then represented using equi-X (strain) interpolated X–Y pairs as defined by the polynomial fit, before carrying out the Y (stress)—averaging. In Figure 9.4, the two representative curves for the quasi-static and higher rate tests are shown juxtaposed with the overall experimental data spread for all the tensile tests in 1-1 as well as 2-2 directions put together.

9.4.1.1   QUASI-STATIC FORMULATION

For development of the quasi-static version of the piecewise-linear plasticity formulation, the quasi-static representative curve was then used to calculate the Young's modulus (E), the yield stress (SIGY) and the DYNA effective true-stress plastic strain parameters which are common for both *MAT_024 and *MAT_081–082 material models. Young's modulus was calculated between the strain limits of 0.0005 mm/mm and 0.0025 mm/mm as prescribed in the ISO-527 standard.[24] The magnitude of true stress corresponding to the strain of 0.0025 (upper limit of elastic regime calculation) was then assigned to the parameter SIGY (Yield stress). The true-stress plastic strain curve was then calculated from the representative engineering stress–strain curve using the following relationships:

True stress, $\sigma = S * (1 + e)$
True strain, $\varepsilon = \ln (1 + e)$
Plastic strain, $\varepsilon_p = \varepsilon - \sigma / \varepsilon$,

where S is the Engineering stress and e is the Engineering strain
    The true stress plastic strain curve was identified in the material model input file with the LCID parameter (refer to Fig. 9.6). PR was calculated from the tensile tests and was found to be 0.41. These parameters complete the definition of the quasi-static piecewise linear plasticity model definition.

9.4.1.2   RATE-DEPENDENT FORMULATION

A similar method was also followed to determine the E and SIGY parameters and the true-stress plastic strain curve corresponding to the representative engineering stress–strain curve at the higher velocity. However, the combination of this information with quasi-static information to arrive at a rate-dependent formulation is not straightforward. This is because the initial slope of the representative tensile stress–strain curve for the higher-velocity test was significantly higher compared to that of the quasi-static curve; *MAT_024 on the other hand, allows for assignment of only one value for E and SIGY. To overcome this limitation of piecewise linear plasticity formulation, the highest recorded modulus (i.e., the initial slope corresponding to the higher velocity test) was assigned to the parameter E. The quasi-static curve then would follow the initial slope of the higher velocity curve (refer to Appendix A). However, to ensure that the true-stress true-strain curve for the quasi-static scenario is not overestimated, the SIGY parameter was assigned a value of the true stress of the higher velocity curve, at a strain significantly lower than 0.0025; the parameter SIGY now signifies a branch point for the curves at the two velocities (again refer to Appendix A). After assignment of E and SIGY which bounds the overlapping elastic regime for both the test curves, the true-stress plastic strain curves were then calculated separately for each velocity as described earlier.

The resulting rate-dependent formulation involves assignment of a Table ID in the LCID parameter, which identifies the average strain rates at each test velocity. These numeric entries for strain rates then become curve identifiers for the true-stress plastic strain curves at each respective strain rate. Assignment of the nominal or representative strain rate corresponding to each test velocity was arrived at using trial and error. Iterations involving tensile simulation of an ASTM D-638[19] test-coupon discretized using quadrilateral shell elements (see Fig. 9.7a) were carried out by systematically changing the nominal strain-rate entries in LCID corresponding to the true-stress—effective-strain curve for each test velocity, until a perfect match was seen between the experimental and simulated curves. Using this method, the nominal strain rates corresponding to tests carried out under quasi-static and higher velocity conditions were identified as 3×10^{-6} and 1×10^{-4} 1/s, respectively. As seen in Figure 9.7b the perfect overlap of the simulated stress–strain curves with the representative stress–strain test curves at the two velocities for the substrate validates the rate-dependent plasticity model.

(a)

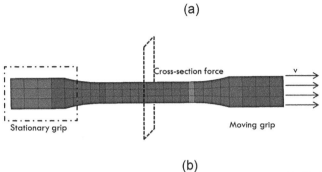

(b)

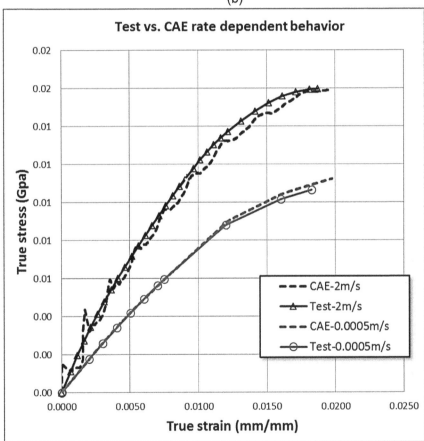

FIGURE 9.7 (a) The ISO-527 shell simulation geometry that is employed for LS-DYNA simulation of tension, to validate the rate-dependent piecewise-linear plasticity material mode (DYNA *MAT_024); and (b) the comparison of the simulated stress–strain curves at quasi-static and higher strain rate tension with the averaged representative stress–strain curves.

The resulting quasi-static and rate-dependent versions of the piece-wise-linear plasticity models (hereafter referred to as *MAT_024_QS and *MAT_024_RD, respectively) were employed in the dynamic localized out-of-plane impact (punch) simulations without any further modifications, and without any assignment to the FAIL parameter (refer to Fig. 9.6a). Figure 9.8a displays the geometry and boundary conditions employed for the punch simulation. It should be noted that for punch simulations, the LD-GMT substrate specimen was discretized using hexahedral solid elements, with two elements along the thickness direction to better capture the penetration and failure as observed during the impact test. The estimates obtained using *MAT_024_QS and *MAT_024_RD are compared with the experimental punch load–displacement curves in Figure 9.8b.

It is clear from an inspection of Figure 9.8b that the quasi-static version *MAT_024_QS provides an accurate estimate of the initial loading transient during the impact. The rate dependent version *MAT_024_RD; on the other hand, over predicts the initial slope during impact loading. This over prediction can be associated with the assignment of higher magnitude of E (Young's modulus) in the rate-dependent formulation. However, beyond the initial loading transient, both models provide very poor representation of the substrate material in impact loading. Neither *MAT_024_QS nor *MAT_024_RD predicts any softening during the course of impact. While in tests, peak load of around 500 N with significant post-peak reduction of loads was observed, the simulations with *MAT_024 only predict a monotonic increase of forces for the entire duration of impact all the way up to 2600 N in case of *MAT_024_RD. In the absence of any assignment to the FAIL parameter, *MAT_024 is unable to predict penetration damage and rupture as observed in the test.

9.4.2 DAMAGE MODEL PARAMETERS

Two versions of *MAT_081–082 models for the LD-GMT substrate were developed, one each corresponding to the quasi-static and rate-dependent versions respectively of *MAT_024 discussed earlier. The damage rule was tailored with the aim of getting a better representation of the substrate characteristics in dynamic impact, particularly the softening and post-peak behavior which was not at all captured by *MAT_024 (refer to Fig. 9.8b). In order to arrive at the appropriate damage curve, the parameters EPPF, EPPFR, and the shape of the damage function identified by LCDM were systematically iterated by trial and error so as to get the best possible representation of the experimental load–displacement curve.

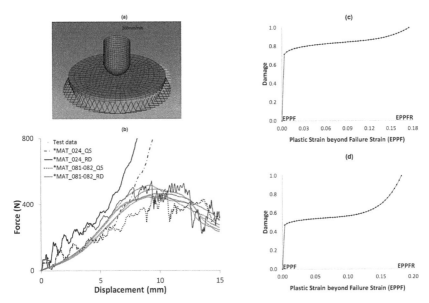

FIGURE 9.8 (a) Punch simulation geometry, (b) comparison of experimental punch load–displacement curves with estimates obtained using quasi-static formulations and rate dependent formulations of *MAT_024 (without assignment of FAIL) and *MAT_081–082 material model input files, (c) the damage curve specification in *MAT_081–82_QS—specifically tailored for dynamic out-of-plane loading—that is overlaid on the quasi-static piecewise-linear plasticity formulation, and (d) the damage curve specification in *MAT_081–82_RD—specifically tailored for dynamic out-of-plane loading—that is overlaid on the rate-dependent piecewise-linear plasticity formulation.

The estimated load–displacement curves obtained with the quasi-static and rate-dependent versions of the *MAT_081–082 models (hereafter referred to as *MAT_081–082_QS and *MAT_081–082_RD respectively) in the impact (punch) simulation are compared with the experimental punch load–displacement curves in Figure 9.8b. To arrive at *MAT_081–082_QS, the quasi-static piecewise-linear plasticity formulation (*MAT_024_QS) was extended with the damage evolution function shown in Figure 9.8c (with EPPF = 0.025 and EPPFR = 0.2). For *MAT_081–082_RD, correspondingly, *MAT_024_RD was extended with the damage evolution function shown in Figure 9.8d (with EPPF = 0.0175 and EPPFR = 0.2).

As seen in Figure 9.8b, the initial slope of the load–displacement curves estimated by *MAT_081–082_QS and *MAT_081–082_RD follow trajectories similar to those estimated by the corresponding piecewise-linear plasticity formulations (*MAT_024_QS and *MAT_024_RD, respectively). Clearly, this initial slope is governed by the magnitude of E (Young's modulus) alone. Just as in case of the *MAT_024 models, the higher magnitude of E assigned in *MAT_081–082_RD results in a stiffer response compared to the test. *MAT_081–082_QS is more accurate in predicting the initial loading transient. Both versions of *MAT_081–082 perform significantly better than their equivalent *MAT_024 formulations in terms of predicting the peak load, the softening and the significant post-peak drop in loads. Clearly, the additional detail provided by the damage evolution function allows *MAT_081–082 to more systematically address the softening, penetration, and rupture behavior observed in the test. However, when inspecting the overall quality of prediction over the entire range of displacements, no clear choice emerges between *MAT_081–082_QS and *MAT_081–082_RD. While the rate-dependent version provides a more accurate estimate of peak load, the displacement corresponding to the peak load, and the load transient after the peak load, it significantly over-predicts the initial slope of the response. On the other hand, *MAT_081–082_QS, while providing a more accurate prediction of the initial stiffness, does not provide a good match at subsequent displacements, although the prediction is still significantly better than either of the *MAT_024 versions.

It needs to be noted here that in subsystem-level scenarios, involving the impact of, e.g., a head form with the headliner in which the LD-GMT substrate has been incorporated, the overall response is not only governed by the LD-GMT substrate but also by the outer foam fabric layer of the headliner assembly, and the countermeasures placed behind the headliner that come in the load path. Thus the pristine response of LD-GMT headliner substrate governs only the initial transient of accelerations during the impact; therefore the choice of material model for headliner substrate would be governed more by the quality of match between simulations and tests in the initial loading regime. With this consideration, it would appear that the quasi-static formulations, i.e., *MAT_024_QS and *MAT_081–082_QS provide adequate representation of the LD-GMT material in the dynamic out-of-plane loading scenarios. However, in impact scenarios in the absence of countermeasure influence—where the LD-GMT substrate softening characteristics become more important—*MAT_081–082_QS would be a better choice compared to *MAT_024_QS. Of course this aspect will be investigated further in the next section.

Interestingly, *MAT_024 employed with the FAIL parameter can also result in estimates of pre-peak softening and post-peak load drops as seen in Figure 9.9. However, when employing the FAIL parameter, while the peak loads are matched and the forces do not keep on increasing, the post-peak signals are much noisier compared to those estimated with *MAT_081–082.

It is hypothesized that this noise may be reduced by increasing the number of elements along the thickness direction of the substrate specimen.

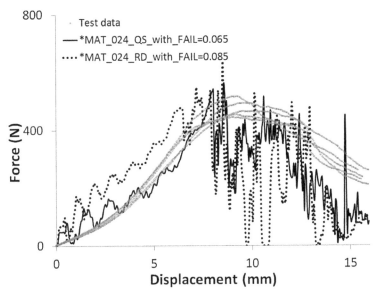

FIGURE 9.9 Comparison of experimental punch load–displacement curves with estimates obtained using quasi-static and rate-dependent formulations of *MAT_024 with assignment of FAIL parameter (the magnitudes of latter specifically optimized to obtain best representations of experimental punch data).

9.5 VALIDATION

The various versions of *MAT_024 (with and without FAIL parameter) and *MAT_081–082 (with damage curves tailored based on dynamic impact loading) for representing the LD-GMT substrate, discussed in the foregoing, were next verified in a subsystem-level headliner head form impact simulation. We showed that the estimates obtained from LS-DYNA simulations of head form impact on a carefully selected location in the headliner, where the influence of countermeasures was expected to be minimum and the initial acceleration (g) response was expected to be predominantly influenced by headliner characteristics alone.

In Figure 9.10a, the estimates obtained from headliner head form impact simulations—with the LD-GMT substrate represented using *MAT_024_QS and *MAT_024_RD—are compared with the experimentally measured acceleration-time response. The test results show a steady rise in acceleration till about 8 ms after which there is a drop followed by another steady rise up to the peak acceleration. At this point two key differences should be noted with respect to the punch test. Even though both tests show a drop following the peak, the post-peak drop in loads during punch tests was due to progressive material rupture associated with penetration of the impactor, while the post-peak drop in acceleration during head form impact tests was associated with withdrawal of the head form following impact (in the head form impact tests, no material breach or rupture was observed). Also, in punch tests and simulations, the entire response, all the way up to peak loads and post-peak behavior is ascribable only to the LD-GMT substrate, while in case of head form impact test and simulations, the peak acceleration is the result of cumulative energy absorption response of the headliner substrate as well as the trims, countermeasures, and the vehicle structural members in the load path. Of course, the initial slope of the acceleration, for the first 6–8 ms, is predominantly governed by the LD-GMT substrate behavior and influences the subsequent magnitude of the peak acceleration. Therefore an acceptable material model for the headliner substrate is expected to provide adequate representation of this initial portion of the acceleration transient.

As seen in Figure 9.10a, this initial response is predicted significantly better when the LD-GMT substrate is represented using *MAT_024_QS. As in case of punch simulations, *MAT_024_RD results in stiffer behavior in the initial loading regime compared to the experiment; clearly this mismatch is also governed by the assignment of the higher magnitude for the Young's modulus E in the model. Neither model however captures the intermediate drop in acceleration around 8 ms; in the test response, this intermediate drop may be associated with the softening response of the headliner before the influence of other components along the load-path drives up the accelerations further. The peak load is better predicted when the headliner substrate is modeled using *MAT_024_RD; however, the estimate of peak load has contributions from several material models in the load-path and therefore the match (or lack thereof) cannot be taken as an indication of the accuracy (or lack thereof) of the material model employed for the headliner substrate alone.

With greater weightage given to the accuracy of predicting the initial slope of, and intermediate drop in, the acceleration transient when evaluating the material model for the headliner substrate, the quasi-static versions *MAT_081–082_QS as well as *MAT_024-QS with the FAIL parameter assigned perform superiorly as shown in Figure 9.10b. Both models provide excellent estimates of the initial loading transient, and also seem capable of predicting an initial softening, albeit at times shorter than those observed in the test. Both models present a qualitatively accurate representation of the response all the way up to 10 ms. It is clear that greater accuracy in predicting the headliner substrate behavior not only results in better estimates of the pristine headliner behavior (in the first 6–8 ms) but also superior estimates of the trajectory leading to peak acceleration following the intermediate softening (between 6 ms and 10 ms). Beyond 10 ms, however, an abrupt drop in accelerations is observed; in the simulations, this is associated with element deletion. This prediction is obviously inaccurate since no breach or rupture was observed in the head form impact test. In addition to the material models themselves, several other factors such as contact settings, element formulation, and non-optimal element size can contribute to premature element deletion. All these factors need to be systematically investigated to eliminate the unexpected element deletion before *MAT_081–082_QS or *MAT_024_QS with FAIL may be employed at subsystem levels. However, it needs to be noted that despite the obvious drawback associated with element deletion, these two models provide the best estimates of the subsystem-level head form impact acceleration transients in the initial loading regime that is dominated by the LD-GMT substrate response.

Figure 9.10c clubs the predicted responses obtained with material models for the headliner substrate that provide the poorest performance at the subsystem-level simulations. Both *MAT_081–082_RD and *MAT_024_RD with FAIL not only estimate a stiffer response in the initial loading regime, but also result in element deletion beyond 10 ms, thus combining the drawbacks discussed in the context of Figures 9.10a and 9.10b.

Based on the foregoing, it is clear that *MAT_024_QS offers the most suitable candidate for predicting the LD-GMT substrate response in headliner subsystem-level simulations, both in terms of predicting the initial stiffness as well as overall stability. However, *MAT_024_QS does not predict the intermediate drop in accelerations. Material models with a framework for representing damage—i.e., MAT_024_QS with FAIL and *MAT_081–082_QS—overcome this limitation, but with an added drawback of unexpected element deletion. If the latter instability can be systematically eliminated, these models, especially *MAT_081–082_QS can offer significant improvement in accuracy of predicting the headliner response. Further, from a fundamental perspective, we need to note that the damage evolution for these models is tailored based on a localized out-of-plane impact test that involves penetration and rupture. The FAIL parameter in case of *MAT_024_QS with FAIL and the EPPFR parameter in case of *MAT_081–082_QS are assigned based on coupon-level localized penetration and failure response, whereas at the subsystem-level head form impact, the deformation is more diffuse and no penetration occurs. This therefore brings up an unmet need for better

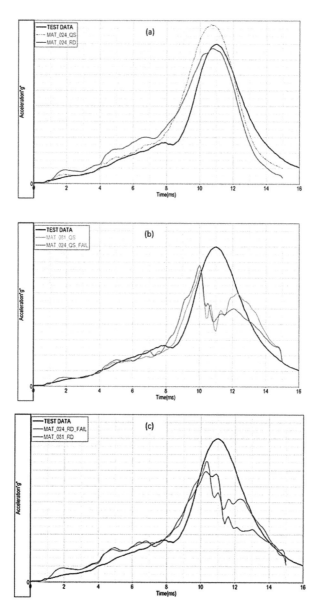

FIGURE 9.10 Comparison of measured acceleration-time responses obtained during head form impact with subsystem-level headliner assembly, with estimates from LS-DYNA head form impact simulations carried out with the headliner substrate represented using (a) *MAT_024 (quasi-static and rate dependent formulations, without assignment of FAIL parameter), (b) quasi-static formulations of *MAT_081–082 and *MAT_024, the latter with FAIL assigned, and (c) rate dependent formulations of *MAT_081–082 and *MAT_024, the latter with FAIL assigned.

tailoring the damage evolution parameters in these promising material models through the use of a customized test setup that offers a more distributed dynamic out-of-plane deformation without penetration.

9.6   CONCLUSIONS

Three key learnings may be extracted from considering the performance of the material models for the LD-GMT substrate both in coupon-level dynamic punch simulations and in subsystem-level headliner head form impact simulations:

Firstly, the initial stiffness response at both coupon-level punch tests and subsystem-level impact tests seems to be primarily governed by the Young's modulus of the substrate. Quasi-static versions of both *MAT_024 and *MAT_081–082, in which the parameter E is assigned a value obtained from quasi-static tensile tests, perform better in predicting this response. This would imply that a rate-dependent model need not be inherently better in predicting the dynamic response of the substrate; of course, a better piecewise-liner plasticity formulation which allows for rate-dependent variation of modulus (which is not accounted for in either *MAT_024 or *MAT_081–082) or a more careful selection of strain rates at which the substrate is tested, can improve the predictability of the rate-dependent formulations.

Secondly, LD-GMT material models which allow a mechanism for representing damage evolution—irrespective of whether it is instantaneous as in the case of *MAT_024_QS with FAIL or progressive as in the case of *MAT_081–082_QS—perform better in predicting both the post-peak load-transient at the coupon level and the pre-peak softening behavior in the case of subsystem-level simulations.

Thirdly, the localized out-of-plane dynamic dart impact test offers a good coupon-level qualitative verification mechanism for screening and choice of material models for representing the LD-GMT substrate in subsystem-level simulations. The models which provided best estimates of the substrate response in punch (i.e., *MAT_081–082_QS and *MAT_024_ QS with FAIL), also provided an adequate representation in subsystem-level head form impact simulations especially in the initial loading regime that is dominated by the headliner substrate response. *MAT_024_QS without the assignment of the FAIL parameter provided a good match with the initial loading transient both at the coupon-level and subsystem-level simulations; however, the inadequacy of this model in predicting failure and post-peak load-drop in coupon-level simulations is translated to the inability of predicting pre-peak softening behavior at the subsystem-level simulations.

Despite its shortcomings in predicting the pristine material response (in punch simulations) accurately, however, *MAT_024_QS without assignment of FAIL emerges as the best compromise in terms of accuracy and stability (at subsystem-level simulations). This is because *MAT_081–082_QS and *MAT_024_QS with FAIL, even though superior in terms of capturing the damage behavior of the substrate, cause instability at the subsystem-level simulations by inducing unexpected element deletion. It should be noted that the damage evolution function for these models is tuned based on a localized out-of-plane impact test that involves penetration and rupture. Therefore, there emerges the need to develop a novel coupon-level test setup that would allow better tailoring of damage evolution parameters using a load case involving distributed dynamic out-of-plane deformation without penetration. Such a load case would be more representative of head form impact scenario. Development of such tests is the focus of ongoing work and will be described in future reports.

ACKNOWLEDGMENTS

The authors acknowledge useful discussions with Keshavlal Rathi, Vesna Savic, Robert Bingham, and Srinivasan Velusamy. Helpful comments from the reviewer of this manuscript are also highly appreciated.

KEYWORDS

- **plasticity**
- **headliner**
- **material**
- **manufacturing**
- **thermoplastics**

REFERENCES

1. Bedford, A.; Wilshaw, T. *Development of High Strength Substrate for Use in Modular Headliner Systems;* SAE Technical Paper 910783: 1991.
2. Harper, C.; Kunal, K. *SPE Automotive and Composites Divisions - 12th Annual Automotive Composites Conference and Exhibition 2012*, ACCE 2012, 79–84, 2012.
3. Hipwell, J.; Proceedings of SPE Automotive Conference, Troy, MI, USA, 2005.
4. Savic V., et al., *Effects of Thickness on Headliner Material Properties*; SAE Technical Paper 2011-01-0463: 2011.
5. Laboratory Test Procedure For FMVSS 201U Occupant Protection in Interior Impact Upper Interior Head Impact Protection, Office of Vehicle Safety Compliance, TP201U-01 (1998)
6. Federal Motor Vehicle Safety Standard (49 CFR PART 571) MVSS 201 OCCUPANT PROTECTION IN INTERIOR IMPACT; ORIGINAL: F.R. Vol. *36* No. (232) - 02.12.1971 and amended to May 1991
7. Haque, E.; Kamarajan, J.; Yang, G. *Development and Characterization of New Headliner Material to meet FMVSS 201 Requirements;* SAE Technical Series 2000-01-0624: 2000.
8. LS-DYNA Version 5.1 Keyword Manual reference, http://www.ls-dyna.com
9. Thomsen, O. T. Theoretical and experimental investigation of local bending effects in sandwich plates. *Compos. Struct.* **1995,** *30*, 85–101.
10. Abrate S. Localized Impact on Sandwich Structures with Laminated Facings. *Appl. Mech. Rev.* **1997,** *50*(2), 69–82.
11. Davalos, J. P.; Qiao, P.; Xu, X. F.; Robinson, J.; Barth, K. E. Modeling and Characterization of Fiber-reinforced Plastic Honeycomb Sandwich Panels for Highway Bridge Applications. *Compos. Struct.* **2001,** *52*, 441–452.
12. Sriram, R.; Vidya, U. K. Blast Impact Response of Aluminum Foam Sandwich Composites. *J. Mater. Sci.* **2006,** *41*, 4023–4039.
13. Qiao, P.; Yang, M. Impact Analysis of Fiber Reinforced Polymer Honeycomb Composite Sandwich Beams. *Composites: Part B.* **2007,** *38*, 739–750.
14. Heimbs, S. Virtual Testing of Sandwich Core Structures using Dynamic Finite Element Simulations. *Comput. Mater. Sci.* **2009,** *45*, 205–216.
15. Heimbs, S.; Pein, M. Failure Behavior of Honeycomb Sandwich Corner Joints and Inserts. *Compos. Struct.* **2009,** *89*, 575–588.
16. Feraboli, P.; Deleo, F.; Wade, B.; Rassaian, M.; Higgins, M.; Byar, A.; Reggiani, M.; Bonfatti, A.; DeOto, L.; Masini, A. Predictive Modeling of an Energy-Absorbing Sandwich Structural Concept Using the Building Block Approach. *Composites: Part A.* **2010,** *41*, 774–786.
17. Heimbs S.; Cichosz, J.; Klaus, M.; Kilchert, S.; Johnson, A. F. Sandwich Structures with Textile-Reinforced Composite Fold cores under Impact Loads. *Compos. Struct.* **2010,** *92* 1485–1497.
18. Gu, G.; Xia, Y.; Lin, C. H.; Lin, S.; Meng, Y.; Zhou, Q. Experimental Study on Characterizing Damage Behavior of Thermoplastics. *Mater. Design.* **2013,** *44*, 199–207.
19. D638-03, Standard Test Method for Determination of Tensile Properties. *ASTM.*
20. D3763, Standard Test Method for High Speed Puncture Properties of Plastics Using Load and Displacement Sensors. *ASTM.*

21. LS-DYNA Version 5.1 Theory Manual reference, http://www.ls-dyna.com
22. Du Bois, P. A.; Kolling, S.; Koester, M.; Frank, T. Material Behavior of Polymers under Impact Loading. *Int. J. Impact Eng.* **2006,** *32*, 725–40.
23. Kolling, S.; Haufe, A.; Feucht, M. Du Bois, P. A. SAMP-1: A Semi-Analytical Model for the Simulation of Polymer. 4th German LS-DYNA Forum. 2005.
24. ISO 527-1:2012 Plastics. Determination of Tensile Properties. Part 1: *General Principles*

APPENDIX A
Incorporating Rate-Dependent Formulation in DYNA Piecewise Linear Plasticity Model

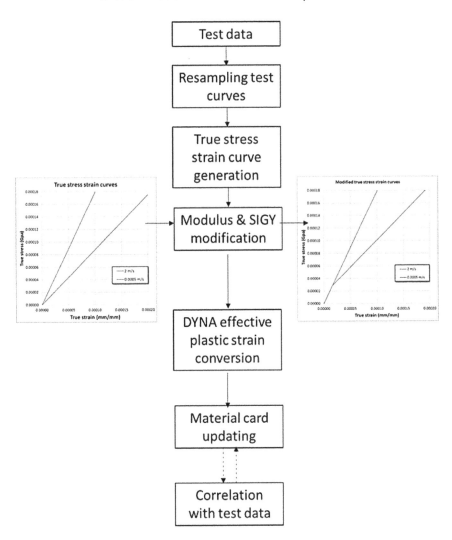

CHAPTER 10

REVIEW OF STRENGTH-CHARACTERIZING METHODOLOGIES IN CONCRETES

GEORGE OKEKE[1], S. JOSEPH ANTONY[1*], and NESIBE G. OZERKAN[2]

[1]*Institute of Particle Science and Engineering, School of Chemical and Process Engineering, University of Leeds, LS2 9JT, Leeds, United Kingdom*

[2]*Center for Advanced Materials, Qatar University, Doha, Qatar*

Corresponding author. E-mail: S.J.Antony@leeds.ac.uk

CONTENTS

ABSTRACT

This chapter presents a review on the recent advancements on the strength characteristics of conventional and composite concretes within the framework of fracture mechanics. Failure behavior of concrete materials under mechanical loading can be studied using fracture mechanics approaches. Using a cracking and energy based criterion, the failure of concrete structures can be described in the context of fracture mechanics. The principles of LEFM have been presented and stress field around a crack-tip is been described using a stress function approach.

10.1   INTRODUCTION

Concrete is a composite-based construction material formed from mixing aggregates (mostly fine and coarse granular material), cementitious materials (such as cement, fly ash, slag, etc.), water, and additives (whenever necessary). Concrete sometimes contains additional ingredients such as chemical admixtures including those in cementitious materials and other products. Concrete is very important in civil engineering applications, as it is used in making a wide variety of civil structures such as architectural structures, foundations, pavements, bridges, brick/block walls, highways, dams, fences, and so on. These civil structures are known to have different design requirements based on mechanical and durability properties, and, hence, concrete mixtures can be designed to meet these requirements.

The performance of concrete depends on factors such as the quality of the ingredients, their mixing proportions, placement, and exposure conditions.[88] Also, the amount and type of cement, water, temperature of mixing, type of fine and coarse aggregate as well as admixture, and the environment in which the concrete is exposed to, will determine its physicochemical and durability behavior. The effect of these parameters can be investigated using various analytical techniques.

The quality of freshly prepared concrete can be determined by the ease and homogeneity with which it can be mixed, transported, compacted, and finished. The quality and workability of concrete can be related to the amount of internal work required to produce adequate compaction.[75] In assessing the quality and workability of concrete, properties such as flowability, moldability, cohesiveness, and compactibility also need to be considered. The workability of concrete can be influenced by various factors, including plasticity of the cement paste, size and grading of the aggregates, shape and surface characteristics of the aggregate, quantities of paste and aggregates, and water content in the concrete mix.[88] Freshly prepared concrete can also be described based on its consistency or fluidity. This describes the ease with which a material flows and, also, the degree of wetness of concrete. Workability characteristic may differ across concrete samples. For example, concretes with similar consistency, may have different workability characteristics. Workability of concrete can be quantified by measuring its consistency. Various tests that can be used in measuring consistency include: the slump test (as described by ASTM C143), compacting factor test (as described in the BS1881 and by AC1 211), ball penetration test (as described in ASTM-C360), remolding test, and Vebe test.[58,88,89]

There has been recent advancement in the development of high performance concrete to meet the ever growing need of advanced superstructures. Most of these advancements have been in all areas of concrete production, including materials, durability, recycling, mixture proportioning, and environmental quality. Examples of such advancements include the development of chemical admixtures such as superplasticizers and supplementary mineral admixtures. These are known to enhance the performance, durability, strength, chemical resistance, water and cement reduction, as well as sustainability of concrete.[4,15,90,120] Macro-defect-free cements[32] as well as chemically bonded ceramics[35,83] are also used to achieve concrete with low porosity and high strength. Also, the addition of corrosion-inhibiting admixtures can achieve high durable concrete with enhanced service life that is mostly exposed to corrosive environments.[29,47] Apart from these admixtures, the inclusion of recycled waste materials such as fly ash, slag, and silica fumes, as well as fiber, is becoming increasingly popular. For example, the inclusion of fiber to concrete can improve the tensile and flexural strength, splitting resistance, impact strength, toughness, shock resistance, as well as plastic shrinkage resistance of concrete.[27]

Concrete is a heterogeneous material and can hence, exhibit some unique fracture characteristics such as tensile cracking. Concrete failure mostly involves the stable growth of large cracking zones and the development of large fractures before the maximum load is reached.[13] The application of fracture mechanics in concrete technology is receiving a growing level of attention as it involves the study of the response and failure of structures as a result of crack initiation and propagation.[101] For quasi-brittle materials such as concrete, fracture mechanics describes the failure in them using a strength, and energy-based failure criterion in the form of elastic and inelastic energies.[21]

In this report, various classical strength characterizing methodologies and techniques in the field of concrete technology within the framework of fracture mechanics are presented. These will provide more understanding into extending theories of fracture mechanics in describing the mechanisms also in advanced and complex heterogeneous materials such as concrete composite materials.

10.2   MECHANICAL PROPERTIES

Concrete possess certain mechanical properties in which they are required to attain specific or optimum specifications, to be regarded as good quality. These properties include: compressive strength, flexural strength, Poisson's ratio, splitting tensile strength, static modulus of elasticity, properties under triaxial loads, creep under compression, abrasion resistance, bond development with steel, pull out strength, penetration resistance, and so on.[88]

Concrete can be regarded as a multiphase or composite material, with regards to its mechanical behavior. The factors that influence the mechanical behavior of concrete are: size and distribution of particles, concentration, shape of particles, topology, composition of the disperse and continuous phases, composition between the continuous and disperse phases, and the pore structure.[58] As a composite material, it constitutes at least two chemically and mechanically distinct materials, with an interface separating the components. This material will possess different properties from the original components.[75,88] The water-to-cement (w/c) ratio is an important factor that determines the strength of a concrete. The relationship between this ratio and strength was initially developed by Abrams in 1918[1] and is presented below:

$$S = A / B^w \tag{10.1}$$

where S is the strength of concrete, and w is the w/c ratio. The constants A and B depend on the age, curing regime, cement type, and the method of testing. This relationship is known as Abram's law. This law mostly applies, provided the concrete is fully compacted, as below to a certain w/c ratio, further reduction with not yield the expected strength. This is because, for low w/c ratios, the concrete is not workable enough to allow full compaction. Also, the compressive strength of concrete can be reduced by air entrainment, and should be considered when applying the law.[2] Compressive strength is inversely proportional to the workability of concrete. Hence, when workability of concrete increases its compressive strength decreases. Air entraining agents act as concrete admixture to increase the workability of concrete, without reducing its compressive strength. The aim is to maintain the desired compressive strength and workability of concrete so as to achieve higher strength concrete.

Furthermore, the strength of concrete also depends on the strength of the paste, coarse aggregate, and the paste-aggregate interface.[88] The paste-aggregate interface is the weakest region of concrete and is where failure occurs before it occurs on the aggregate or the paste. Its weakness is due to the weak bonding and the development of cracks which tend to develop due to bleeding and segregation, and volume changes of the cement paste during setting and hydration. In a freshly prepared concrete, bleeding refers to the settlement of solids followed by the formation of a layer of water on the surface. The seepage of water to the surface transports some particles via localized channels. This can lead to the formation a layer of weak concrete containing diluted cement paste and fines from the aggregate. Upon the development of cracks due to bleeding, a transition zone which has a higher porosity and permeability is created which extends to about 50 μm from the surface of the aggregate. Typical test program for determining most of the fundamental mechanical properties of concrete on a single specimen is shown in Table 10.1.

TABLE 10.1 Test Program for Determining Most of the Fundamental Mechanical Properties of Concrete.[84]

Material property		Specimen
1	Dynamic Young's modulus	
2	Fracture energy	
3	Tensile strength	
4 a	Flexural tensile strength	
4 b	Young's modulus	
5	Compression strength	

10.3   RHEOLOGICAL CHARACTERISTICS OF CONCRETE

Rheology can be defined as the study of flow behavior, and is usually applied to materials that exhibit time-dependent response to stress, such as fluid materials. Flow can be measured using shear, with parameters, stress (τ) and strain ($\dot{\gamma}$). Its parameters can be obtained from measurements of torque and flow rate. These parameters can be used in defining viscosity (η) as presented below:

$$\eta = \tau / \dot{\gamma} \qquad (10.2)$$

Viscosity can be defined in other ways, such as: plastic viscosity (defined as the slope of stress versus strain rate for a plastic material) and differential viscosity (defined as the slope of the curve relating stress and strain rate). There are several types of flow behavior, with the Newtonian behavior recognized as the simplest (Fig. 10.1). It constitutes of a linear relationship between stress and strain rate, and zero stress at zero strain rate.[88] Newtonian behavior is that of an ideal fluid, and is analogous to the Hookean behavior in a solid. Plastic behavior also known as Bingham is also exhibited in a wide variety of fluids. In the case of this type of fluid behavior, flow only

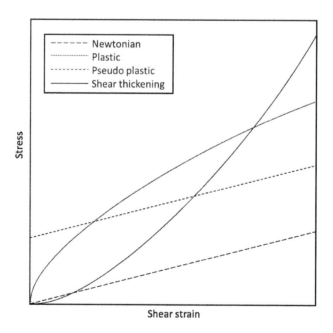

FIGURE 10.1   Types of rheological behavior.[65]

initiates above some stress level, and is known as the yield stress. Following this, the relationship between the stress and strain rate becomes linear. A behavior, in which viscosity decreases as strain rate increases, is referred to as pseudo plastic or shear thinning. An uncommon type of fluid behavior is the thickening behavior.

Two main factors affect the rheological behavior of concrete, and they are: the volume fraction of solid particles in the suspension and the extent of particle flocculation or dispersion.[88] Consequently, an increase in the volume fraction of solids (ϕ) will lead to an increase in the viscosity. Einstein proposed a relationship between viscosity of the suspension and the volume fraction of solids, as presented below:[110]

$$\eta = \eta_c (1 - 2.5\phi) \tag{10.3}$$

where η is the viscosity of the suspension and η_c is the viscosity of the fluid phase. Equation 10.3 represents the relationship at low volume fractions. At higher volume fractions, an extensive formula has been proposed by Krieger and Dougherty and is presented below:[69]

$$\eta = \eta_c \left(1 - \frac{\phi}{\phi_m} \right) \tag{10.4}$$

where ϕ_m represents the maximum possible volume fraction for the particles, which for randomly close-packed spheres is usually taken as 65%.

In the case of the second factor, flocculated particles can form discrete aggregates or a gel. Flocculation is important for colloidal particles, which are usually smaller than 1 μm. With the application of a considerable amount of stress, a disruption of the flocculated network can occur, leading to a flow of the suspension. This is because the forces are mostly fairly weak and can be broken easily by shear. The yield stress is the stress at which this breakdown occurs. Hence, flocculation exhibits plastic behavior with the yield stress, reflecting the forces holding particles together.[88] Pseudo plastic (shear thinning) behavior can occur where the breakdown is incomplete at the yield stress so that the suspension is still considered as being flocculated. However, it is possible that the suspension could still flow and the remaining flocculation is disrupted as the strain rate is increased. In such a case, the pseudo plastic behavior could be accompanied by thixotropy, which is a progressive and reversible decrease in viscosity upon the application of stress at a constant level.

Freshly prepared concrete can be regarded as a fluid material, and there-fore its rheological behavior can influence the way it is being processed.

Hence, the measurement and control of rheological parameters is paramount in the preparation of good quality concrete. A rheological empirical measurement technique such as the slump test only describes a part of the behavior of concrete. However, there are other measurement techniques that can describe the workability of concrete. For example, an insight into the flow behavior of concentrated suspensions in fresh concrete is of great importance.

Concrete is well known to show plastic behavior, and can be deduced from a typical concrete flow curve (measured using a rheometer) as shown in Figure 10.2. Using a rheometer, the measurement of torque (Γ) and angular velocity (Ω) at various applied loadings can be used to predict the behavior of the concrete under shear. For this, the following equations can be used:[44]

$$\tau(\dot{\gamma}) = \frac{h\dot{\gamma}}{2\pi R_2^4} \sum_{i=0}^{\infty} \left[\left(\frac{R_1}{R_2} \right)^{4i} F \left(\frac{R_1}{R_2} \right)^i \frac{h\dot{\gamma}}{R_2} \right] \tag{10.5}$$

$$F(\Omega) = (3/\Omega)\, \Gamma(\Omega) + \partial\, \Gamma(\Omega)/\partial\, \Omega \tag{10.6}$$

$$\dot{\gamma} = \Omega R_2 / h \tag{10.7}$$

where τ is the shear stress, γ is the shear rate, R_1 and R_2 are the inner and outer radii of the sample, h is the effective height of the sample, Γ is the torque applied to the sample, and Ω is the angular velocity of the rotating part. The shear yield stress, τ_0, and plastic viscosity, η_{pl}, which are the two characteristics of Bingham materials, can be deduced from the following equations[44]:

$$\tau_0 = \frac{3\Gamma_0}{2\pi\left(R_2^4 - R_1^3\right)} \tag{10.8}$$

$$\eta_{pl} = \frac{2h(\partial\Gamma/\partial\Omega)}{\pi\left(R_2^4 - R_1^4\right)} \tag{10.9}$$

The flow behavior of concrete is largely influenced by the water-to-cement (w/c) ratio, as adjusting the mixture proportions can result in the desired degree of flow and segregation resistance in concrete. An increase in the w/c ratio results in a lower yield stress and plastic viscosity. Furthermore, increasing the yield stress will result in harder concrete, while decreasing both yield stress and plastic viscosity will result in wetter concrete. An example of the relationship between flow behavior and w/c ratio is shown in (Fig. 10.3).

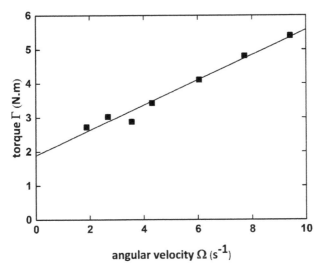

FIGURE 10.2   Concrete flow curve.[88]

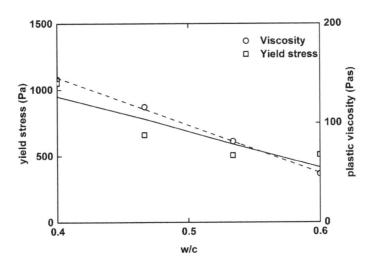

FIGURE 10.3   Effect of w/c on yield stress (grave, 60% total aggregate, 40% sand).[88]

Similarly, an increase in concentration of aggregate results in an increase in plastic viscosity. Also, the type of aggregate could affect concrete flow. For example, concrete with river gravel is expected to have lower viscosity and yield stress as a result of its much rounder and finer surface. The relationship between flow behavior and sand content is presented in (Fig. 10.4).

The sand ratio is seen to greatly influence the yield stress. It can be seen that the lowest yield stress occurred when the sand ratio was about 38%, which coincided approximately with the sand content that provided for the highest packing density for the coarse aggregates. This behavior is similar to that of the observed plastic viscosity.

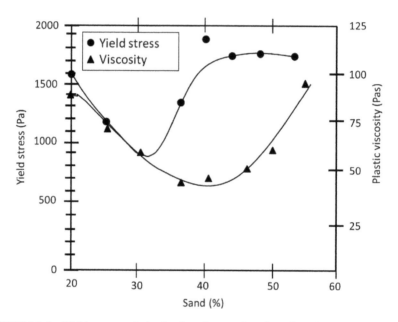

FIGURE 10.4   Yield stress and plastic viscosity as a function of sand content (gravel, w/c 0.45).[88]

Furthermore, admixtures can also influence the flow of concrete. The influence of the replacement of cement by fly ash to reduce both yield stress and plastic viscosity has been investigated by Ramachandran and Beaudoin.[88] They observed that yield stress increased at low fly ash contents and decreased mostly at higher values and is presented in (Fig. 10.5).

Since the flow of concrete is known to be plastic, it is important to describe its flow behavior using the two values of yield stress and plastic viscosity. Rheometer design can pose challenges as the suspension contains particles with a wide range of sizes, from submicron-sized cement to centimeter-sized coarse aggregates. Ideally, the larger particles will tend to settle due to gravity. This could lead to segregation during shearing. The plastic flow behavior of concrete caused by the large size of coarse aggregates, leads to a slip condition which is a common problem. Several rheometers are available for testing freshly prepared concrete and mortar. These rheometers include: coaxial cylinder,[6,7,70] Tattersall apparatus,[44,107] BML viscometer,[118] and BTRHEOM.[44]

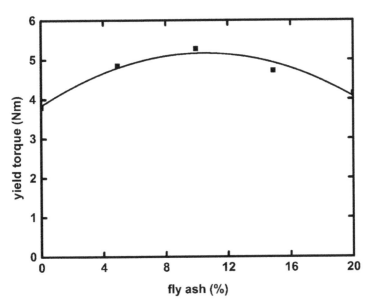

FIGURE 10.5   Yield torque as a function of fly ash content (river gravel, w/c 0.45, 40% sand).[88]

10.4   TESTING FOR CONCRETE STRENGTH

Strength generally indicates the measurement of the stress required for a material to yield. The influence of applying load on concrete induces internal mechanical changes within a concrete, leading to fracture. A very distinctive change is the failing ability of the concrete material to transfer stress during increased deformation. The resistance to such changes is known as the strength of concrete, and this is developed from: resistance to cracking on the interface of cement paste (or mortar) and coarse aggregates; resistance to cracking in the hardened paste; resistance to cracking in the coarse aggregate particles; and aggregate interlocking and friction between the broken internal surfaces.[86] Generally, the test methods which cause the least damage require correlation for assessment of material strength that tends to be complex. For example, surface hardness and pulse velocity tests cause little damage, are cheap, quick, and well suited for comparative and uniformity assessments;

however, their correlation for absolute strength prediction may pose many difficulties.[20] On the other hand, core tests cause the most damage and are slow and expensive; however, they provide the most reliable in situ strength assessment. Partially destructive methods mostly require calibration with fewer details, for carrying out strength tests. They cause damage to the surface of the material as they test only the surface zone, and could be associated with high variability.

10.4.1 COMPRESSIVE STRENGTH

A common test conducted on concrete is the compression test. It is easy to carry out and cost-efficient. The compressive strength of a concrete is one of its most important mechanical properties which characterizes the quality of the concrete.[10] Its importance in measuring the performance of concrete makes it widely used by engineers in designing structures and buildings. In structural designs, concrete is mostly exposed to compression loading, since its tensile strength is low. Therefore, in terms of structural design, concrete is directly related to compressive strength.[10] Compressive strength is also directly and indirectly related to other properties of concrete such as impermeability, modulus of elasticity, and resistance to weathering agents.[74] Other physical properties can be measured during compression testing including density, water absorption, indirect tensile strength and movement characteristics such as expansion due to alkali-aggregate reactions.

10.4.1.1 PREPARATION OF CONCRETE SPECIMEN FOR COMPRESSIVE STRENGTH TEST

Compressive strength can be determined using 150 × 300 mm cylinders and 150 mm cubes. However, there is no strict limitation on the size of the specimen as standards (ASTM C 470-02a) allow the use of smaller specimens depending on the maximum size of the aggregate.[76] The test cylinder can be prepared either in a reusable (made of e.g., steel, cast iron, brass, and a wide variety of plastic), or non-reusable mould (made of e.g., sheet metal, plastic, and waterproof products). To prevent bonding between the concrete and the mold, the inside surfaces of the molds, is to be applied with a thin layer of mineral oil. The concrete should be placed within the mold in layers and compacted in layers by 25 strokes of a 16 mm diameter steel rod with a rounded end, to achieve compaction of a high-slump concrete in three layers. Internal or external vibration can be used to achieve low-slump concrete with compaction in two layers (ASTM C 192-06). The top surface of a cylinder is to be plane and smooth for uniform loading across the surface during testing (as required by ASTM C 617-98). This can be obtained using two methods: grinding and capping. Three materials can be used for the purpose of capping: stiff Portland cement paste on freshly-cast concrete, and either a mixture of sulfur and a granular material, or a high-strength gypsum plaster on hardened concrete.[76] Steel caps can also be used for the purpose of capping. The cap should be thin (preferably 1.5–3 mm), with a strength comparable to that of the concrete being tested. A most suitable capping material would be the sulfur-clay mixture which can accommodate concrete strengths of up to 100 MPa. Furthermore, the test cylinder should have ends that are normal to its axis so that the end planes are also parallel to each another. A small tolerance of 6 mm is allowed and is usually at an inclination of the axis of the specimen to the axis of the testing machine.

ASTM C 192-06 provides the standard curing conditions for the test cylinders. The molded specimens are to be stored for not less than 20 h and not more than 48 h, when cast in the laboratory. The temperature in this case should be $23 \pm 1.7°C$ to prevent the loss of moisture. Following this, the demolded cylinders are stored at similar temperature and under moist conditions or in saturated lime water for the recommended period of testing age.

10.4.1.2 DETERMINATION OF COMPRESSIVE STRENGTH

ASTM C 39-05 recommends a procedure for the determination of compressive strength of concrete test cylinders. This can be performed by using the testing machine at a constant stress rate of 0.25 ± 0.05 MPa/s. This can be higher when applying the first half of the projected loading range. The compressive strength can be obtained from the ratio of the maximum recorded load to the cross sectional area of the specimen and is reported to the nearest 0.05 MPa.

Failure of concrete subjected to compression test is associated with lateral forces which are developed between the end surfaces of the concrete specimen and the adjacent steel platen of the testing machine. The restraint of the concrete induces these forces, and expands laterally due to the effect of the stiffer steel with a much smaller expansion. The friction developed at the concrete platen interface, as well as the distance from the end surfaces of the concrete, influence the degree of platen restraint on the concrete. This leads to an increase in the compressive strength as a result of the lateral shearing stress. The satisfactory failure modes of the test cubes shown in Figure 10.6, illustrate the influence of the platen restraint on the cubes. The effect of shear is seen to decrease toward the center of the cube in such a way that the sides of the cube possess cracks that are almost vertical. The cube could also break to form an undamaged central core as seen in Figure 10.6(a).

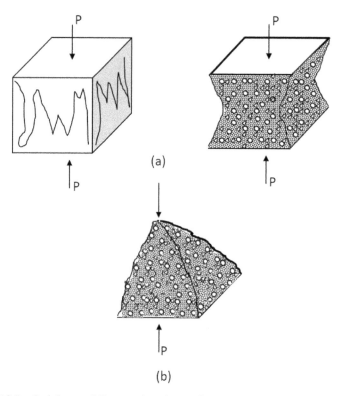

FIGURE 10.6   Satisfactory failure modes of test cubes according to BS EN 12390-3: 2002: (a) non-explosive and (b) explosive.[76]

The other types of failure modes not in this category are classed as unsatisfactory and suggest a problem with the testing machine. Furthermore, the influence of shear reduces with increase in the ratio of height to width of the specimen. This means that the specimen could fail just by lateral splitting at its central area. This case occurs in a standard cylinder test (e.g., test cores) where the height-to-diameter ratio is 2, as shown in Figure 10.7.[76] The cylinder or core diameter depends on the core-cutting tool, while its height depends on the thickness of the slab.

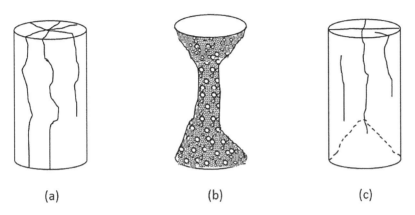

(a)   (b)   (c)

FIGURE 10.7 Typical failure modes of standard test cylinders: (a) splitting, (b) shear, and (c) splitting and shear.[76]

A correction factor can be used to estimate the strength which would have been obtained using a height-to-diameter ratio of 2. The correction factor is given by ASTM C 42-04, and depends on the level of strength of the concrete, as shown in Figure 10.8. The strength of the cube is about 1.25 times that of the cylinder since the influence of platen restraint on the mode of failure in a cube is greater than that in a standard cylinder.[76] Generally, the strengths of both the cube and cylinder concrete specimen will depends on the moisture condition of concrete at the time of testing.

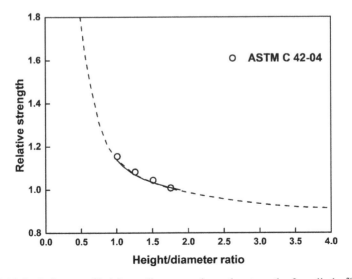

FIGURE 10.8 Influence of height-to-diameter ratio on the strength of a cylinder.[76]

10.4.2   TENSILE STRENGTH

Flexural strength provides a measure of the tensile strength of concrete. Conventionally, concrete is used in structures due to its compressive strength, without paying much attention to its tensile strength. However, it is important not to completely ignore the tensile strength of concrete, especially for crack control undergoing non-mechanical loads.[57] Flexural strength is the measure of an unreinforced concrete beam or slab to resist failure or deformation in bending. Knowledge of the flexural strength of concrete is important in structural designing, for example, designers of pavements use a theory based on flexural strength. The flexural strength is usually expressed as *Modulus of Rupture* (MR) in psi (MPa), and is determined by standard test methods such as the ASTM C 78 (third-point loading) or ASTM C 293 (center-point loading). It is the theoretical maximum tensile stress reached in the bottom fiber of a test beam.[76] So far, experimental methods have been suggested to determine tensile strength, and they have been classified into two categories: direct and indirect methods.[57] Due to the difficulties in gripping and alignment of a specimen, the direct tensile test method is not often used for plain concrete. It is mostly used to determine tensile behavior of high-performance fiber-reinforced cementitious composites[71] and engineered cement composites.[62] The indirect tensile test method are often used to determine the tensile strength of concrete due to the simple procedures involved, and ease of preparing the specimen. This is because it is difficult to apply uniaxial tension to a concrete specimen as the ends are required to be gripped while avoiding bending.[76] These methods include the splitting (or Brazilian) test and the three or four-point bending tests.[57]

10.4.2.1   PREPARATION OF CONCRETE SPECIMEN FOR TENSILE STRENGTH TEST

The size of the beam for a tensile strength test is required to be $150 \times 150 \times 750$ mm. However, for a maximum aggregate size less than 25 mm. BS EN 12390-2:2000 provides information regarding the fabrication and curing of standard test beams. Similar guideline has been provided by ASTM C 78-02, although the beam dimension in this case is $152 \times 152 \times 508$ mm, and the loading rate is between 0.0143 and 0.020 MPa/s. The modulus of rupture (f_{bl}) is calculated to the nearest 0.1 MPa when the fracture occurs within the middle one-third of the beam.[76] This is based on the elastic theory, and is given as:

$$f_{bl} = \frac{Pl}{bd^2},$$ (10.10)

where P is the maximum total load, l is the span, d is the depth of the beam, and b is the width of the beam. Based on ASTM C 78-02, a case where fracture occurs in the exterior of the middle on-third, the modulus of rupture can be taken as:

$$f_{bl} = \frac{3Pa}{bd^2}$$ (10.11)

In a case where failure occurs at an area where $(l/3 - a) > 0.05\ l$, then the result can be disregarded.[76]

Furthermore, in the case of a splitting test, a concrete cylinder similar to that used in compressive strength testing is placed horizontally between platens of a testing machine. Following this, the load is increased until failure occurs by splitting in the plane of the specimen with the vertical diameter.[76] A standard compression test with jigs required for supporting the test specimens, is shown in Figure 10.9, and is suggested by BS

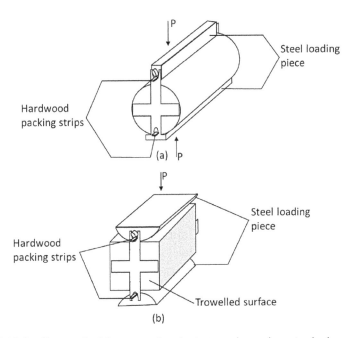

FIGURE 10.9   Jigs required for supporting the test specimens in a standard compression test machine for determination of splitting strength in: (a) cylinder and (b) cube or prism.[76]

EN 12390-6: 2000 and ASTM C 496-04. A high horizontal compressive stress at the top and bottom of the cylinder is expected under these conditions. This is also associated with a vertical compressive stress of similar magnitude and some biaxial compression so that failure is initiated by the horizontal uniform tensile stress acting over the remaining cross section of the cylinder. In tensile strength, the load is applied at a constant rate of increase in tensile strength of 0.04–0.06 MPa/s, and 0.011–0.023 MPa/s in accordance to BS EN 12390-6: 2000, and ASTM C 496-04, respectively.

Following the above, the tensile splitting strength f_{st} can be calculated as:[76]

$$f_{st} = \frac{2P}{\pi Ld} \tag{10.12}$$

where P is the maximum load, L is the length of the specimen, and d is the diameter or width of the specimen.

10.4.3   DIRECT AND INDIRECT TESTS FOR DETERMINING THE TENSILE STRENGTH OF CONCRETE

Direct and indirect test methods have been developed for evaluating the tensile strength and characteristics of concrete. Although direct tensile tests are preferable, a major challenge in conducting these tests on a concrete specimen is the difficult in achieving a uniform tensile stress across a section of the concrete without causing stress concentrations of too high a magnitude on other sections of the concrete.[84] For example, when a specimen is fitted in a testing machine by clamping grips, these grips induce stress concentrations and multiple axial stresses. Also, when subjected to uniaxial stress, lateral strains could emerge. Another approach is to use indirect tests such as bend or splitting tests. These tests involve loading a cylindrical specimen with compressive loads, and results in a relatively uniform tensile stress perpendicular to and along the plane comprising the applied load. Following this, the indirect tensile strength can computed using information on the ultimate load and specimen dimensions. Direct and indirect tests for determining the tensile strength of concrete are shown in Figure 10.10.

The subsequent section will focus mainly on indirect tensile test methods as these are widely used.

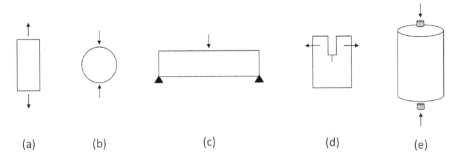

FIGURE 10.10   Direct and indirect tests for determining the tensile strength of concrete: for direct test, (a) uniaxial tension test, and for indirect test, (b) Brazilian splitting test, (c) three-point bending test, (d) wedge-splitting test, and (e) double punch test.[115]

10.4.3.1   INDIRECT TENSILE TESTS

The indirect tensile test is a substitute for the direct tensile test which requires extensive specimen preparation and a more rigorous testing procedure.[77] Splitting and bending tests such as: the Brazilian splitting test and (three-point or four-point) bending test, can be used to carry out indirect tensile tests. Indirect tests are relatively simple and the type of specimen and equipment used in these tests are similar to those used for compression testing. For this test, failure is initiated in a region of relatively uniform tensile stress, and is only fairly damaged by surface conditions.

10.4.3.1.1   Splitting Test: Brazilian Test

The tensile strength of brittle materials can be obtained using the Brazilian test in a compression machine. Here, a disc specimen is compressed diametrically with opposite and symmetric line loads.[52] It is calculated using an equation which assumes isotropic or transverse isotropic material properties, and is based on the analytical solutions of these materials under concentrated loads, and loads that are distributed over a small arc of the disc's circumference.[24,41,52] A major shortcoming of the Brazilian test is the fact the stress state at the center of the testing disc is not a purely tensile mode.[22] Moreover, in the case of an isotropic disc under diametrical loading, the absolute value of the compressive stress at the center of disc is three times larger than that of the tensile stress.[41] This test was initially proposed by Japanese Akazawa in 1943, and later by Brazilian Carneiro in 1949.[78] Results from the Brazilian test are usually used in computing the roof stability in underground excavations, especially those bedded horizontally in a strata, since the failure plane in the test specimen is relatively perpendicular to the bedding planes.[77] The validity of the Brazilian test was discussed by,[34] where he mentioned that although the failure of the test specimen initiates at the center of the disc, it however occurs at the loading points, sometimes. It has been observed that the tensile strength obtained from the Brazilian test might be much lower than the true tensile strength, as indicated by the bi-axial state of stress. This observation is due to the large compressive stress acting perpendicular to the maximum tensile stress.[79,119] However, the tensile value at failure obtained during a Brazilian test has proven to be a useful measure of the tensile strength.[14]

The theory of elasticity[109] can be used to evaluate the state of stress in a cylinder loaded between two opposite line loads. Based on this theory, loading of the cylinder produces a nearly uniform maximum principal tensile stress along the diameter, which allows the cylinder to fail by splitting. The failure in a split tensile test is size dependent similar to most brittle failures of concrete. Hondros,[41] Lundborg,[63] Sabnis and Mirza,[98] Chen and Yuan,[23] and Torrent and Brooks[111] performed various research and observed that the split-cylinder tensile strength depends on the diameter of the cylinder. For example,[41] observed that the splitting strength increases with diameter. However, Sabnis and Mirza[98] and Chen and Yuan[23] observed that the splitting tensile strength decreases with the diameter of the cylinder. Furthermore, it was observed that for small diameters, the strength decreases as the diameter increases, but begins to increase after a certain diameter is exceeded.[14] In a larger structure, the release of the stored energy is at the same nominal stress, which means that a lower nominal stress produces the required energy to break the material. This is the reason for the size effect on the failure in a split tensile test.[11,12] The failure mode in the Brazilian test depends mainly on a wide range of parameters including: the material properties, stiffness of the load platens, and the size of the specimen.[14] Experiments have shown that splitting cracks are initiated at a location in the uniform tensile zone[80] after which, secondary cracks begin to open at a distance from the load platen. A wedge formation and possible collapse of the specimen occurs when one of the secondary cracks continues to propagate. Thus, two peak loads (one due to splitting, and the other due to wedge formation) may occur in a well-controlled test.[14] Rocco et al.[94,95,96] and Olesen et al.[81] performed experiments to investigate the effect of size and boundary conditions in a Brazilian test. They observed that when the diameter of the cylinder increases (or width of the loading strips decreases), an enhanced estimate of the tensile strength can be achieved. Furthermore, in the case where plastic materials such as fibers, are added to the cement-based matrix, the split-test cylinder is not appropriate for determining the tensile strength.[81]

The stress distribution between the line loads for three different widths of load-distributing strips from the theory of elasticity is shown in Figure 10.11.

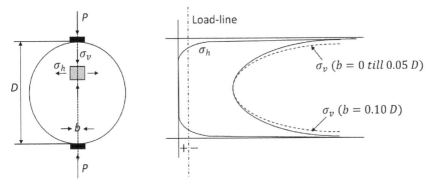

FIGURE 10.11   Stresses between the two line loads in a Brazilian splitting test for two different widths b of the load-bearing strips.[78]

The stresses are seen to act perpendicular to the line linking both line loads P. Horizontal tension (σ_h) is seen to be more prominent at the centre of the cylinder, and a confined compressive state of stress develops below the loading strips. The horizontal tensile stresses between the loading strips can be calculated using the following expression:[115]

$$\sigma_h = \frac{2P}{\pi Ld} \approx 0.64 \frac{P}{Dd} \qquad (10.13)$$

where D is the diameter of the cylinder, d is the depth, and P is the line load. The plotted vertical stress σ_v is also a compressive stress. The resulting state of the stress in the crack plane is also a biaxial tension compression.

10.4.3.1.2   Bending: Three-point Bend Test

Difficulties in a uniaxial tensile test can be avoided by carrying out flexural tests and can be performed with a simple compressive machine. The three-point bend test is a classical test generally performed on beams and involves loading a beam under three-point bending with certain dimensions so that interlaminar shear failure can occur.[102] The beam specimen in this test is subjected to a compressive stress in the upper section so that it experiences a tensile stress in its lower section. The lower face of the specimen absorbs the highest tensile stress which results in failure.[66] This test consists of two parallel supports for the specimen, and a single load-point in the middle, between the supports, which is where the force is introduced (Fig. 10.12). The supports are placed in way that will enable them rotate freely on their axes. This will help to minimize the influence of friction on the measurement. The supports should also individually rotate about an axis perpendicular to and parallel to the axes of the specimen so that the specimen can align itself when subjected to stress. Frictionless roller systems such as steel roller supports can be specially designed for this purpose.[85] However a slight plastic deformation at the point of contact of the roller bearings could significantly affect the frictional restraint.[115]

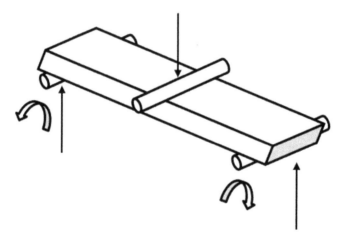

FIGURE 10.12   Schematic illustration of a typical three-point bending test.

The bending strength can be determined by dividing the maximum bending moment by the section modulus, and for a simple three-point bending test is given as:[115]

$$\sigma_{fl} = \frac{6Pl^2}{4bh^2}$$ (10.14)

where P is the maximum load the specimen can sustain, l is the span between the supports, and $A = b \times h$ is the sectional area.

Three-point bend tests can be used to determine flexural tensile strength and fracture energy. When used to determine the former, the support conditions are less important and the test is only indicative since the gradients of the stress and strain of the whole loading history are not exactly suitable. Hence, the deviations on the order of 5–10% are negligible.[115] Similarly, if used to determine fracture energy, the test is quite unreliable. For example, the prism beams used in a standard three-point bending tests are quite short and thick (with dimensions ranging from $100 \times 100 \times 600$ mm^2 to $150 \times 150 \times 700$ mm^2), and may not meet the requirement from the Bernoulli beam theory.[115] Thus, the results from this test are mostly indicative and can only be compared to results obtained from similar test method. This test method is however, quick, simple, and convenient, hence its wide popularity. Performing this test on a fiber-reinforced composite is simpler compared to a direct tensile test method due to the less significance of the effects of flaws and geometrical stress concentrations.

10.4.3.2 NOTCHES AND BEAMS

Notches can be used to perform experiments with a stable displace-controlled fracture.[115] The location of the fracture zone is relatively fixed, using a notch. The reason for fixing the fracture zone could be a case where a specimen is tested for crack growth at different places. The loading configuration of notched beams using test methods such as three-point bending, allows for simple stress states and convenient test control. A notched beam test can be used to investigate mixed-mode fracture characteristics by offsetting the mechanical notch location (Fig. 10.13).[56] Notches can be used to evaluate post cracking behavior and toughness measurements of concrete specimen by measuring the displacement at the opening of the crack mouth. Fracture properties of concrete such as the critical stress intensity factor and critical crack tip opening displacement of concrete, can be determined by using three-point bend tests on notched beams.[103] Notched beam tests can be performed in such a way that the deflection in the specimen are localized at the crack mouth while the rest of the beam is restrained from undergoing any inelastic deformation. This helps to minimize the energy dissipated over the entire volume of the specimen thereby absorbing and directing all the energy toward the fracture along the notch plane.[9,37] Notches can be sawn or cast unto a concrete sample, so that the location of a crack is relatively fixed. The relative notch depth can be defined as the ratio of the notch depth to the height of the beam specimen.[121]

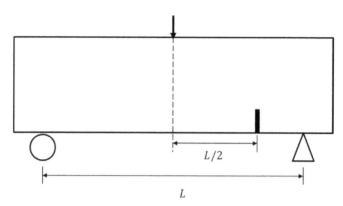

FIGURE 10.13   Schematic representation of a notched beam using an offset notch method.

10.5   FRACTURE MECHANICS APPROACH

Fracture mechanics involves the study of stress and strain behavior of homogeneous, brittle materials (such as concrete).[76] Overall, it provides an understanding on the mechanism of concrete failure. Fracture theory gives the opportunity to utilize the strengths of concrete more efficiently. This theory of concrete fracture provides quantitative predictions of the behavior of concrete under load. Concrete fracture theory includes numerical relationships: between the strength and composition of a concrete; between the various strengths of concrete, including its strengths under sustained, repeated, and multiaxial loads; for the deformations of a concrete as a function of the composition and type of testing.[86] As concrete are highly heterogeneous composite materials, the main reason for their failure may include weakness of the paste, the aggregate, the paste-aggregate interface, or any combination of these factors.

10.5.1   LINEAR ELASTIC FRACTURE MECHANICS

Fracture mechanics theories have been developed for various types of nonlinear material behavior (such as plasticity and viscoplasticity), and dynamic effects.[5] However, to fully understand linear elastic fracture mechanics (LEFM) provides more insight into more advanced concepts in fracture mechanics. There are three common modes of fracture (Fig. 10.14), namely: the opening mode (or mode I), in-plane shear (or mode II), and out-of-plane shear (or mode III).[115]

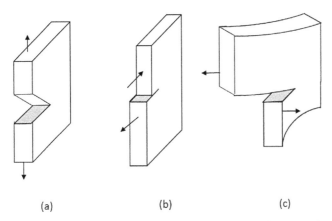

FIGURE 10.14   Three fracture modes: (a) mode I—opening mode, (b) mode II—in-plane shear, and (c) mode III—out-of-plane shear.[115]

The three modes can be combined to form mixed-modes, so that, for example, tension and shear can either be referred to as mode I+II or mode I+III. Mode I can be regarded as the most important mode of fracture, in engineering, however, the other modes are still of importance.[115] The stress intensity factor K was derived by Irwin,[49] using crack-tip stresses. Consider the near-tip stresses in a Cartesian coordinate system shown in the tip of slit-like crack, in Figure 10.15, the main equations for the three modes are given below:[18,61,105,115]

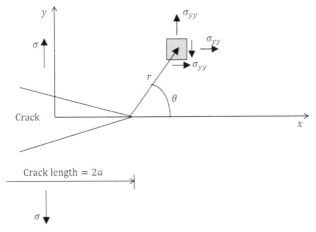

FIGURE 10.15   Tip of slit-like crack showing near-tip stresses in a Cartesian coordinate system. The length of the crack is $2a$, and is in an infinite plate subjected to tension. The local stresses σ_{xx}, σ_{yy}, τ_{xy}, are of interest and are defined by polar coordinates (r, θ) at point P.[115]

For mode I, the three stresses σ_{ij} and displacements u_i at point P are[115];

$$
\begin{Bmatrix} \sigma_{xx} \\ \sigma_{yy} \\ \sigma_{xy} \end{Bmatrix} = \frac{K_I}{\sqrt{2\pi r}} \begin{Bmatrix} \cos\dfrac{\theta}{2}\left[1-\sin\dfrac{\theta}{2}\sin\dfrac{3\theta}{2}\right] \\[2mm] \cos\dfrac{\theta}{2}\left[1+\sin\dfrac{\theta}{2}\sin\dfrac{3\theta}{2}\right] \\[2mm] \sin\dfrac{\theta}{2}\cos\dfrac{\theta}{2}\cos\dfrac{3\theta}{2} \end{Bmatrix} \tag{10.15}
$$

For plane stress, $\sigma_{zz} = 0$, and for plane strain, $\sigma_{zz} = v\,(\sigma_{xx} + \sigma_{yy})\ \sigma_{xz} = \sigma_{yz} = 0$, and;

$$
\begin{Bmatrix} u_x \\ u_y \end{Bmatrix} = \frac{K_I}{2E}\sqrt{\frac{r}{2\pi}} \begin{Bmatrix} (1+v)\left[(2\kappa-1)\cos\dfrac{\theta}{2}-\cos\dfrac{3\theta}{2}\right] \\[2mm] (1+v)\left[(2\kappa+1)\sin\dfrac{\theta}{2}-\sin\dfrac{3\theta}{2}\right] \end{Bmatrix} \tag{10.16}
$$

For plane stress, $u_z = -\dfrac{vz}{E}(\sigma_{xx} + \sigma_{yy})$, and for plane strain, $u_z = 0$

In eq 10.16, for plane stress, $\kappa = (3 - v)/(1+v)$, and for plane strain, $\kappa = (3 - 4v)$. K_I is the stress intensity factor for mode I and is given as $K_I = \sigma_{yy}^{\infty}\sqrt{\pi a}$ and has dimension, $MPa\sqrt{m}$.

Similarly, for mode II,

$$
\begin{Bmatrix} \sigma_{xx} \\ \sigma_{yy} \\ \sigma_{xy} \end{Bmatrix} = \frac{K_{II}}{\sqrt{2\pi r}} \begin{Bmatrix} -\sin\dfrac{\theta}{2}\left[2+\cos\dfrac{\theta}{2}\cos\dfrac{3\theta}{2}\right] \\[2mm] \sin\dfrac{\theta}{2}\cos\dfrac{\theta}{2}\cos\dfrac{3\theta}{2} \\[2mm] \cos\dfrac{\theta}{2}\left[1-\sin\dfrac{\theta}{2}\sin\dfrac{3\theta}{2}\right] \end{Bmatrix} \tag{10.17}
$$

For plane stress, $\sigma_{zz} = 0$, and for plane strain, $\sigma_{zz} = v\,(\sigma_{xx} + \sigma_{yy})\ \sigma_{xz} = \sigma_{yz} = 0$, and;

$$
\begin{Bmatrix} u_x \\ u_y \end{Bmatrix} = \frac{K_{II}}{2E}\sqrt{\frac{r}{2\pi}} \begin{Bmatrix} (1+v)\left[(2\kappa+3)\sin\dfrac{\theta}{2}+\sin\dfrac{3\theta}{2}\right] \\[2mm] -(1+v)\left[(2\kappa-3)\cos\dfrac{\theta}{2}+\cos\dfrac{3\theta}{2}\right] \end{Bmatrix} \tag{10.18}
$$

For plane stress, $u_z = -\frac{vz}{E}(\sigma_{xx} + \sigma_{yy})$, and for plane strain, $u_z = 0$

In eq 10.18, κ is similar to the definition given for eq 10.16. K_{II} is the stress intensity factor for mode II and is given as $K_{II} = \sigma_{yy}^\infty \sqrt{\pi a}$ with similar dimension as in the case of mode I.

For mode III,

$$\sigma_{xx} = \sigma_{YY} = \sigma_{zz} = 0$$

$$\sigma_{xY} = 0$$

$$\begin{Bmatrix} \sigma_{xz} \\ \sigma_{yz} \end{Bmatrix} = \frac{K_{III}}{\sqrt{2\pi r}} \begin{Bmatrix} -\sin\dfrac{\theta}{2} \\ \cos\dfrac{\theta}{2} \end{Bmatrix} \qquad (10.19)$$

$$u_x = u_y = 0$$

$$u_z = -\frac{4K_{III}}{E}\sqrt{\frac{r}{2\pi}}\left[(1+v)\sin\frac{\theta}{2}\right] \qquad (10.20)$$

The stress intensity factor, K_{III}, for mode III is $K_{III} = \sigma_{yz}^\infty \sqrt{\pi a}$

10.5.2 PLASTIC CRACK-TIP MODEL

Barenblatt[8] developed a cohesive crack model for concrete fracture mechanics. The model was developed based on the atomic potential for crystalline solids, and hence the cohesion between atoms induced by this potential.[115] He proposed that the cohesive zone at the crack tip should be smaller than the size of the entire crack. In other work on plastic metals done by,[33] the interatomic cohesive forces were substituted with the yield stress of the metal. The principle of plastic crack-tip is shown in Fig. 10.16, where the size of the plastic zone is t, and progresses just to stress-free crack with length, $2a$.[115]

From Figure 10.16 it can be assumed that the total crack length is $2(a+1)$, and a closing pressure σ_{ct} acts equal to σ_y over the tip region of the longer crack which is over the segments of length t. Thus, two stress intensities are taken into account: one from the crack-tip stresses σ_y, and the other is from the tensile stress σ_{yy}^∞, and hence,[18,115]

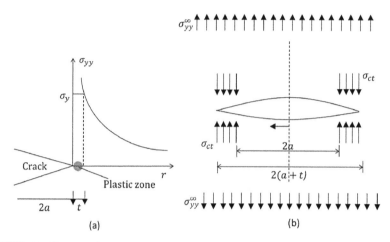

FIGURE 10.16 (a) Principle of the plastic crack-tip model and (b) closing pressure at crack-tips.[115]

$$k_{\sigma_{yy}^{\infty}} = \sigma\sqrt{\pi(a+1)} \tag{10.21}$$

$$k_{\sigma y} = 2\sigma_y\sqrt{\frac{(a+1)}{\pi}}\,arc\cos\frac{a}{a+1} \tag{10.22}$$

Based on the condition that when superimposed, two stress intensities should cancel each other, the size of the plastic crack-tip zone t is:

$$t = \frac{\pi^2\left(\sigma_{yy}^{\infty}\right)^2 a}{8\sigma_y^2} \tag{10.23}$$

10.5.3 FRACTURE PROCESS IN UNIAXIAL TENSION

The process of fracturing at the particulate level of concrete can be classified into four stages, namely (Fig. 10.17)[115]: (0) elastic stage, (A) stable microcracking, (B) unstable macrocracking, and (C) bridging.

10.5.3.1 STAGE (0): ELASTIC STAGE

Stage (0) begins from the origin. This would usually be the point just after loading commences, and the material behavior may be elastic in nature.

Here, the material can initially respond either in a linear-elastic (the most common) or nonlinear-elastic way, depending on the composition of the material.[115] Without any initial defects on the material, loading and unloading will behave identical.

10.5.3.2 STAGE (A): STABLE MICROCRACKING

Small microcracks tend to develop along the interface between aggregates and the cement matrix, even before the peak of the stress is reached. Stage (A) occurs at the initial part (prepeak) of the stress-deformation diagram (Fig. 10.17) following the nucleation and growth of the first microcracks. The acoustic emission technique has been used[82,99] to measure microcracking in single-notched tensile specimens. They observed that the acoustic emission count was significant at the peak stress. Also, early microcracking was discovered in a work on Serena Sandstone carried out by[67] using electronic speckle pattern interferometry. However, the earlier than normal microcracking he observed may have been as a result of his machined sandstone specimen notches.

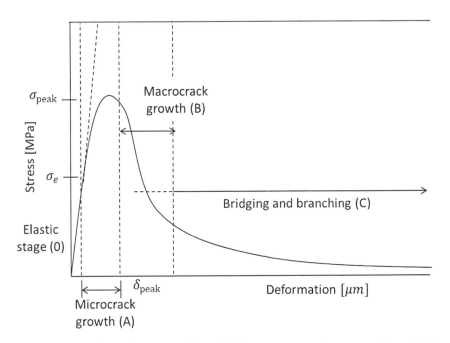

FIGURE 10.17 Schematic representation of the fracture process of concrete under uniaxial tension.[115]

Furthermore, when small cracks grow, elastic energy is released (taken as the deviation from ideal linear elastic behavior) and the sum of this energy leads to an increasing curvature of the stress-deformation curve,[115] as seen in Figure 10.18.

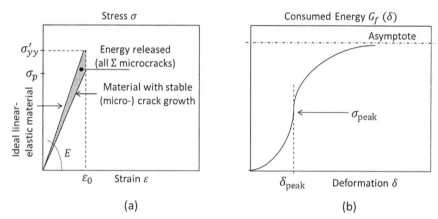

(a)   (b)

FIGURE 10.18   Stress-deformation response in stage (A) caused by (a) the energy release from nucleation and stable growth of microcracks and (b) total energy $G_f(\delta)$ consumed in the fracture process.[115]

The amount of energy required to create the cracks can be regarded as the area below the stress-deformation diagram, and is similar to the fracture energy, G_f. In order to eliminate the crack initiation stage, the prepeak curve is usually linearized, and its energy, subtracted from the total energy.[115]

10.5.3.3   STAGE (B): UNSTABLE MACROCRACKING

Stage (B) occurs at the peak of the stress and results in a steep curve due to an unstable macrocrack growth. Macrocracks usually have lengths and depths that are of similar order of magnitude as the dimensions of the specimen being considered.[115] Their crack widths are substantially larger than those of microcracks. Although they can be seen with the naked eyes, these cracks can also be easily spotted with tools such as impregnation, dyeing, photo-elastic coatings, and so on.[60] carried out test on charcoal granite, which has similar behavior to concrete in tension. This test was performed on a double-edge notched plate which was loaded between freely rotating loading platens in deformation-control. Their results are shown in Figure 10.19, where some level of irreversible deformation is observed in the postpeak regime where several loading cycles were applied. Along the softening curves, and between the bracket, the optical crack length, a_m, which also includes the initial notch depth of 13 mm, is shown. The crack length from the tip of the notch is 28.9 mm (i.e., the difference between depths 41.9 and 13 mm), at a load of 1 kN where the ligament between the two notches is for the largest broken part.

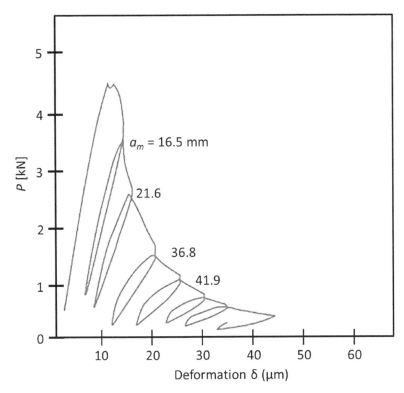

FIGURE 10.19   Optical crack length, a_m measurements in an uniaxial tension test on granite between freely rotating loading platens.[60,115]

Results of deformation controlled uniaxial tensile tests are shown in Figure 10.20. These tests were carried out on a double-edge notched specimen of size 200×200 mm², notch depth of 25 mm, and a thickness of 25 mm. The front and back of the specimen where the crack was expected to grow, was coated with a thin photoelastic coating of 1 mm thickness. It was observed that the main crack started to propagate just after the peak, with two overlapping crack branches which was as a result of the fixed boundaries which were used. At crack opening of 22 μm, the steep part of the softening curve becomes like a shallow tail.

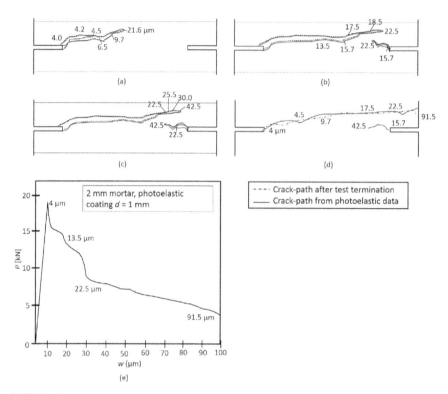

FIGURE 10.20   Photoelastic coating technique in monitoring crack propagation in a single-edge notched specimen subjected to uniaxial tension between fixed (nonrotating) loading platens.[116]

Furthermore, impregnation experiments (results in Fig. 10.21) carried out at subsequent crack openings shows that the macrocrack begins from the notch and extends farther along the specimen surfaces. This leaves a core in an uncracked form, in the specimen.[112,113]

10.5.3.4   STAGE (C): CRACK-FACE BRIDGING

At stage (C), the unstable macrocrack is reasonably stabilized by bridging. It is possible for bridging to begin almost after the macrocrack propagates, leading to an overlapping of stages (B) and (C). This crack overlap as can be seen in Figure 10.21(d), is as a result of structural effects.[115] A load-eccentricity gradually increases when the crack begins to propagate from the left notch, so that a bending moment which hinders the crack from propagating occurs. This hindrance is as a result of the nonrotating loading platens which were used in the test. The mechanism of overlapping of cracks also known as handshake crack is seen to occur in many materials as well as a wide variety of structural conditions. Impregnation tests[112] were also carried out to study overlapping cracks, and results are shown in Figure 10.21. Here, examples of overlapping cracks, as well as effect on the carrying capacity in the tail of the softening curve can be deduced. The tail is seen to increase with increasing aggregates size, and seems to have ligament size between two overlapping crack-tips that clearly explains the reason for the increase in the carrying capacity in the tail.[115]

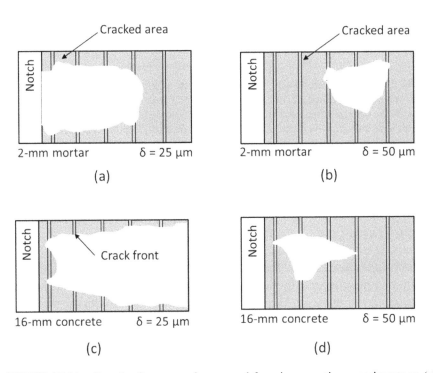

FIGURE 10.21   Growth of macrocrack measured from impregnation experiments on (a) and (b) for single-edge notched specimens of 2 mm mortar, and (c) and (d) for 16 mm concrete.[115]

Crack overlaps in concretes can also be seen in Figure 10.22(c) where the main crack propagates through the porous lytag particles in the lytag concrete, and produces crack-face bridges around the harder sand particles.

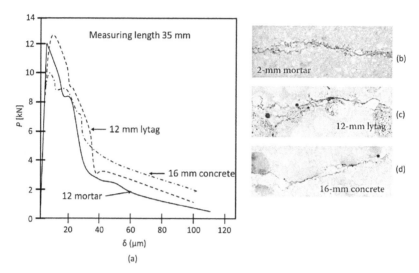

FIGURE 10.22 Load-deformation diagrams for three different concretes: 2 mm mortar, 16 mm gravel concrete, and 12 mm lytag concrete. (b)–(d) shows three examples of crack overlaps in the corresponding three materials. The cracks are located at the interfaces between aggregate and cement matrix in the case of the 2 mm mortar and 16 mm concrete. In the lytag concrete, cracks propagate through the weak and porous lytag particles but along the interface of the larger sand grains.[115] (Source for b–d: Reprinted with permission from van Mier, J. G. M. *Concrete Fracture: A Multi-Scale Modeling Approach*; CRC Press, © 2012 Taylor and Francis.)

10.5.3.5 FOUR-STAGE FRACTURE MODEL FOR CONCRETE AND OTHER MATERIALS

The four stages discussed in the previous sections form a basis for the development of a macroscopic model for tensile fracture of concrete. This model as illustrated in Figure 10.23 uses the principle associated with the physical mechanisms that surround most materials that fail in a quasi-brittle way upon the propagation of one or more cracks.

In the prepeak regime of stage (0) and (A), a continuum model approach can be used (assuming uniformity in the distribution of stress and strains) and is given as:[115]

$$\sigma = \frac{F}{A}$$
$$\varepsilon = \frac{\Delta l}{l}$$

(10.24)

where A and l are the cross section and length of a tensile test specimen, respectively.

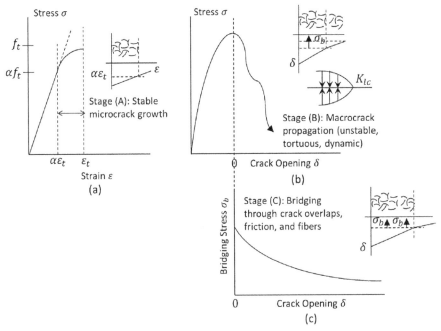

FIGURE 10.23 Four-stage fracture model for concrete and other materials such as glass, metal, rock, fiber-reinforced composites, and so on.[114,115]

When the peak strength is reached, a different approach is required as a change occurs which can be considered as a phase-transition. Since a macroscopic crack begins to propagate at this peak, a continuum approach may not be suitable for this purpose. This is because new boundaries are formed in the specimen when the macrocracks grows hence it will be difficult to average the crack-width to retrain the classical state variable strain.[115] The best approach would be to use the classical criterion from LEFM for describing crack propagation, and can be given as:

$$K_I = \sigma \sqrt{\pi a} f(a.\theta) = K_{Ic} \qquad (10.25)$$

Initially, cracks propagate with several smaller cracks and a distinct homogeneity, and therefore, the stress intensity is required to be obtained under same condition. Since the main microcrack is associated with several branching and bridging, it is necessary to include a bridging stress to eq 10.25.[115] The bridging stress is equal to the tail of the softening curve and can be measured under fixed boundary conditions where the crack propagation stage (B) is distinguished from the bridging stage (C). The specimen used in this case is required to be wide in dimension, which will allow the macrocrack to extend with been prevented. However, it should not be too wide to lead to instabilities when a second crack from the opposite notch develops, during test control. The composition of the material is very important as it influences the bridging stress. The composition includes; the aggregate size, the strength and stiffness of the matrix and aggregates in the concrete, and the fibers in the mixture.[115] Two cases of bridging can be seen in Figure 10.24: (a) normal concrete where bridging is initiated from only strong aggregates, and (b) fiber-reinforced concrete, where bridging develops from both fibers and aggregates. The model for tensile fracture can be described using three main components: prepeak nonlinear elasticity, postpeak macrocrack propagation,

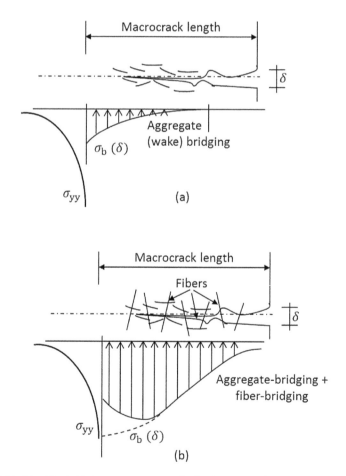

FIGURE 10.24   The bridging stress concept in the four-stage fracture model for (a) plain gravel concrete and (b) fiber-reinforced cement composites.[114]

and bridging. In normal concrete, the bridging stress is uniformly distributed. However, for fiber-reinforced concrete, the stress is uniform until a total pull-out of the fibers occurs. The bridging stress decreases to zero rapidly at mostly large deformations, after pull-out. For fiber-reinforced concrete, the bridging stress may also decrease gradually, depending on the material composition. The cracking process is controlled by the stress-singularity from the main propagating crack.[115]

10.5.3.6 RELATIONSHIP BETWEEN TENSILE AND COMPRESSIVE FRACTURE

Similar to fracture in uniaxial tension, a model for compression is available for fracture process under compression.[115] Assuming that damage from initial defects can be neglected, and considering the linear elastic stage (0), early stages of microcracking can be observed at a particular stress (may be between 30% and 50% of peak strength). Figure 10.25 shows the model for compression. At stage (A), the stability of a microcrack growing under compression is much improved; hence, the stable microcracking is larger

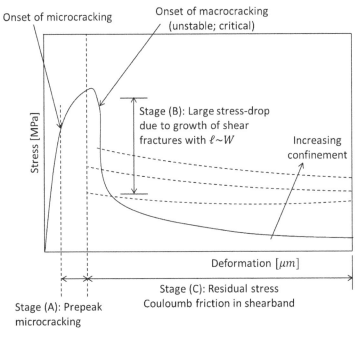

FIGURE 10.25 Four-stage fracture model for concrete under compression.[115]

and more evident compared to that in tension. A phase transition occurs at the point where peak-stress is reached, similar to the model in tension. A mode II (in-plane shear) crack is formed as the most critical crack is propagated, and will advance during the steep part of the softening curve.

At stage (C), the residual stress will increase, subject to the magnitude of the confining stress. Comparing the model for compression to that in the case of tension, there is an extended stage (A), and different mechanisms in the bridging stage (C). In stage (C), the effect of frictional restraint on the shear band begins just after peak stress.[115]

The direction of shear crack in a compressive test can be influenced by several factors such as: the compressive loading σ_a, the confining stress σ_c, and the frictional stress τ_b, along the boundaries of the specimen in direct contact with the loading platen. In Figure 10.25, the shear band and external stresses are shown. The local equilibrium can be used in deriving the normal and shear along the inclined plane, and can be given as:[115]

$$\tau = \sigma_c \sin \alpha + \sigma_a \cos \alpha + \tau_b \sin \alpha \qquad (10.26)$$

where α is the angle in the axial loading direction.

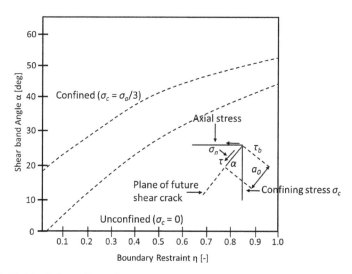

FIGURE 10.26   Effect of boundary restraint ($\eta = \tau_b/\sigma_a$) on the inclination angle α of a shear crack in uniaxial compression ($\sigma_c = 0$) and under confined conditions ($\sigma_c = \sigma_a/3$).[115]

$$\sigma_n = \sigma_c \cos \alpha - \sigma_a \sin \alpha + \tau_b \sin \alpha \qquad (10.27)$$

A coupling between the boundary shear and the axial stress (i.e., $\tau_b = \eta\sigma_a$ ($\eta > 0$)) can be assumed, and eqs 10.26 and 10.27 can further be simplified as follows:[115]

$$\tau = \sigma_c + \sigma_a (\cos \alpha + \eta \sin \alpha) \tag{10.28}$$

$$\sigma_n = \sigma_c \cos \alpha + \sigma_a (-\sin \alpha + \eta \cos \alpha) \tag{10.29}$$

It can be assumed that when $\sigma_n = 0$, the shear crack would propagate under pure mode II. Substituting in eqs 10.26–10.29, the direction of the shear band can be deduced. The outcome of this is seen in Figure 10.25 for uniaxial compression ($\sigma_c = 0$), and for confined compression $\sigma_c = \sigma_a/3$, assuming there is a transfer of friction from the confining stress to the concrete).[115] The inclination of the shear band is seen to increase when the boundary restraint η, increases. A higher inclination angle is predicted for an increase in the curve in Figure 10.25, under confined conditions (i.e., $\sigma_c = \sigma_a/3$), but is most likely to be limited to the regime of lower-confinement. A transition from brittle to ductile behavior is observed for a large increase in σ_c. Here, it is possible that no macroscopic crack can develop as the fracture process is advanced at stage (A).

To compute the four-stage model for compressive fracture, the length of the initial crack can be taken as σ_0, and is the result of the pre-peak microcracking. A micromechanical model is required to compute the crack-size distribution in compression. The residual carrying capacity σ_1 at crack length $\sigma_1 > \sigma_0$ is given as:[115]

$$\frac{\sigma_1}{\sigma_p} = \sqrt{\frac{\sigma_1}{\sigma_0}} \cdot \frac{f\left(\dfrac{a_1}{W}\right)}{f\left(\dfrac{a_0}{W}\right)} \tag{10.30}$$

Where σ_p is the stress at peak, and $f(a/W)$, is the geometrical function for this case. The propagation criterion can now be taken as the mode II case, where $K_{II} = K_{II,c}$ = constant. In stage (C), bridging will be caused mostly by coulomb or rolling friction, and will delay the macrocrack growth. The effect of this can be derived from the summing of the two contributions to the stress-intensity factor, and is given as:[115]

$$K_{II} = K_{II,loa} + K_{II, Cou. frict} \tag{10.31}$$

The above information provides the concept of a model for compressive fracture. However, further experiments are required to investigate the microcracking process in stage (A), as well as the role of friction on the stress intensity. Also, the transition from a brittle to ductile behavior also requires investigation.

10.5.4   FRACTURE MECHANICS IN CONCRETE

Materials fracture upon the application of sufficient amount of work and stress at the atomic level. This breaks the bonds that hold the atoms together.[5] For fracture to occur, the stress at the atomic level will need to exceed the cohesive strength of the material, and the flaws that lead to fracture must reduce the global strength by increasing the local stress. Concrete is more complex than an ideal brittle material, and therefore the classical theories of fracture developed for brittle materials do not fully describe the failure of concrete under loading.[86]

The strength of concrete is known to be mostly influenced by the presence of flaws, discontinuities and pores rather than on the chemical composition or physical structure of the products of hydration of cement, and on their relative volumetric proportions.[76] To well understand the influence of the flaws on concrete strength, it is important to take into account the mechanics of fracture of concrete under stress. These flaws could also be caused by the presence of aggregates contained within the concrete, with their individual flaws which lead to microcracking at the interface with the cement paste.

10.5.4.1   INITIATION AND PROPAGATION OF CRACK IN CONCRETE

Information about the mechanism of a fracture can be obtained from the way a crack is propagated. Fracture mechanisms include: shearing,[40] cleavage,[19] fatigue,[19,40] and crazing[39]. Shearing occurs when the origin and growth of cracks is instigated by shear stress. Cleavage occurs when the crack is propagated along grain boundaries, and can be found mostly in materials with weak or damaged grain boundaries. Fatigue occurs when a crack is subjected to cyclic loading so that the crack tip increases only gradually (may be better captured under a microscope) for each loading cycle, provided that the stress is high enough, but not too high to lead to a sudden fracture. Crazing occurs when sub-micrometer voids (also known as crazes) develop as a critical load level is exceeded.

The failure of concrete under load usually occurs through advanced internal cracking caused by internal tensile stresses. The process of cracking results in a final disintegration of the concrete specimen under load.[26] In his research, Brandtzaeg[17] who first observed internal cracking found that under uniaxial compressive loading, the volume of concrete specimens decreased and was approximately proportional to the load applied.[93] The concrete decreases further, upon a load increase above the critical stress (also known as compressive stress). However, a point is reached where a small increase in load will yield no further change in the volume of the concrete specimen. Furthermore, at much higher loads, an increase in volume is observed and the actual volume of the tested specimen at the ultimate load becomes greater than it was before the compressive load was applied. Following his findings, Brandtzaeg concluded that the expansion and subsequent failure of the specimen was as a result of a gradual development of internal microcracking and tension within the specimen parallel to the direction of the applied compressive stress.

Complete failure occurs when the cracks start to form continuous patterns.[86] Further observations for compressive, tensile, including other loadings show the gradual development of microcracking in concrete under loads that are less than the ultimate load. For example, it was observed by Jones that the ultrasonic pulse velocity in the direction of loading remains constant as the load is increased to failure, for cubes in compression.[53] However, there is a decrease in the pulse velocity at loads lower than the ultimate, in the transverse direction. Overall, these findings show that internal cracking begins at a fraction of the ultimate load and the cracks are oriented parallel to the direction of loading.[86] Microcracking can be further investigated using X-ray photography, microscopy,[36,108] optical microscopes,[73] fluorescent-dye techniques reflective photoelasticity,[106] laser interferometry for investigating surface cracks, acoustic emission for investigating internal cracks by monitoring the noise generated from cracking.[25,97]

Microcracks can exist in concrete at the interface between coarse aggregate and mortar, even before an external load is applied. These bond cracks could be as a result of difference in change of volume of matrix and aggregate during hydration and drying,[43,108] including bleeding and segregation. The presence of these cracks shows that the interface is the weakest link within the concrete and this mostly occurs in composite materials which compose of components of different stiffness.[86] Generally, bond cracks initially develop around the larger aggregates or particles, which means that concrete strength will decrease with an increase in aggregate size. A continuous crack pattern is formed by linking between neighboring bond cracks when most (about 70–90%) of the ultimate load and cracks that occur through the mortar and/ or aggregates start to increase. The load carrying capacity of the concrete decreases when the continuous crack pattern extends further. Crack development is significantly influenced by the properties of the coarse aggregates within the concrete, as larger size and amount of aggregate will lead to an increase in interfacial cracking. The pattern of crack propagation can be influenced by the rate of loading.[86] For example, fewer cracks are developed before failure than when the strain rate of the specimen is kept constant, when the rate of stress is kept constant during testing.

10.5.5 STRESS CONCENTRATIONS

Materials usually contain imperfections which are the source of stress concentrations, and may lead to failure of materials well below their theoretical strengths. Stress concentrations are caused by pores, inclusions, interfaces between distinct material phases, and so on. Surface defects usually lead to early failure at stress levels well below the maximum possible limit.[115]

Inglis[48] initially provided quantitative evidence for the stress concentration effect of flaws by analyzing elliptical holes in flat plates. This provided the background for describing the internal cracking of concrete under load. His approach suggested that flaws in materials in the form of pores, voids, cracks can potentially perform as tensile stress concentrators, irrespective of whether the load is under compression or tension. This means that the stresses at the flaw tip could be high enough to exceed the average stress and ultimate tensile strength of the material.[86] Furthermore, Inglis observed that for flaws with narrow and elliptically shaped tips, the stress concentration σ_m/σ_u in a linearly elastic material that is of a high strength and non-cracking will be approximately given as[48]:

$$\frac{\sigma_m}{\sigma_u} = 2\left(\frac{a}{r_0}\right)^{0.5} \tag{10.32}$$

Where σ_m is the maximum tensile stress at the elliptic tip of the flaw, σ_u is the applied tensile stress a distance away from the flaw tip and is uniform and normal to the plane of the flaw, α is the depth of the flaw, and r_0 is the minimum radius of curvature at the tip. This can be further illustrated in Figure 10.26, using a flat plate with a circular hole with radius $r = \alpha$.[115]

Stretching the plate to infinity will lead to non-uniformity in the distribution of the stresses in the surrounding of the hole. The highest stress concentrations will show at the sides of the hole at points A and B.

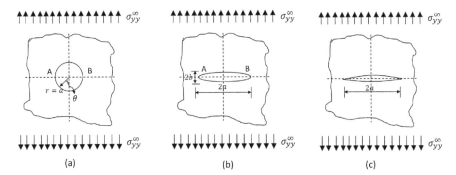

FIGURE 10.27 Flat plate with (a) circular hole, (b) elliptical hole, and (c) slit.[115]

The radial and shear stress components (σ_{rr} and $\tau_{r\theta}$) are equal to zero at the edge along the hole. Also, the tangential stress $\sigma_{\theta\theta}$ can be derived from the linear elasticity theory[109] and is given as:[115]

$$\sigma_{\theta\theta} = \sigma_{yy}^\infty - 2\sigma_{yy}^\infty \cos 2\theta \tag{10.33}$$

Where σ_{yy}^∞ is the externally applied stress. It can be observed that the stresses along the sides of the hole (i.e., $\theta = \pi/2$ and $3\pi/2$) are three times the external stress. Also, compressive stresses are equal to $-\sigma_{yy}^\infty$, above and below the hole (i.e., $\theta = 0$ and π). In the event of tensile failure, the actual measured stress is three times lower than the material could withstand without the circular hole.[115] In the case of the elliptical shape of the hole, the stress concentrations become more pronounced as the more stretched the ellipse, the higher the stress concentrations, and the lower the measured failure stress of the plate. The tangential stresses at A and B increase when the semi-axes of the ellipse are equal to a and b, to:[115]

$$\sigma_{\theta\theta} = \sigma_{yy}^\infty \left(1 + \frac{2a}{b}\right) \tag{10.34}$$

In the case of a slit, the tangential stress at the tip will become infinitely large ($\sigma_{\theta\theta} \to \infty$).

10.5.6   FRACTURE TOUGHNESS

Fracture toughness is the resistance of a material to crack propagation, and can be measured by testing the strength of specimens (mostly brittle materials) with notches. Notch sensitivity refers to the point at which the stress concentration at the end of the notch reduces the strength. The notch sensitivity of a concrete can be defined as the ratio of its flexural strength (calculated from the net cross section of the notched beam) to the flexural strength of a similar but unnotched beam.[100] This ratio would have the size of unity if the notched beam had the same fracture load in bending as a similar unnotched beam with the same depth as the residual depth beneath the notch of the notch beam.[86] This ratio can be used as a measure of the notch sensitivity of a material. For instance, it is less than unity for a brittle material, and for a ductile material, the ratio may approach unity due to relaxation of the stress concentration by plastic flow.

10.5.7   FRACTURE UNDER SUSTAINED AND REPEATED LOADS

The failure of concrete under sustained and repeated loads is also associated with the principle of crack propagation.[16,31,87] The prolonged time under load and the repetition of load gives rise to the propagation of cracks under sustained and repeated loads, respectively. Also, failure is bound to occur earlier under sustained load, for a high amount of induced stress. Failures with less repetition are produced by high loading. Larger deformations can be observed on concrete under sustained load, compared to under short-term loading. With this, stress is redistributed within the concrete. This means that cracks will be propagated under sustained load irrespective of the constant stress due to the fact that the prolonged period of time under the load will create more weak spots, as well as provide more opportunity for the cracks to locate the weakest paths for propagation.[86] Furthermore, in terms of repeated load, greater deformations occur in concrete compared to when standard static loading is applied. This enables cracks to grow under repeated cycles of loading even when the maximum load remains unchanged.

10.5.8   FRACTURE UNDER MULTIAXIAL LOADING

The testing condition in which a concrete specimen is subjected to could influence its load-bearing ability under multiaxial or triaxial loading, as its ability to bear load could be greater or lesser compared to when under uniaxial loading. Overall, the load-bearing ability will be increased accordingly, if the combined load (multiaxial and triaxial) were to obstruct the crack propagation,[59] as compared to uniaxial loading. For example, concrete will have lower strengths when subjected to tensile stress in one direction and compressive stresses in the other two directions, compared to when subjected to uniaxial compressive strength. However, concrete subjected to triaxial compressive stresses, will have higher strength. Two distinct modes of failure observed for paste, mortar and concrete specimens, under triaxial compression. The failure mode was tensile under low confining pressures, and a slitting form of fracture, similar to that under uniaxial compression, was observed. This form of fracture is associated with relatively large axial compression and lateral tensile strains, as well as significant internal microcracking. They observed that the axial strains at failure were smaller, lateral strains were compressive, very little microcracking at failure occurred, and specimens remained in a good condition.

10.5.9   THE J CONTOUR INTEGRAL

The J contour integral is a fracture characterising parameter for nonlinear materials.[5] The basis for using the fracture mechanics approach using the elastic-plastic deformation as nonlinear elastic, was provided by reference [91]. The uniaxial stress–strain behavior of elastic-plastic and nonlinear elastic materials is shown in Figure 10.27.[5] The materials responses to loading are different even though their loading behavior is similar. The elastic-plastic has a linear unloading path with the slope similar to Young's modulus. The nonlinear elastic material follows an unloading path similar to the path as it was loaded. A given strain in an elastic-plastic material may be associated with multiple stress values provided the material is subjected to cyclic loading or unloaded, and therefore it is much easier to examine an elastic material compared to a material that exhibits irreversible plasticity.[5] However, there is a distinct relationship between stress and strain in an elastic material. The mechanical response of the two materials in Figure 10.27 is identical provided the stresses in both materials increase monotonically. A theory similar to nonlinear elasticity and which relates total strains to stresses in a material is the deformation theory of plasticity. This theory has been applied to study crack propagation in a nonlinear material[91] where it was shown that the nonlinear energy release J could be given as a path-independent line integral. It has been found that J well characterize crack-tip stresses and strains in nonlinear materials.[46,92] Hence, the J integral can be considered as both energy and a stress intensity parameter.[5]

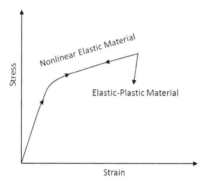

FIGURE 10.28   Schematic comparison of the stress–strain behavior of elastic plastic and nonlinear elastic materials.[5]

10.5.9.1   J INTEGRAL AS AN ENERGY PARAMETER (NONLINEAR ENERGY RELEASE RATE)

The energy release rate can be defined as the potential energy released from a structure when the crack grows in an elastic material.[5] However, when the crack grows or the specimen is unloaded in an elastic-plastic material, most of the strain energy absorbed by the material, is not recovered as a growing crack in an elastic-plastic material leaves a plastic wake.

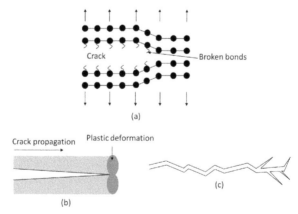

FIGURE 10.29   Crack propagation in various types of materials: (a) ideally brittle material, (b) quasi-brittle elastic-plastic material, and (c) brittle material with crack branching.[5]

A path-independent contour integral for the analysis of cracks has been provided by.[91] Here, the J integral is equal to the energy release rate in a nonlinear elastic body that contains a crack. The energy release rate nonlinear elastic materials with regards to J is given as:[5]

$$J = -\frac{d\Pi}{dA} \qquad\qquad (10.35)$$

where Π is the potential energy, and A is the crack area. The potential energy is given as:

$$\Pi = U - F \qquad\qquad (10.36)$$

where U is the strain energy stored in the body and F is the work done by external forces. For a cracked plate which exhibits a nonlinear load-displacement curve (Fig. 10.30) with plate of thickness, $A = a$, the potential energy for load control will be:[5]

$$\Pi = U - P\Delta = -U^* \qquad\qquad (10.37)$$

where U^* is the complimentary strain energy and is given as:

$$U^* = \int_0^P \Delta \, dP \qquad\qquad (10.38)$$

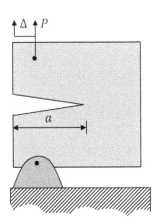
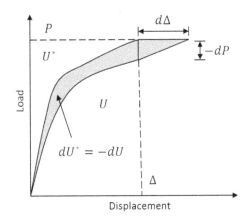

FIGURE 10.30   Nonlinear energy release rate.[5]

Therefore, if the plate is subjected to load control, J is given as:

$$J = \left(\frac{dU^*}{da}\right)_P \tag{10.39}$$

In a case where the crack advances at a fixed displacement, $F = 0$, then the J is given as:

$$J = -\left(\frac{dU}{da}\right)_\Delta \tag{10.40}$$

J for load control is equal to that for displacement control, and can be expressed in terms of load and displacement as shown below:

$$J = \left(\frac{\partial}{\partial a}\int_0^P \Delta dP\right)_P \tag{10.41}$$

$$= \int_0^P \left(\frac{\partial \Delta}{\partial a}\right)_P dP \tag{10.42}$$

or

$$J = -\left(\frac{\partial}{\partial a}\int_0^\Delta P d\Delta\right)_\Delta \tag{10.43}$$

$$= -\int_0^\Delta \left(\frac{\partial P}{\partial a}\right)_\Delta d\Delta \tag{10.44}$$

In the case of a linear elastic material (mode I—opening mode), J is given as:

$$J = -\frac{K_I^2}{E'} \tag{10.45}$$

10.5.9.2   J INTEGRAL AS A STRESS INTENSITY PARAMETER

Hutchinson[46] and Rice and Rosengren[92] showed that J can be used to evaluate crack-tip conditions in a nonlinear elastic material by assuming a power law relationship between plastic strain and stress. The relationship for uniaxial deformation for the case of the addition of elastic strains is known as the Ramberg–Osgood equation and is given as:[5]

$$\frac{\varepsilon}{\varepsilon_0} = \frac{\sigma}{\sigma_0} + \alpha\left(\frac{\sigma}{\sigma_0}\right)^n \tag{10.46}$$

where σ_0 is the reference stress value which is equal to the yield strength, $\varepsilon_0 = \sigma_0/E$, is a dimensionless constant, and n is the strain-hardening exponent. There are two conditions in this case: the first applies when the stress–strain varies as $1/r$ to remain path independent, and the second applies when the stress–strain behavior reduces to a simple power law as a result crack tips which are very close to the plastic zone and which lead to small elastic strains in relation to the total strain. These conditions show that a variation of stress and strain exists ahead of the crack tip, and can be described by the following equations:[5]

$$\sigma_{ij} = k_1 \left(\frac{J}{r} \right)^{\frac{1}{n+1}} \tag{10.47}$$

$$\varepsilon_{ij} = k_2 \left(\frac{J}{r} \right)^{\frac{n}{n+1}} \tag{10.48}$$

where k_1 and k_2 are proportionality constants. The stress and strain distributions are derived by applying the appropriate boundary conditions:[5]

$$\sigma_{ij} = \sigma_0 \left(\frac{EJ}{\alpha \sigma_0^2 I_n r} \right)^{\frac{1}{n+1}} \bar{\sigma}_{ij}(n,\theta) \tag{10.49}$$

$$\varepsilon_{ij} = \frac{\alpha \sigma_0}{E} \left(\frac{EJ}{\alpha \sigma_0^2 I_n r} \right)^{\frac{1}{n+1}} \bar{\varepsilon}_{ij}(n,\theta) \tag{10.50}$$

where I_n is an integration constant that depends on n, $\bar{\sigma}_{ij}$ and $\bar{\varepsilon}_{ij}$ are the dimensionless functions of n and θ. Equations 10.49 and 10.50 above are known as the Hutchinson, Rice, and Rosengren (HRR) singularity.[46,92] Similar to the way the stress intensity factor defines the amplitude of the linear elastic singularity, the J integral also defines the amplitude of the HRR singularity. Therefore, it can be considered that J well describes the conditions within the plastic zone.

10.5.10   FRACTURE UNDER TENSILE STRENGTH

The presence of flaws or cracks leads to very high stress concentrations at their tips under load which makes localized microscopic fracturing to occur when the average or nominal stress in the entire material is low (Fig. 10.31).

The orientation of a crack in a direction normal to the applied load, leads to increased weakness of concrete strength.

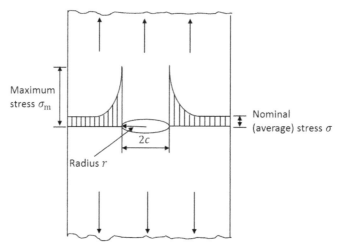

FIGURE 10.31 Stress concentration at the tip of a crack in a brittle material under tension.[76]

From Figure 10.31, it can be seen that the greater the maximum stress the longer and sharper the crack. This means that the greater the value of c, the smaller the value of r. This is also shown in the following expression:[76]

$$\frac{\sigma_m}{\sigma} = 2\left(\frac{c}{r}\right)^{\frac{1}{2}} \tag{10.51}$$

where σ_m is the maximum stress, σ is the nominal stress, c is the crack.

Also, when the external load increases, the maximum stress, σ_m, increases to a point where the failure stress of the material containing the crack is reached. This is known as the brittle fracture strength of the material, σ_f, and is given by:[76]

$$\sigma_f = \left(\frac{WE}{\pi c}\right)^{\frac{1}{2}} \tag{10.52}$$

where W is the work required to cause a fracture, and E is the modulus of elasticity. At this point, there is a release of energy stored in the material, which is associated with the formation of new surfaces and extension of cracks. An impending failure of the entire material occurs if the energy is not enough to continue the propagation of the crack. Also, if the energy released is too low, the crack is prevented until the external load is increased.

From the brittle fracture theory, failure is usually initiated by the largest crack which is oriented in the direction normal to the applied load.[76] Hence, size and shape of the concrete could possibly affect concrete strength, as there is a higher probability that a large concrete will contain a large number of critical cracks, enough to initiate failure. The energy released upon the initial propagation of a crack is enough to continue the propagation as a result of the increase of maximum stress and the decrease of brittle fracture strength upon an extension of the crack.

10.5.11   FRACTURE UNDER COMPRESSIVE STRENGTH

Compressive strength can be considered under bi- and triaxial stress, and uniaxial compression. In a case where two unequal principal stresses are compressive, the stress along the edge of an internal flaw is tensile at some points in order for fracture to occur.[76] For a combination of two principal stresses, P and Q, the fracture criteria are shown in Figure 10.32.

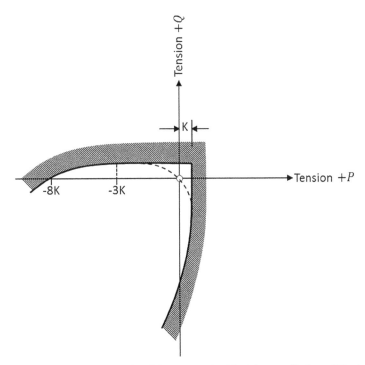

FIGURE 10.32   Orowan's criteria of fracture under biaxial stress,[76] where K is the tensile strength in direct tension.

From Figure 10.32, fracture is seen to occur under a combination of two principal stresses P and Q so that the point representing the state of stress crosses the curve outwards into the shaded area. Upon the application of uniaxial compression, the compressive strength is eight times the direct tensile strength (i.e., $8K$).

Furthermore, fracture patterns of concrete under different stress states, are shown in Figure 10.33.

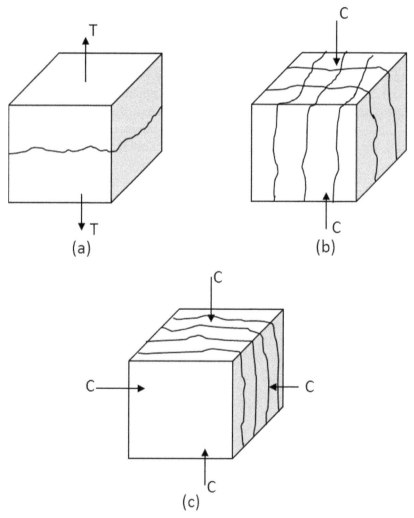

FIGURE 10.33 Fracture patterns of concrete under: (a) uniaxial tension, (b) uniaxial compression, and (c) biaxial compression.[76]

Fracture is seen to occur mostly in a plane normal to the direction of the load, under uniaxial tension. Under uniaxial compression, the cracks are mostly parallel to the applied load, and are caused by a localized tensile stress in a direction normal to the compressive load. However some cracks under this compression form at an angle to the applied load, and are caused by the collapse as a result of the development of shear planes. It is important to note that cracks are formed in two planes parallel to the load, hence the specimen are split into column-type fragments,[76] as seen in Figure 10.33(b). Similarly, for biaxial compression, failure is seen to occur in one plane parallel to the applied load which leads to the formation of slab-type fragments, as seen in Figure 10.33(c). In the case of triaxial compression, failure is seen to occur by crushing.

10.5.12   MICROCRACKING AND STRESS–STRAIN RELATION

Microcracking occurs due to differential volume changes between the cement paste and the aggregate. It is also as a result of differences in stress–strain behavior, and in thermal and moisture movement.[76] In concrete, very fine bond cracks can exist at the interface between coarse aggregate and hydrated cement paste irrespective of if load is applied. The stability of such cracks can be retained and will not develop further under stress up to about 30% of the ultimate strength of the concrete. Stress–strain relations for cement paste, aggregate, and concrete, are shown in Figure 10.34. Linearity is observed for the stress–strain relations for the aggregate and cement paste, while that for concrete becomes curvilinear at higher stresses. This can be described by the development of bond cracks at the interfaces between the two phases caused by microcracking at stresses above 30% of the maximum strength. At this stage, an increase in size and number of the microcracks begins to occur. The cracks at this point are stable under sustained loading, and can be referred to as slow propagation of microcracks. To avoid early localized failure, the concrete is able to redistribute local high stresses to regions of lower stress as a result of the development of microcracks as well as creep. Cracks expand through the mortar matrix (i.e., cement paste and fine aggregate) and therefore bridge the bond cracks in such a way that the crack pattern is formed continuously. This occurs at 70 to 90% of the maximum strength. At this stage (which can be referred to as fast propagation of cracks), sustaining the load might lead to failure as time advances. This failure regime over time can be referred to as static fatigue. Increasing the load will lead to rapid failure at the nominal maximum strength.

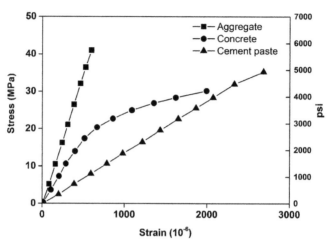

FIGURE 10.34   Stress–strain relations for cement paste, aggregate, and concrete.[76]

During the measurement of axial or compressive strain, the application of load at a constant rate of stress until failure occurs at the maximum stress can lead to an extension of the lateral strain (Fig. 10.34). For stress levels below 30% of the maximum strength, the ratio of lateral strain to axial strain (i.e., Poisson's ratio) is constant. Beyond this point, this ratio slowly increases, and further increases rapidly at 70–90% of the ultimate strength as a result of the formation of mostly vertical unstable cracks.[76] As can be seen in the volumetric stain curve of Figure 10.34, the specimen is hardly a continuous body at this stage, and a change from slow contraction to a rapid increase in volume, occurs.

Furthermore, cracks can be detected using ultrasonic and acoustic emission tests. For these, the transverse ultrasonic pulse-velocity decreases as cracking develops. The sound level emitted increases, and in both cases, shows large changes before failure.

In standard compression tests on cylinders or cubes, stress is expected to increase at a constant rate when the concrete is loaded in uniaxial compression as can be seen in the strain-strain curve shown in Figure 10.33. However, in a situation where the specimen is loaded at a constant rate of strain, the stress–strain curve is seen to decline prior to failure. This trend can be seen in Figure 10.35.[117] In this case, the displacement has to be controlled rather than the load, and the test requires a testing machine with a rigid frame. The declining trend implies that concrete has a capacity to withstand some load after the maximum load has been exceeded. This is because the linking of microcracks is delayed prior to complete breakdown.

Also, a steeper descending curve for concrete made of lightweight aggregate indicates that it exhibits a more brittle nature compared to that made with normal weight aggregate. A more brittle type of behavior is the case for high-strength concrete, as both the parts of the curve that increase and decrease, are steeper. The area bounded by the entire stress–strain curve represents the work required to cause failure or fracture toughness.

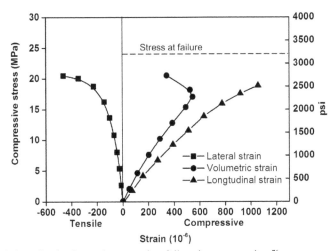

FIGURE 10.35   Strains in a prism tested to failure in compression.[76]

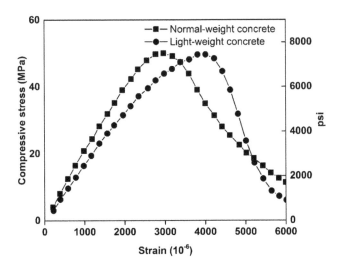

FIGURE 10.36   Stress–strain relation for concretes tested at a constant rate of strain.[76,117]

10.5.13   FACTORS THAT INFLUENCE MICROCRACKING

Porosity and the type of aggregate can influence microcracking. However it is difficult to measure porosity as the degree of hydration is not easily measured. Also, the influence of aggregate cannot be quantified easily. Consequently, the main factors influencing microcracking and therefore the strength of concrete are: degree of compaction, age, temperature, w/c ratio, aggregate/cement ratio, quality of the aggregate (i.e., grading, surface texture, shape, strength, and stiffness), the maximum size of the aggregate, and transition zone.

10.5.14   RELATIONSHIP BETWEEN TENSILE AND COMPRESSIVE STRENGTH

Theoretically, compressive strength is about eight times larger than the tensile strength. This indicates that a fixed relationship between both strength types exists. The ratio of the strengths is dependent on the overall strength level of the concrete. The ratio of tensile to compressive strength is lower with increasing compressive strength. This means that tensile strength increases with age at a lower rate than the compressive strength. Other factors that influence this relationship include: the method of concrete in tension testing, the size of the specimen, the shape and surface texture of coarse aggregate, and the moisture condition of the concrete. Direct or uniaxial tension tests are difficult to perform due to the problem faced with the complete gripping of the specimen in such a way that premature failure close to the edge of the attachment does not occur. Alternative methods for measuring tensile strength in flexure (modulus of rupture) and indirect tension (splitting) have been suggested. These methods reveal different numerical results, and can be arranged in the following order: direct tension < splitting tension < flexural tension. The discrepancy in results could be as a result of: firstly, the size of the laboratory specimen resulting in the decreasing of the volume of concrete that undergoes tensile stress. This implies that there is a higher probability of a weak element which will lead to failure in a larger than in a smaller volume. The second reason is that both the splitting and flexural test methods involve non-uniform stress distributions which influence crack propagation and leads to delay of maximum failure. In the direct test, a uniform stress distribution exists and a crack is quickly propagated through the section of the specimen once it is formed. Typical values of tensile strength as a function of compressive strength for different test methods are shown in Figure 10.37.

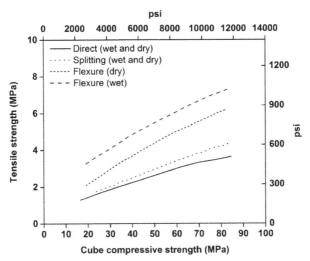

FIGURE 10.37 Relation between tensile and compressive strengths of concrete with rounded coarse normal and light weight aggregates. Flexural test: 100 × 100 × 500 mm, splitting test: 150 × 300 mm, direct test: 75 × 355 mm, and compression test: 100 mm.[76]

It can be seen that the moisture condition of concrete influences the relationship between the flexural and compressive strengths. Continuously wet-stored concrete have been compared with wet cured concrete, and stored in a dry environment. The compressive strength of the drying concrete is greater than the continuously wet-stored concrete. Also, the splitting and direct tensile strengths are not influenced. However, the flexural strength of the drying concrete is lower than that of wet concrete due to the sensitivity of this test, and in the presence of shrinkage cracks.

Tensile strength (f_t) can be related to compressive strength (f_c) using the expressions[76,117]

$$f_t = k f_c^n \qquad (10.53)$$

Where k and n are coefficients which depend on the key factors as discussed earlier, and on the shape of the compression specimen (i.e., cube or cylinder). The coefficients are influenced by the properties of the mix used.[76]

10.5.15 FATIGUE STRENGTH

Fatigue failure in concrete can be classified under two types: those that occur under a sustained load or gradually increasing load (also known as static fatigue or creep rupture), and those that occur under cyclic or repeated loading.[76] For a particular threshold value, a time-dependent failure can occur at stresses greater than this value, but smaller than the temporary static strength. This applies to both types of fatigue failure.

When testing for fatigue strength, the duration of test is important as strength depends on the rate of loading. Standard loading rates occur for the determination of compressive strength, and have been suggested by ASTM C 39-05 and BS EN 12390-3: 2002. Figure 10.38 shows the influence of the duration of test or rate of loading, on the strength and strain capacity in compression. It can be observed that for a decrease in the rate of loading (or increase in duration of test time), applying a steadily increasing load will result in a lower strength compared to the case of a standard test. Also, applying the load more rapidly will result in a higher strength with a smaller strain at failure. It has been observed that concrete behaves more brittle when load is applied more rapidly than at lower rates, when the strain at failure is increased by creep and microcracking.[76] In a standard test, the compressive strength is obtained from a test of a short duration (approximately 2–4 min).[76]

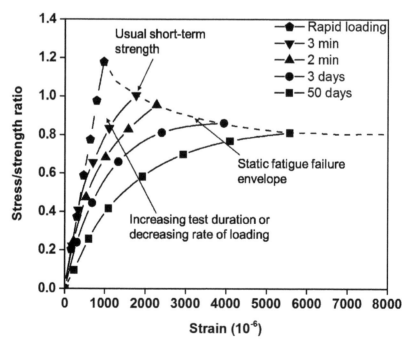

FIGURE 10.38 Influence of test duration or rate of loading on strength and strain capacity, in compression.[45,76]

10.5.16 ELASTICITY

Concrete is a structural material that exhibits an almost elastic behavior when load is initially applied to it. However, under sustained loading, concrete is seen to exhibit strain which increases with time under a constant stress. This is also known as creep, and occurs even at very low stresses and under normal environmental conditions of temperature and humidity.

The concept of pure elasticity involves the appearance and disappearance of strains upon application and removal of stress. Figure 10.39 illustrates four categories of stress–strain responses that which describe elasticity. These categories are: (a) linear and elastic, (b) non-linear and elastic, (c) linear and non-elastic, and (d) non-linear and non-elastic. Steel correspond to (a), while certain plastics and timber correspond to (b), and these describe pure elasticity. Categories (c) and (d) best describe brittle materials such as glass and rocks, and concrete respectively. The stress–strain diagram in Figure 10.40 (c) shows separate linear curves which represent loading and unloading. Following loading, a permanent deformation occurs when the load is completely withdrawn. For case (d), a permanent deformation also occurs, and can be mostly identified in concrete that undergo moderate and high stress levels of compression and tension loads.[76]

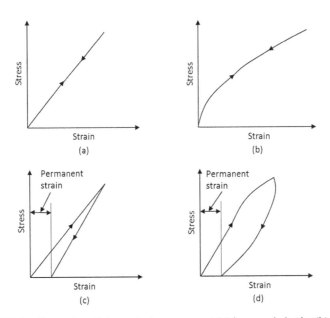

FIGURE 10.39 Categories of stress–strain response: (a) Linear and elastic, (b) Non-linear and elastic, (c) Linear and non-elastic, and (d) Non-linear and non-elastic.[76]

The modulus of elasticity can be derived by obtaining the slope of the relationship between stress and strain. The linear cases (i.e., (a) and (b)) can be used to obtain the Young's modulus. To obtain the Young's modulus of a typical concrete as illustrated in (d), the linear section of the curve at the origin, will have to be considered. For the other section of the curve which non-linear, the tangent of the curve can be measured to obtain the initial tangent modulus (for the curve at the origin), or tangent modulus (at any point on the curve). The non-linearity which exists in concrete at normal stresses is usually due to creep.

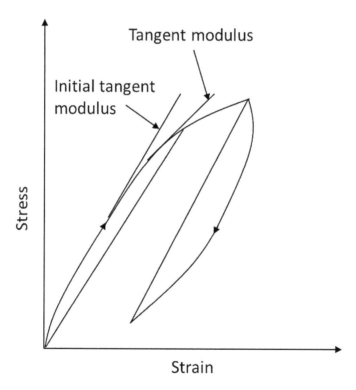

FIGURE 10.40   Typical stress–strain curve for concrete.[76]

10.6   STRENGTH CHARACTERIZATION OF COMPOSITE REINFORCED CONCRETE

Concrete is weak in tension, hence, it is usually reinforced with continuous-bar elements in the tensile zone in order to sustain large transverse loading.

However, such reinforcement may not be able to fully resolve the issue of micro- and macro-cracking. The idea behind including these reinforcements is to replace the function of the tensile zone of a section of the concrete, and assume the tension equilibrium force in that section.[72] Consequently, randomly spaced discontinuous fiber elements can aid in reducing the propagation of the microcracks which are known to develop at the early stages of loading. Reinforced concrete containing composites usually of high strength and modulus, such as fiber, is becoming widely used in civil infrastructure as the inclusion of these composites may improve the toughness, flexural strength, tensile strength, impact strength, and failure mode of the concrete.[50] Studies show that the inclusion of Fiber-based Reinforced Composites (FRC) can increase the compressive strength and the deformability of the concrete composite when subjected to vertical and lateral loads.[42] Natural and synthetic fiber-reinforced concrete are currently been used around the world for mostly civil engineering applications. Natural fiber-reinforced composites are cheap and can be recycled, and are suitable for developing countries. Synthetic based fibers are difficult to recycle, and are not as environmentally-friendly as natural fiber-based composites. A wide range of organic polymer matrices (e.g., epoxy-based) are also used due to their availability and durability under aggressive environments, as well as mechanical properties that meet the requirements for structural applications.[28] The concept of incorporating fiber-based reinforcement in concrete fabrication is based on the fact that the fibers are the principal load-carrying members, while the surrounding matrix (e.g., polymer, metal, ceramic, etc.) restrict the fiber components to the desired location and orientation. These matrix act as a load transfer medium and protects the fiber materials from environmental damages due to, for example, temperature and humidity.[64]

Mobasher[68] has summarized a review based on current trends of research and development with regards to sustainability of construction materials, and categorized this under the following points:

- Durability—deals with tackling the demand for more durable construction materials by enhancing design efficiency, proper engineering, and increased service life through more durable construction.
- Quality—deals with improving quality control measures and determination of early-age and long-term properties of concrete materials. This may be achieved via life cycle cost modeling combined with statistical quality control measures.

- Innovations—deals with innovating alternative concrete-based materials with higher specific strength, ductility, and stiffness, by accounting for these in preliminary design.
- Cross-disciplinary efforts—deals with the development of software tools that integrate various design considerations that account for sustainability, intelligence, and economics during the design and engineering of concrete structures.
- Education—deals with reducing the gap between research and development using tools for carrying our analysis, design, and technology transfer.

10.6.1 MECHANICAL TESTING AND STRENGTH CHARACTERISTICS OF FIBER-REINFORCED COMPOSITE

As with pure concrete, the macrostructural properties of Fiber-reinforced composite such as: strength, stiffness, and ductility can be determined. Also micromechanical properties such as cracking, including fracturing and failure mode can be detected for these types of concrete composites. By carrying out mechanical testing and observing the decrease in the load-carrying capacity with an increase in the strain or widening of cracks in a tensile stress field, a stress–strain curve can be obtained for these concretes. Similar to pure concrete, the strength of this concrete type can be studied by evaluating stress–strain curves, stiffness, toughness, fracture toughness, ultimate tensile strength, crack width, crack spacing, and so on.

10.6.1.1 COMPRESSION TEST

Compression test is commonly used for characterizing concrete, and can be used to obtain the maximum stress (i.e., strength), modulus of elasticity, and post-peak regime to determine the entire response for stress–strain.[68] This test can be conducted under a load or actuator displacement control conditions, based on standard ASTM test requirement. At the point of maximum load, a resistance to increasing load by the specimen is reached. The effect of AR glass fibers on the compressive strength and ductility of concrete in compression is shown in Figure 10.41[30] in the form of a stress versus circumferential strain for concrete containing two fibers of lengths 12 and 24 mm. These stress–strain results are provided for various ages at 3, 7, and 28 days.

An increase in strength from 3 to 7 days was observed and higher compared to 7 to 28 days. Also, the postpeak energy absorption at 28 days is much higher than at 3 days. Furthermore, a better toughening of the system is observed after 28 days of curing.

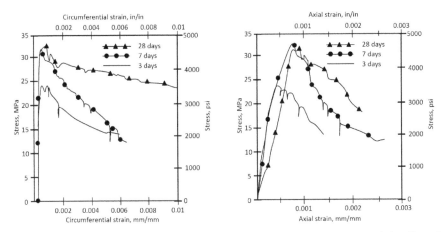

FIGURE 10.41 Effect of AR glass fibers on the compressive strength and ductility of concrete, $V_f = 10$ kg/m^3, $w/c = 0.4$.[30,68]

Zhou et al.[122] investigated the compressive behavior of Polyvinyl alcohol (PVA) fiber-reinforced engineered cementitious composites in uniaxial compression for compressive strengths ranging from 35 to 60 MPa. They further studied compressive parameters, such as the elastic modulus, engineering strain at the peak stress, Poisson's ratio, and toughness index. A servo-hydraulic testing machine with a capacity of 4600 kN was used for compression test, as well as two compressometers and one circumferential extensometer to evaluate elastic modulus and Poisson's ratio. To stress–strain relationship was obtained by measuring strain using compressometers or LVDT's at the central region of the specimen.

They observed that the addition of PVA fibers made the cementitious composites more ductile than mortar under uniaxial compression, and the failure mode changed from brittle splitting to ductile shear failure. They further observed an increase in the elastic moduli of the composite with the compressive strength. In their case, the addition of the PVA fiber had little effect on the strain at peak stress. A slight increase in their Poisson's ratio with the compressive strength was observed for strength lower than 50 MPa. However, above this strength, the Poisson's ratio remains fairly constant.

FIGURE 10.42   Failure mode of cylinder specimens under compression.

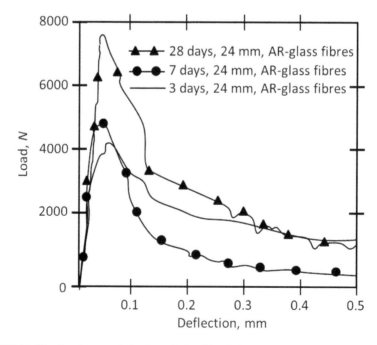

FIGURE 10.43   Load versus deflection relationship of Fiber-Reinforced Composite materials representing a relatively low-volume fraction for fiber loading, $V_f = 0.6$ kg/m^3, $w/c = 0.55$.[30]

10.6.1.2   FLEXURE TEST

A three-point bending test can potentially eliminate deformations which may arise as a result of specimen rotations or support stability. The Crack Mouth Opening Deformation (CMOD) can be measured across the face of the notch using extensometers. Direct tensile tests have been performed on cement composites like AR glass and fabric composites and it was observed that the presence of fibers reduces the potential for localization. The shapes of the curves in Figure 10.44 show the increase in the load-carrying capacity of the composite with age. It can be observed that the response is quasi-linear up to the peak load, while the specimens retain a major section of their tangential stiffness. The effect of low fibers is more obvious in the early ages as the strength of the matrix material had not significantly increased. It can also be seen that the postpeak deformation is less significant, but decreases significantly as the specimen age increases.

10.6.1.3   FRACTURE TOUGHNESS AND STRAIN ENERGY RELEASE RATE

Cracking of concrete is complex since it is an anisotropic and heteroge-nous material. The complex process of cracking involves micro-cracking shielding, aggregate bridging and crack branching, which also dissipate energy during the fracture process. The cracking strength of concrete is mainly evaluated using the concept of fracture mechanics.[3] The propaga-tion of a crack is initiated when the crack tip stress intensity factor reaches the fracture toughness value (defined in terms of critical stress intensity factor K_{Ic}, or fracture energy G_c. To determine fracture toughness, notches or sharp cracks are introduced to the specimens. The test characterizes the resistance of a material to fracture in the presence of a sharp crack under severe tensile constraint, such that the state of the stress near the front of the crack approaches plane strain. In this situation, the crack-tip plastic or non-linear viscoelastic region is smaller than the crack size and dimensions of the specimen in the constraint direction. A typical specimen configuration for fracture toughness test is shown in Figure 11.45.

The sample width, W, is the $W = 2B$. The crack length, a, is usually within the region of $0.45 < a/W < 0.55$.

To establish a valid K_{Ic} has been achieved, a conditional result, K_Q, is calculated. This involves the construction on the test record to determine the result is consistent with the size of the specimen. Using the size criteria, and calculating a quantity using, $2.5(K_Q/\sigma_y)^2$, where σ_y is the yield stress, then K_Q is equal to K_{Ic} if this quantity is less than the specimen thickness B, the crack length, a, and the ligament $W - a$. If this is not the case, then the test is not valid.

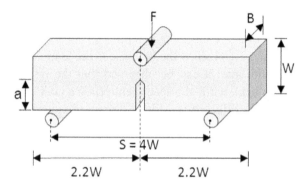

FIGURE 10.44   Schematic diagram of the specimen for the fracture toughness test.

A general expression for K_Q, is:

$$K_Q = \left(\frac{P_Q}{BW^{1/2}}\right) f(x) \qquad (10.54)$$

where $(0 < x < 1)$, then:

$$f(x) = 6x^{1/2} \frac{\left[1.99 - x(1-x)(2.15 - 3.93x + 2.7x^2)\right]}{(1+2x)(1-x)^{3/2}} \qquad (10.55)$$

where, P_Q is the load.

Some fracture toughness values are shown in Table 10.2.

TABLE 10.2   Fracture Toughness Values from Various Investigators.

Size of specimen (mm)	Test method	K_{Ic} (MPa.m$^{1/2}$)	Investigator
508 × 152 × 152	3-pt and 4-pt	0.67–1.13	54
48 × 12 × 12–800 × 100 × 200	3-pt	0.6–1.1	104
600 × 80 × 76–1800 × 80 × 300	3-pt	0.867 ± 0.063	55
305 × 29 × 76, 609 × 57 × 152, and 914 × 86 × 229	3-pt	0.787–1.127	51
760 × 150 × 150	3-pt	1.014–1.076	38

10.7 CONCLUSION

In this book chapter, we have reviewed the recent advancements on the strength characteristics of conventional and composite concretes within the framework of fracture mechanics. Fracture mechanics approaches are helpful in understanding the failure behavior of concrete materials under mechanical loading. Using a cracking and energy based criterion, the failure of concrete structures can be described in the context of fracture mechanics.

Here, the principles of LEFM have been presented and stress field around a crack-tip is been described using a stress function approach. The stress component in the direction perpendicular to the crack was found to approach infinity at the crack tip. The concept of the J contour integral (a fracture characterizing parameter for non-linear materials) was discussed. Here, J contour integral is used to characterize crack-tip stresses and strains, as well in describing the nonlinear energy release rate and stress intensity parameter. Fracture process in uniaxial tension has been presented, and this comprises of an elastic stage, stable microcracking, unstable macrocracking, and crack-face bridging. These four stages form a basis for the development of a macroscopic model for tensile fracture of concrete, and uses the principle associated with the physical mechanisms associated with most materials that fail in a quasi-brittle manner upon the propagation of one or more cracks. As microcrack is associated with several branching and bridging, this model takes into account a bridging stress which is influenced by the composition of the concrete specimen. In normal concrete, the bridging stress is seen to be more uniformly distributed. In fiber-reinforced concrete, the stress is uniform until a total pull-out of the fiber occurs. However, this stress may decrease gradually for fiber-reinforced concrete, depending on the material composition.

Furthermore, the failure behavior of standard and reinforced concrete composite beginning from the fundamental understanding of crack propagation and damage progression under several loading types including uniaxial, multiaxial, sustained and repeated, is reviewed using fracture mechanics. It was observed that the failure of concrete under load usually occurs as a result of gradual internal progression of microcracks and tension within the specimen parallel to the direction of the applied compressive stress. Complete failure occurs when the cracks begin to form continuous patterns. Also, microcracks at the interface between for example, coarse aggregate and mortar, show that the interface is the weakest link within the concrete and mostly occurs in composite materials which comprise of components of different stiffness. Bond cracks tend to initially develop around the larger aggregates or particulate inclusions, which means that the strength of concrete would decrease with an increase in aggregate size. For concrete under sustained load, it was observed that lager deformations can occur, compared to under short-term loading. The prolonged period of time under the load will create more weak spot within the concrete, as well as provide more opportunity for the cracks to locate the weakest paths for propagation. More deformation occurs in concrete under repeated load than when under standard static loading. Hence cracks will grow under repeated cycles of loading even for a fixed maximum load. For concrete subjected to multiaxial such as triaxial compression, two failure modes were observed. The failure was tensile under low confining pressures, and a slitting form of fracture, similar to that under uniaxial compression. This form of fracture has been associated with relatively large axial compression and lateral tensile strains, including significant internal microcracking.

Fracture mechanics is a powerful design tool for researchers and designers to develop advanced structures with improved safety and structural reliability. The growing need for such advanced structures has motivated engineers to develop concrete reinforced with innovative particulate inclusions. These composites such as with recycled waste materials, as well as fiber, are known to improve the overall performance and durability of the resulting concrete material. A major challenge with such composite materials is their complex heterogonous nature which makes their failure behavior difficult to describe. Some factors that contribute to the complexity of these materials include cracking at the interface between dissimilar materials, interactions between inclusions, development of microcracks, and voids. Though a comprehensive level understanding of such particulate composites is not yet understood, fracture theories are useful in this regard as they can be extended to describe the mechanisms of failure in such materials in future.

Information provided in this book will help engineers, research and designers in developing advanced concrete structures with enhanced structural performance, durability as well as reliability. This will also help the graduate students by providing a coherent reference source of materials in the field of mechanics of concrete.

KEYWORDS

- **concrete**
- **composite material**
- **rheology**
- **plastic viscosity**
- **compressive strength**

REFERENCES

1. Abrams, D. A. *Design of Concrete Mixtures*; Structural Materials Research Laboratory, Lewis Institute: Chicago, 1919.
2. Abrams, D. A. *Flexural Strength of Plain Concrete*; Structural Materials Research Laboratory, Lewis Institute: Chicago, 1922.
3. Alam, M. R.; Azad, M. A. K.; Kadir, M. A. Fracture Toughness of Plain Concrete Specimens Made with Industry-burnt Brick Aggregates. *J. Civ. Eng.* **2010,** *38*(1), 81–94.
4. Albayrak, G.; Canbaz, M.; Albayrak, U. Statistical Analysis of Chemical Admixtures Usage for Concrete: A Survey of Eskisehir City, Turkey. *Procedia Eng.* **2015,** *118*, 1236–1241.
5. Anderson, T. L. *Fracture Mechanics: Fundamentals and Applications*, 3rd Edition; Taylor & Francis: Boca Raton, 2005.
6. Banfill, P. F. G. Feasibility Study of a Coaxial Cylinders Viscometer for Mortar. *Cem. Concr. Res.* **1987,** *17*(2), 329–339.
7. Banfill, P. F. G.; Saunders, D. C. On the Viscometric Examination of Cement Pastes. *Cem. Concr. Res.* **1981,** *11*(3), 363–370.
8. Barenblatt, G. The Mathematical Theory of Equilibrium Cracks in Brittle Fracture. *Adv. Appl. Mech.* **1962,** *7*, 55–129 (Elsevier).
9. Barr, B.; Gettu, R.; Al-Oraimi, S. K. A.; Bryars, L. S. Toughness Measurement—The Need to Think Again. *Cem. Concr. Comp.* **1996,** *18*(4), 281–297.
10. Başyiğit, C.; Çomak, B.; Kilinçarslan, Ş. A New Approach to Compressive Strength Assessment of Concrete: Image Processing Technique. *AIP Conf. Proc.* **2012,** *1476*(1), 65–69.
11. Bažant, Z. P. Size Effect in Blunt Fracture: Concrete, Rock, Metal. *J. Eng. Mech.* **1984,** *110* (4), 518–535.
12. Bažant, Z. P. Fracture Energy of Heterogeneous Materials and Similitude. In *Fracture of Concrete and Rock*; Shah, S., Swartz, S., Eds.; Springer: New York, 1989; p 229.
13. Bazant, Z. P. In *Fracture Mechanics of Concrete Structures,* International Conference on Fracture Mechanics of Concrete Structures (FraMCoS1), 1–5 June 1992, Beaver Run Resort, Breckenridge, Colorado, USA, Society, A. C., Ed.; Taylor & Francis: Beaver Run Resort, Breckenridge, Colorado, USA, 1992; p 146.

14. Bazant, Z. P.; Kazemi, M. T.; Hasegawa, T.; Mazars, J. Size Effect in Brazilian Split-cylinder Tests. Measurements and Fracture Analysis. *ACI Mater. J.* **1991,** *88*(3), 325–332.

15. Bedard, C.; Mailvaganam, N. P. The Use of Chemical Admixtures in Concrete. Part 1: Admixture-cement Compatibility. *J Perform. Constr. Facil.* **2005,** *19*(4), 263–266.

16. Béres, L. Failure Process of Concrete Under Fatigue Loading. *Rheol. Acta* **1974,** *13*(3), 639–643.

17. Brandtzaeg, A. *Failure of a Material Composed of Non-isotropic Elements*; Trondhjem: Bruns i Komm., 1927; p 2.

18. Broek, D. *Elementary Engineering Fracture Mechanics*; Nijhoff: The Hague; London, 1982.

19. Broek, D. *The Practical Use of Fracture Mechanics*; Springer Netherlands: Dordrecht, 1989.

20. Bungey, J. H.; Grantham, M. G.; Millard, S. *Testing of Concrete in Structures*, 4th Edition; Taylor & Francis, London, 2006.

21. Carpinteri, A. *Applications of Fracture Mechanics to Reinforced Concrete*; Taylor & Francis, 1992.

22. Chen, C. S.; Hsu, S. C. Measurement of Indirect Tensile Strength of Anisotropic Rocks by the Ring Test. *Rock Mech. Rock Eng.* **2001,** *34*(4), 293–321.

23. Chen, W. F.; Yuan, R. L., Tensile Strength of Concrete: Double-Punch Test. *J. Struct. Div. ASCE* **1980,** *106*(8), 1673–1693.

24. Claesson, J.; Bohloli, B. Brazilian Test: Stress Field and Tensile Strength of Anisotropic Rocks Using an Analytical Solution. *Int. J. Rock Mech. Min. Sci.* **2002,** *39*(8), 991–1004.

25. Dai, Q.; Ng, K.; Zhou, J.; Kreiger, E. L.; Ahlborn, T. M. Damage Investigation of Single-edge Notched Beam Tests with Normal Strength Concrete and Ultra High Performance Concrete Specimens Using Acoustic Emission Techniques. *Constr. Build. Mater.* **2012,** *31*(0), 231–242.

26. Damgaard Jensen, A.; Chatterji, S. State of the Art Report on Micro-cracking and Life-time of Concrete—Part 1. *Mater. Struct.* **1996,** *29*(1), 3–8.

27. Das, S.; Iyer, P.; Kenno, S. Y. Mechanical Properties of Fiber-Reinforced Concrete Made with Basalt Filament Fibers. *J. Mater. Civ. Eng.* **2015,** *27*(11), 4015015.

28. De Caso y Basalo, F. J.; Matta, F.; Nanni, A. Fiber Reinforced Cement-based Composite System for Concrete Confinement. *Constr. Build. Mater.* **2012,** *32*, 55–65.

29. De Schutter, G.; Luo, L. Effect of Corrosion Inhibiting Admixtures on Concrete Properties. *Constr. Build. Mater.* **2004,** *18*(7), 483–489.

30. Desai, T.; Shah, R.; Peled, A.; Mobasher, B. *Mechanical Properties of Concrete Reinforced with AR-glass Fibers*, Proceedings of the 7th International Symposium on Brittle Matrix Composites (BMC7), Warsaw, October 13–15 2003, pp 223–232.

31. Desayl, P. Strength of Concrete Under Combined Compression and Tension-Determination of Interaction Curve at Failure from Cylinder Split Test. *Mater. Constr.* **1969,** *2*(3), 179–185.

32. Donatello, S.; Tyrer, M.; Cheeseman, C. R. Recent Developments in Macro-defect-free (MDF) Cements. *Constr. Build. Mater.* **2009,** *23*(5), 1761–1767.

33. Dugdale, D. S. Yielding of Steel Sheets Containing Slits. *J. Mech. Phys. Solids* **1960,** *8*(2), 100–104.

34. Fairhurst, C. On the Validity of the 'Brazilian' Test for Brittle Materials. *Int. J. Rock Mech. Min. Sci. Geomech. Abstr.* **1964,** *1*(4), 535–546.

35. Ferrari, V. J.; Arquez, A. P.; Hanai, J. B.; de Souza, R. A. Development of High Performance Fiber Reinforced Cement Composites (HPFRCC) for Application as a Transition Layer of Reinforced Beams. *Revista IBRACON de Estrut. e Materiais* **2014,** *7,* 965–975.

36. Floyd, O. S.; Stanley, O. X-Rays for Study of Internal Structure and Microcracking of Concrete. *J. Proc.* **1963,** *60*(5), 575–588.

37. Gopalaratnam, V. S.; Gettu, R. On the Characterization of Flexural Toughness in Fiber Reinforced Concretes. *Cem. Concr. Compos.* **1995,** *17*(3), 239–254.

38. Hamoush, S. A.; Abdel-Fattah, H. The Fracture Toughness of Concrete. *Eng. Fract. Mech.* **1996,** *53*(3), 425–432.

39. Hertzberg, R. W. *Deformation and Fracture Mechanics of Engineering Materials*; Wiley: New York, London, 1976.

40. Hertzberg, R. W. *Deformation and Fracture Mechanics of Engineering Materials*; Wiley & Sons: New York, 1995.

41. Hondros, G. The Evaluation of Poisson,s Ratio and the Modulus of Materials of a Low Tensile Resistance by the Brazilian (Indirect Tensile) Test with Particular Reference to Concrete. *Aust. J. Appl. Sci.* **1959,** *10*, 243–268.

42. Nanni, A.; Bradford, N. M. FRP Jacketed Concrete Under Uniaxial Compression. *Constr. Build. Mater.* **1995,** *9*(2), 115–124.

43. Hsu, T. C. Fatigue and Microcracking of Concrete. *Mater. Struct.* **1984,** *17*(1), 51–54.

44. Hu, C.; de Larrard, F.; Sedran, T.; Boulay, C.; Bosc, F.; Deflorenne, F. Validation of BTRHEOM, the New Rheometer for Soft-to-fluid Concrete. *Mater. Struct.* **1996,** *29*(10), 620–631.

45. Hubert, R. Researches Toward a General Flexural Theory for Structural Concrete. *J. Proc.* **1960,** *57*(7), 1–28.

46. Hutchinson, J. W. Singular Behaviour at the End of a Tensile Crack in a Hardening Material. *J. Mech. Phys. Solids* **1968,** *16*(1), 13–31.

47. Hwee Ling, L. *Handbook of Research on Recent Developments in Materials Science and Corrosion Engineering Education*; IGI Global: Hershey, PA, USA, 2015; pp 1–493.

48. Inglis, C. E. Stresses in a Plate Due to the Presence of Cracks and Sharp Corners. *Trans. Inst. Nav. Archit.* **1913,** *55*, 219–241.

49. Irwin, G. R. Fracture. In *Handbuch der Physik;* Flugge, S., Ed.; Springer-Verlag: Berlin, 1958; Vol. 6, pp 551–590.

50. Islam, S. M.; Hussain, R. R.; Morshed, M. A. Z. Fiber-reinforced Concrete Incorporating Locally Available Natural Fibers in Normal- and High-strength Concrete and a Performance Analysis with Steel Fiber-reinforced Composite Concrete. *J. Compos. Mater.* **2012,** *46*(1), 111–122.

51. Jenq, Y.; Shah, S. Two Parameter Fracture Model for Concrete. *J. Eng. Mech.* **1985,** *111*(10), 1227–1241.

52. Jianhong, Y.; Wu, F. Q.; Sun, J. Z. Estimation of the Tensile Elastic Modulus Using Brazilian Disc by Applying Diametrically Opposed Concentrated Loads. *Int. J. Rock Mech. Min. Sci.* **2009,** *46*(3), 568–576.

53. Jones, R. The Development of Microcracks in Concrete. *Rheol. Acta* **1962,** *2*(1), 34.

54. Kaplan, M. F. Crack Propagation and the Fracture of Concrete. *J. Proc.* **1961,** *58*(11), 591–610.

55. Karihaloo, B. L.; Nallathambi, P. Fracture Toughness of Plain Concrete from Three-point Bend Specimens. *Mater. Struct.* **1989,** *22*(3), 185–193

56. Kim, H.; Wagoner, M.; Buttlar, W. Micromechanical Fracture Modeling of Asphalt Concrete Using a Single-edge Notched Beam Test. *Mater. Struct.* **2009,** *42*(5), 677–689.

57. Kim, J. H.; Chau-Dinh, T.; Zi, G.; Kong, J. S. The Effect of Compression Stresses, Stress Level and Stress Order on Fatigue Crack Growth of Multiple Site Damage. *Fatigue Fract. Eng. Mater. Struct.* **2012,** *35*(10), 903–917.

58. King, A.; Raffle, J. F. Studies on the Settlement of Hydrating Cement Suspensions. *J. Phys. D: Appl. Phys.* **1976,** *9*(10), 1425–1443.

59. Krishnaswamy, K. T. Measurement of Internal Strains in Concrete. *Matériaux et Constr.* **1968,** *1*(4), 361–364.

60. Labuz, J. F.; Shah, S. P.; Dowding, C. H. Fracture Analysis of Subsize, Charcoal Granite Specimens. *Int. J. Rock Mech. Min. Sci. Geomech. Abstr.* **1985,** *22*(6), 192–193.

61. Lawn, B. *Fracture of Brittle Solids*; Cambridge University Press: Cambridge, 1993.

62. Li, V. C.; Wu, H.-C.; Maalej, M.; Mishra, D. K.; Hashida, T. Tensile Behavior of Cement-Based Composites with Random Discontinuous Steel Fibers. *J. Am. Ceram. Soc.* **1996,** *79*(1), 74–78.

63. Lundborg, N. The Strength-size Relation of Granite. *Int. J. Rock Mech. Min. Sci. Geomech. Abstr.* **1967,** *4*(3), 269–272.

64. Mallick, P. K. *Fiber-reinforced Composites: Materials, Manufacturing, and Design*; CRC Press: Boca Raton, FL, 2008.

65. Mascolo, G.; Ramachandran, V. S. Hydration and Strength Characteristics of Synthetic Al-, Mg- and Fe Alites. *Matériaux Constr.* **1975,** *8*(5), 373–376.

66. Mazel, V.; Diarra, H.; Busignies, V.; Tchoreloff, P. Study of the Validity of the Three-Point Bending Test for Pharmaceutical Round Tablets Using Finite Element Method Modeling. *J. Pharm. Sci.* **2014,** *103*(4), 1305–1308.

67. Meda, A. Tensile Behaviour in Natural Building Stone: Serena Sandstone. *Mater. Struct.* **2003,** *36*(8), 553–559.

68. Mobasher, B. *Mechanics of Fiber and Textile Reinforced Cement Composites*; CRC Press Inc: Boca Raton, 2012.

69. Mueller, S.; Llewellin, E. W.; Mader, H. M. The Rheology of Suspensions of Solid Particles. *Proc. Roy. Soc. A: Math. Phys. Eng. Sci.* **2009.**

70. Murata, J. Flow and Deformation of Fresh Concrete. *Matériaux et Constr.* **1984,** *17*(2), 117–129.

71. Naaman, A. E.; Reinhardt, H. W. *High Performance Fiber Reinforced Cement Composites 2: Proceedings of the International Workshop*; Taylor & Francis, 1996.

72. Nawy, E. G. *Concrete Construction Engineering Handbook*; CRC Press: Boca Raton, 2008.

73. Nemati, K. M. Fracture Analysis of Concrete Using Scanning Electron Microscopy. *Scanning* **1997,** *19*(6), 426–430.

74. Nematzadeh, M.; Naghipour, M. Compressive Strength and Modulus of Elasticity of Freshly Compressed Concrete. *Constr. Build. Mater.* **2012,** *34*(0), 476–485.

75. Neville, A. M. *Properties of Concrete*; Longman Scientific & Technical: Harlow, 1994.

76. Neville, A. M.; Brooks, J. J. *Concrete Technology*; Prentice Hall: Harlow, London, 2010.

77. Newman, D. A.; Bennett, D. G. The Effect of Specimen Size and Stress Rate for the Brazilian Test—A Statistical Analysis. *Rock Mech. Rock Eng.* **1990,** *23*(2), 123–134.

78. Nilsson, S. The Tensile Strength of Concrete Determined by Splitting Tests on Cubes. *RILEM Bull.* **1961,** *2*(11), 63–67.

79. Nova, R.; Zaninetti, A. An Investigation into the Tensile Behaviour of a Schistose Rock. *Int. J. Rock Mech. Min. Sci. Geomech. Abstr.* **1990,** *27*(4), 231–242.

80. Ojdrovic, R.; Petroski, H. Fracture Behavior of Notched Concrete Cylinder. *J. Eng. Mech.* **1987,** *113*(10), 1551–1564.

81. Olesen, J. F.; Østergaard, L.; Stang, H. Nonlinear Fracture Mechanics and Plasticity of the Split Cylinder Test. *Mater. Struct.* **2006,** *39*(4), 421–432.

82. Ouyang, C.; Landis, E.; Shah, S. P. Detection of Microcrackiivg in Concrete by Acoustic Emission. *Exp. Tech.* **1991,** *15*(3), 24–28.

83. Parra-Montesinos, G. J.; Reinhardt, H. W.; Naaman, A. E. *High Performance Fiber Reinforced Cement Composites 6: HPFRCC 6,* 1st Edition; Springer Netherlands: Netherlands, 2012; p 560.

84. Petersson, P. E. Direct Tensile Tests on Prismatic Concrete Specimens. *Cem. Concr. Res.* **1981,** *11*(1), 51–56.

85. Planas, J. *Report 39: Experimental Determination of the Stress-Crack Opening Curve for Concrete in Tension;* Final report of RILEM Technical Committee TC 187-SOC. RILEM Publications: 2007.

86. Popovics, S. *Strength and Related Properties of Concrete: A Quantitative Approach;* John Wiley: Chichester, New York, 1998.

87. Raithby, K. D. Flexural Fatigue Behaviour of Plain Concrete. *Fatigue Frac. Eng. Mater. Struct.* **1979,** *2*(3), 269–278.

88. Ramachandran, V. S.; Beaudoin, J. J. *Handbook of Analytical Techniques in Concrete Science and Technology: Principles, Techniques, and Applications;* Noyes Publications: Norwich, NY, 1999.

89. Ramachandran, V. S.; Beaudoin, J. J.; Paroli, R. M. The Effect of Nitobenzoic and Aminobenzoic Acids on the Hydration of Tricalcium Silicate—A conduction Calorimetric Study. *Thermochim. Acta* **1991,** *190*(2), 325–333.

90. Ramanathan, P.; Baskar, I.; Muthupriya, P.; Venkatasubramani, R. Performance of Self-compacting Concrete Containing Different Mineral Admixtures. *KSCE J. Civ. Eng.* **2013,** *17*(2), 465–472.

91. Rice, J. R. A Path Independent Integral and the Approximate Analysis of Strain Concentration by Notches and Cracks. *J. Appl. Mech.* **1968,** *35*(2), 379.

92. Rice, J. R.; Rosengren, G. F. Plane Strain Deformation Near a Crack Tip in a Power-law Hardening Material. *J. Mech. Phys. Solids* **1968,** *16*(1), 1–12.

93. Richart, F. E.; Brandtzæg, A.; Brown, R. L. *The Failure of Plain and Spirally Reinforced Concrete in Compression;* University of Illinois: Urbana Champaign, 1929.

94. Rocco, C.; Guinea, G. V.; Planas, J.; Elices, M. Mechanism of Rupture in Splitting Test. *Mater. J.* **1999a,** *96*(1), 52–60.

95. Rocco, C.; Guinea, G. V.; Planas, J.; Elices, M. Size Effect and Boundary Conditions in the Brazilian Test: Experimental Verification. *Mater. Struct.* **1999b,** *32*(3), 210–217.

96. Rocco, C.; Guinea, G. V.; Planas, J.; Elices, M. Size Effect and Boundary Conditions in the Brazilian Test: Theoretical Analysis. *Mater. Struct.* **1999c,** *32*(6), 437–444.

97. Rossi, P.; Godart, N.; Robert, J. L.; Gervais, J. P.; Bruhat, D. Investigation of the Basic Creep of Concrete by Acoustic Emission. *Mater. Struct.* **1994,** *27*(9), 510–514.

98. Sabnis, G. M.; Mirza, S. M. Size Effects in Model Concretes? *J. Struct. Div. ASCE* **1979,** *105,* 1007–1020.

99. Shah, S. P.; Carpinteri, A. *Fracture Mechanics Test Methods For Concrete*; Taylor & Francis, 1991.

100. Shah, S. P.; McGarry, F. J. Griffith Fracture Criterion and Concrete. *J. Eng. Mech. Div.* **1971**, *97*(6), 1663–1676.

101. Shah, S. P.; Swartz, S. E.; Ouyang, C. *Fracture Mechanics of Concrete: Applications of Fracture Mechanics to Concrete, Rock and Other Quasi-brittle Materials*; Wiley: New York, 1995.

102. Sideridis, E.; Papadopoulos, G. A. Short-beam and Three-point-bending Tests for the Study of Shear and Flexural Properties in Unidirectional-fiber-reinforced Epoxy Composites. *J. Appl. Polym. Sci.* **2004**, *93*(1), 63–74.

103. Sivakumar, A.; Sounthararajan, V. M. Toughness Characterization of Steel Fibre Reinforced Concrete—A Review on Various International Standards. *J. Civ. Eng. Constr. Technol.* **2013**, *4*(3), 65–69.

104. Strange, P. C.; Bryant, A. H. The Role of Aggregate in the Fracture of Concrete. *J. Mater. Sci.* **1979**, *14*(8), 1863–1868.

105. Suresh, S. *Fatigue of Materials*; Cambridge University Press: Cambridge, New York, 1991.

106. Swamy, R. N. Study of the Micro-mechanical Behaviour of Concrete Using Reflective Photoelasticity. *Matériaux et Constr.* **1971**, *4*(6), 357–370.

107. Tattersall, G. H.; Banfill, P. J. G. *The Rheology of Fresh Concrete*; Pitman: Boston [Mass.]; London, 1983.

108. Hsu, T. T. C.; Slate, F. O.; Sturman, G. M.; Winter, G. Microcracking of Plain Concrete and the Shape of the Stress-Strain Curve. *J. Am. Concr. Inst. Proc.* **1963**, *60*(2), 209–224.

109. Timoshenko, S. P.; Goodier, J. N. *Theory of Elasticity*; McGraw-Hill: New York, Maidenhead, 1970.

110. Toda, K.; Furuse, H. Extension of Einstein's Viscosity Equation to That for Concentrated Dispersions of Solutes and Particles. *J. Biosci. Bioeng.* **2006**, *102*(6), 524–528.

111. Torrent, R. J.; Brooks, J. J. Application of the Highly Stressed Volume Approach to Correlated Results from Different Tensile Tests of Concrete. *Magazine of Concrete Research* [Online], 1985, pp 175–184. http://www.icevirtuallibrary.com/content/article/10.1680/macr.1985.37.132.175. Acess date 25/5/2015.

112. van Mier, J. G. M. Mode I Fracture of Concrete: Discontinuous Crack Growth and Crack Interface Grain Bridging. *Cem. Concr. Res.* **1991**, *21*(1), 1–15.

113. van Mier, J. G. M. *Fracture Processes of Concrete*; Taylor & Francis, 1996.

114. van Mier, J. G. M. Framework for a Generalized Four-stage Fracture Model of Cement-based Materials. *Eng. Fract. Mech.* **2008**, *75*(18), 5072–5086.

115. van Mier, J. G. M. *Concrete Fracture: A Multi-Scale Modeling Approach*; CRC Press, 2012.

116. van Mier, J. G. M.; Nooru-Mohamed, M. B. Geometrical and Structural Aspects of Concrete Fracture. *Eng. Fract. Mech.* **1990**, *35*(4–5), 617–628.

117. Wang, P. T.; Shah, S. P.; Naaman, A. E. Stress-strain Curves of Normal and Lightweight Concrete in Compression. *J. Am. Concr. Inst.* **1978**, *75*(11), P603–611.

118. Whorlow, R. W. *Rheological Techniques*; Ellis Horwood: Chichester; New York, 1980.

119. Wijk, G. Some New Theoretical Aspects of Indirect Measurements of the Tensile Strength of Rocks. *Int. J. Rock Mech. Min. Sci. Geomech. Abstr.* **1978**, *15*(4), 149–160.

120. Yazdani, N.; Jin, Y. E. Substitution of Fly Ash, Slag, and Chemical Admixtures in Concrete Mix Designs. *J. Mater. Civ. Eng.* **2003**, *15*(6), 602–608.

121. Zhang, P.; Guan, Q.-Y.; Liu, C.-H.; Li, Q.-F. Study on Notch Sensitivity of Fracture Properties of Concrete Containing Nano-SiO2 Particles and Fly Ash. *J. Nanomater.* **2013,** *2013*, 1–7.
122. Zhou, J.; Pan, J.; Leung, C. Mechanical Behavior of Fiber-Reinforced Engineered Cementitious Composites in Uniaxial Compression. *J. Mater. Civ. Eng.* **2014,** *27*(1), 04014111.

CHAPTER 11

SOLID PARTICLES STABILIZED (PICKERING) EMULSION POLYMERS

CHANDRASHEKARA R. HARAMAGATTI*,
SHRUTI BHATTACHARYA, and SUBHADIP SIKDAR

*Research and Technology Center, Asian Paints Limited,
Plot No. C3-B/1, TTC MIDC Pawane, Thane–Belapur Road,
Navi Mumbai 400703, Maharashtra, India*

*Corresponding author.
E-mail: chandrashekara.haramagatti@asianpaints.com

CONTENTS

ABSTRACT

This review presents the various aspects of pickering emulsions and recent developments in the area. Initial part of this review presents its mechanism. A comparison with conventional emulsion technique is presented to emphasize its role. Pickering emulsion systems with various nanoparticles, which include both modified and unmodified nanoparticles are also presented for the better understanding. Effect of nanoparticles on key factors in emulsion polymerization like particle size distribution and reaction kinetics are also discussed here. A detailed discussion on the properties of pickering emulsions is presented along with their potential applications.

11.1   INTRODUCTION

Emulsion polymerization is a very popular technique to produce the polymer latexes in industrial scale. Though, there are other polymerization techniques available, emulsion polymerization technique has attracted more attention, since one can have good control over the polymerization process,[1] architecture of the latex particles,[2–4] etc. Generally, the emulsion polymerization is carried out in presence of suitable emulsifier/surfactant, which acts as a latex-stabilizing agent. Not just the stabilizing agent, the surfactant micellar structures provide a site for nucleating the polymer particles.[5] Hence, the surfactants are essential part of emulsion polymer particles. However, there are disadvantages associated with surfactant-stabilized polymer particles. The surfactant molecules migrate toward the surface of the polymer film after the film formation.[6] Surfactants are more sensitive to water and hence the polymer films have great tendency to absorb water because of the migrated surfactants.

In early 20th century, pioneering work of Ramsdon[7] and Pickering[8] showed that, the emulsions can also be stabilized by solid particles. Later, the emulsions which are stabilized by solid particles are named after Pickering and are known as Pickering emulsions. In Pickering emulsions, the solid particles of intermediate wettability with nanometer sizes play a crucial role in stabilizing emulsion particles by adsorbing at the monomer–water (liquid–liquid) interface. The schematics of the particle at the interface of the liquids is shown in Figure 11.1.

When the solid particles are adsorbed at oil–water interface (left-hand side) and create a contact angle of <90°, oil in water type of emulsions are generated. For such cases, the arrangement of the solid particles at the interface might look like the schematics shown in Figure 11.1 (left-hand side). Similarly, when the solid particles create a contact angle of >90°, water in oil type of emulsions are generated. The right-hand side of the Figure 11.1 shows the arrangement of the solid particles for such cases.[9] For better understanding on the differences in the conventional surfactant stabilized and solid particles stabilized latexes, a schematic representation of arrangement of the stabilizing species/particles are shown in Figure 11.2.

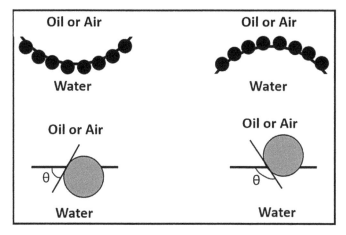

FIGURE 11.1 Schematics of the solid particles at the interface and their corresponding contact angles.

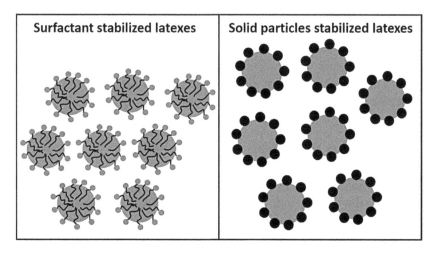

FIGURE 11.2 Schematic representation of the arrangement of surfactants and solid particles to stabilize the polymer latexes.

The inorganic particles are incorporated into the surfactant-stabilized latexes, or inorganic particles along with surfactants used for stabilization of the latexes.[10–15] Such latexes are generally termed as composite/hybrid latexes and the inorganic particles contribute to enhance the properties of the latexes. The composite latexes/films impart enhanced mechanical,[16] thermal,[17] optical,[18] and many other properties which are very essential for many applications.

Inorganic particles such as, SiO_2,[17,19–23] TiO_2,[24] Fe_2O_3,[25] cellulose nano-crystals,[26] clay,[27–29] etc. have been used for stabilization of polymer latexes in Pickering emulsion polymer. However, SiO_2 and clay have attracted more as solid particles for stabilization of latexes due to their unique properties to enhance the properties of the latexes or latex films. In some cases, the surface of the inorganic particles (e.g., SiO_2 and clay) is modified with suitable reagents to enhance the compatibility of the inorganic particles and polymer latexes. The use of macro monomers were also found to be useful in increasing the interaction between clay particles and the monomers during polymerization.[13,29] Though, there have been good number of research articles available on solid particle stabilized latexes, most of them are synthesized in low solid content of <20%. However, there was an exception, where 40% solid content clay particles stabilized latexes were synthesized successfully.[29] The polymer synthesis at high solid content (>40%) would be of great advantage from an industrial point of view because of high productivity. The other important industrial advantages would be the reduction of cycle time either by synthesizing the high solid content polymers or by reducing the polymer processing time without sacrificing the properties/performance of the product.

The following part of this chapter is focused on the use of neat silica, modified silica, neat clay, and modified clay as the stabilizing agent for polymer particles, followed by the possible mechanisms of the solid particles stabilized polymer latexes.

11.2   NANO-SiO$_2$ STABILIZED POLYMER LATEXES

As it was mentioned above, SiO_2 particles at nanometer scale (~20 nm) are one of the widely used solid particles for the Pickering emulsions. Silica particles are a category of nontoxic, stable, and high thermal and mechanical-resistant inorganic material. These excellent properties of SiO_2 allow them to be used in various fields, for example, plastics, coatings, rubbers, etc. Generally, the SiO_2 particles are hydrophilic in nature due to the presence of large number of –OH functional groups attached to its surface. This might be a disadvantageous; because of the presence of large number of hydroxyl groups, the SiO_2 particles tend to agglomerate. Hence, these particles are often modified to reduce the number of hydroxyl groups on the surface. Surfactants,[14] coupling agents,[30] alcohols,[31] etc. are used as surface modifying agents for SiO_2. The importance of modification of silica surface by coupling agents, for example, 3-methacryloxypropyl triethoxysilane on behavior at the interface of water monomer is shown in Figure 11.3.

FIGURE 11.3 Photographs of the two phase system of water monomer, and the behaviors of commercially available SiO_2, and the modified SiO_2 at the interface (unpublished).

Beaker 1 shows the monomer–water two phase system, beaker 2 shows most of the SiO_2 (commercially available) settling in one phase. It clearly indicates that, the added SiO_2 is more of hydrophilic in nature and hence accumulating in water phase (lower phase in beaker) probably because of large number of –OH moieties on the SiO_2 surface. Whereas, the surface modified SiO_2 sits at the interface of water monomer which is clearly demonstrated in beaker 3. Thus, the surface modification of SiO_2 particles enables these particles as potential candidates of stabilizing agents in Pickering emulsions.

Kang et al. have demonstrated the effect of surface modification of commercial SiO_2 by the silane coupling agents on the surfactant-free emulsion polymerization of methylmethacrylate-hydroxyethyl methacrylate.[30] A schematic of the surface modification of SiO_2 is shown in Figure 11.4, and the details of the modification can be found elsewhere.

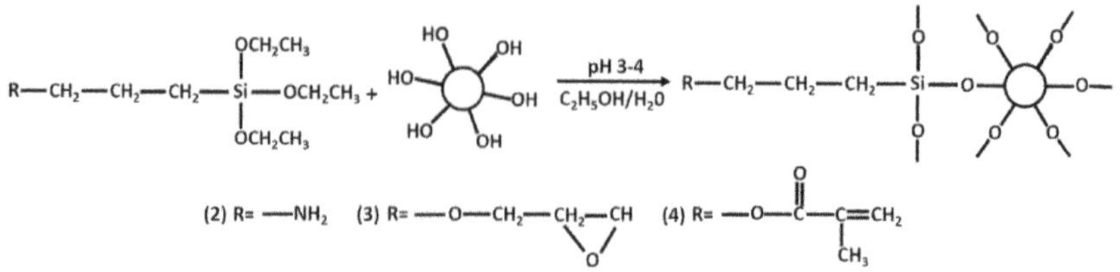

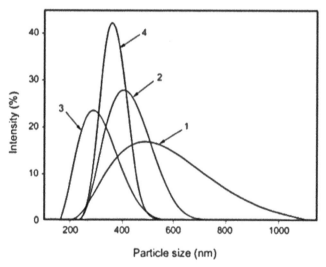

FIGURE 11.4 Schematic illustration of the surface modification of SiO₂ [30] (Reprinted with permission from Kang, J-S.; Yu, C-l.; Zhang, F-A. Effect of Silane Modified SiO₂ Particles on Poly(MMA-HEMA) Soap-Free Emulsion Polymerization. *Iran. Polym. J.* **2009**, *18,* 927–935.)

The emulsion stabilized with surface-modified SiO₂ resulted in lower and narrow distribution of particle size compared with the unmodified SiO₂ as stabilizing agent as shown in Figure 11.5. However, the coagulum content was slightly higher in case of emulsion prepared with surface-modified SiO₂.

FIGURE 11.5 Particle size and its distribution for surfactant-free composite emulsions with (1) SiO₂ particles and (2), (3), and (4) corresponds the SiO₂ particles modified with the reagents mentioned in the Figure 11.4[30] (Reprinted with permission from Kang, J-S.; Yu, C-l.; Zhang, F-A. Effect of Silane Modified SiO₂ Particles on Poly (MMA-HEMA) Soap-Free Emulsion Polymerization. *Iran. Polym. J.* **2009**, *18,* 927–935.)

Commercial aqueous glycerol functionalized silica sols were also used as Pickering agents for the synthesis of film forming styrene–butyl acrylate (ST-BA) copolymer in presence of cationic azo initiator (azobisisobutylni-trile).[32] The resultant latex films were transparent. The transmission electron microscopy (TEM) images suggested the core-shell type of morphology, where the copolymer part forms the core and the silica particles form the shell. Figure 11.6 shows the TEM image of the poly (ST–BA) copolymer latexes.

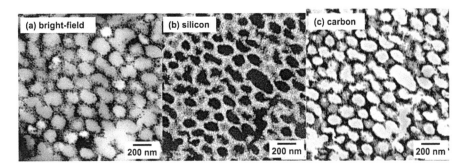

FIGURE 11.6 TEM images of ultramicrotomed poly (ST-*stat-n*-BA)/silica nanocomposite particles. (a) The bright-field image and the two elemental maps obtained for (b) silicon and (c) carbon reveal that these film-forming nanocomposite particles possess a distinct "core-shell" morphology, whereby the copolymer component forms the core, and the silica particles form the shell. This particular nanocomposite was prepared using 50:50 ST/*n*-BA[32] (Reprinted with permission from Schmid, A.; Scherl, P.;Armes, S. P.; Leite, C. A. P.; Galembeck, F. Synthesis and Characterization of Film-Forming Colloidal Nanocomposite Particles Prepared via Surfactant-Free Aqueous Emulsion Copolymerization. *Macromolecules* **2009**, *42*, 3721–3728. © 2009 American Chemical Society.)

Similarly, core-shell type of ST-SiO$_2$ latexes were also prepared using 2,2-azobis(2-methyl- *N*-(2-hydroxyethyl) propionamide and potassium persulfate.[21] The latexes showed polymer particles as the core and the silica particles as a shell. A positively charged silica particle was also a good candidate for the Pickering emulsion. Polystyrene core and silica shell composites were prepared using potassium persulfate as an initiator.[23] It was found that, colloidally stable latexes were obtained by addition of the silica sol 30 min after the initiation and nucleation of the surfactant free emulsion polymerization. The pH, SiO$_2$ solution to monomer volume ratio influences the stability and morphology of the latex particles. The morphology of the polymer latex depends also on the type of monomer system. The TEM images of the latexes with different monomer systems are shown in Figure 11.7. The SiO$_2$ particles stabilized polymethyl methacrylate latexes resulted in spherical latexes as shown in Figure 11.7a.

Hairy outer layer structured latex of poly (acylonitrile), and soft shell of poly (*n*-BA) are some of the other morphology observed as shown in Figures 11.7b and 11.7c, respectively.[20]

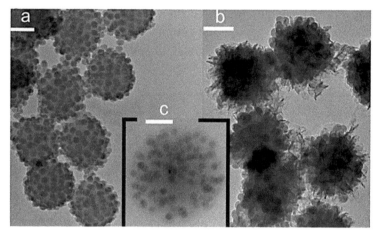

FIGURE 11.7 TEM images (scale bar) 100 nm of (a) poly(methyl methacrylate) latex armored with silica nanoparticles obtained by Pickering emulsion polymerization. Multilayered nanocomposite polymer colloids with, (b) a "hairy" outer-layer of poly(acrylonitrile), and (c) a soft shell of poly(n-BA) [20] (Reprinted with permission from Colver, P. J.; Colard, C. A. L.; Bon, S. A. F. Multilayered Nanocomposite Polymer Colloids Using Emulsion Polymerization Stabilized by Solid Particles. *J. Am. Chem. Soc.* **2008**, *130*, 16850–16851. © 2008 American Chemical Society.)

11.3 CLAY-STABILIZED POLYMER LATEXES

The chemical structure of the montmorillonite (MMT) clay is shown in Figure 11.8. Generally, these clays are of assembly of layers with an individual layer thickness of ~1 nm. Each layer is separated by exchangeable ions and layers of water molecules. Thus, they can be easily dispersible in water at appropriate concentration. Otherwise, at higher amount of clay, they have tendency to form gel phase or house of card structures. MMT clay was used in preparation of polymer-clay nanocomposites in early 1960.[33] Later, in 1990s it was found that, a small amount of MMT clay incorporated into the nylol-6 resulted in tremendous increase in thermal and mechanical properties of the nanocomposite materials.[34]

Clay particles are one of the widely used inorganic materials as stabilizing agent in Pickering emulsion. Laponite/Laponite RD/Laponite XLS are commercial grade clays, and are been used as stabilizing agents in surfactant-free emulsions. Polystyrene latexes were prepared using Laponite RD as a stabilizing agent by mini-emulsion polymerization.[27] The armoring of the Laponite RD clay particle on the polystyrene latexes are clearly observed in field emission gun scanning electron microscopic (FEGSEM) images as shown in Figure 11.9a and the polystyrene latex film formed at 230°C is shown in Figure 11.9b. Apart from the FEGSEM images, the armoring of the clay particles on the surface of the latexes was also evidenced by the atomic force microscopic images.[27]

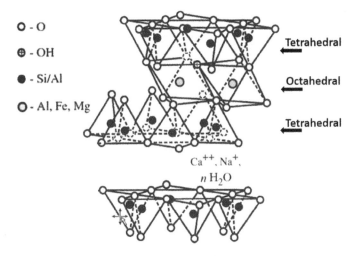

FIGURE 11.8   Chemical structure of montmorillonite clay.

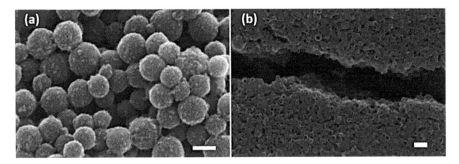

FIGURE 11.9   FEGSEM images of (a) a Laponite-armored polystyrene latex made via Pickering mini emulsion polymerization (scale bar 100 nm) and (b) film formed from a Laponite-armored polystyrene latex at 230°C (scale bar 400 nm)[27] (Reprinted with permission from Bon, S. A. F.; Colver, P. J. Pickering Miniemulsion Polymerization Using Laponite Clay as a Stabilizer. *Langmuir.* **2007**, *23*, 8316–8322. © 2007 American Chemical Society.)

A variety of copolymers of methyl methacrylate, BA, ST, 2-ethylhexyl acrylate (EHA) are successfully prepared with laponite as stabilizing agent.[35] The solid content of these polymers are typically less than 24 wt%. The corresponding transmission electron microscopic (TEM) images are shown in Figure 11.10.

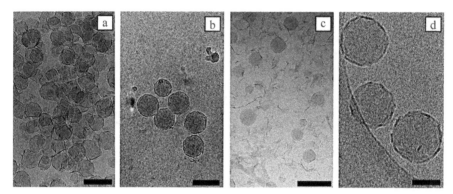

FIGURE 11.10   Cryo-TEM images of polymer latex particles: (a) poly (methyl methacrylate -co-n-(BA)/laponite, (b) poly(Sty-co-BA)/laponite, (c) poly(Sty-co-2- EHA)/laponite, and (d) poly(Sty-co-2-EHA)/laponite with methacrylic acid. Scales bars of 100, 100, 200, and 50 nm, respectively (Reprinted with permission from Teixeira, R. F. A.; McKenzie, H. S.; Boyd, A. A.; Bon, S. A. F. Pickering Emulsion Polymerization Using Laponite Clay as Stabilizer to Prepare Armored "Soft" Polymer Latexes. *Macromolecules.* **2011**, *44*, 7415–7422. © 2011 American Chemical Society.)

The cryo-TEM images clearly evidence the armoring of the clay particles on the surface of the polymer latexes.

The surface modification of clay particles with suitable modifying agents and subsequent use as stabilizing agent also enhances the interaction of clay particles to the polymer latexes. Clay particles can be modified by surfactants,[14] polymer brushes,[17] and oligomeric ST.[36] In some cases a poly(ethylene glycol) monomethylether methacrylate (PEGMA), acrylate-terminated polyethylene glycol macromonomer[13,29,37] have been used to promote the interaction of the clay particles and the polymer latexes. The polymerization of ST in presence of macromonomer PEGMA enhanced the kinetics of the polymerization. The particle size of polystyrene latex in absence of PEGMA was significantly high and the polymerization was very poor. It was also found that the polymer conversion increased in presence of macromonomer with increase in concentration of clay.[13]

The cryo-TEM images of the surfactant-free, Laponite-stabilized polystyrene latexes prepared with 10 wt% of the PEGMA-Laponite complex clearly show the polystyrene latexes of regular spherical shape stabilized by the clay particles in presence of PEGMA macromonomer[(13)]. However, the acrylate-terminated polyethylene glycol macromonomer was found to be more effective in enhancing the clay–latex interaction, and producing the polymer latex free of coagulum.[29] Generally, the solid content of the solid particle-stabilized polymer latexes of ~20% are reported with an exception of 40% solid polymer latex stabilized by Laponite by Lami et al.[29] The lower particle size of the latexes is generally observed when low solid polymer was synthesized. Similarly, the particle size of high solid (40%) Pickering emulsion polymer (copolymer of ST-BA) resulted higher particle size (~628 nm) which was measured by dynamic light scattering technique. However, low (~300 nm) particle size was obtained with appropriate amount of clay; methacrylic acid for the copolymerization of ST-BA system.

11.4   MECHANISM OF STABILIZATION OF POLYMER LATEXES

It is very well-established fact that, in case of conventional surfactant-stabilized emulsion polymer, the surfactant micelles are the nucleating site of polymerization. The free radical reacts with the monomer in the continuous phase (water), which then reacts with few more monomers and results an oligomer, which is generally termed as z-mers. Such z-mers will enter into the micellar core and the polymerization continues.[5] A schematic representation of the micellar nucleation is shown in Figure 11.11.

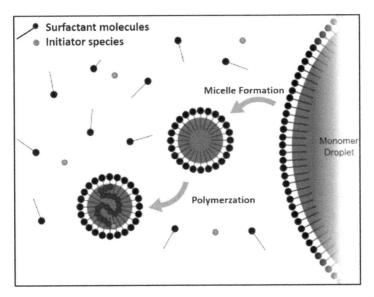

FIGURE 11.11   Schematic representation of the micellar nucleation of surfactant stabilized emulsion polymerization.

The big monomer droplets act as reservoirs of the monomers, and the monomers diffuse from the big droplet to the continuous phase, which then enters into the micellar core as shown in the schematics representation.

Whereas, in case of solid particle-stabilized emulsion polymers, the micellar nucleating mechanism is not possible due to the absence of the surfactants. Few attempts are made to understand the nucleation mechanism,[21,22,38] polymerization kinetics,[36,37,39] and thermodynamic stability[40–42] of the solid particles stabilized emulsion polymers.

The polymerization mechanism depends on the type of initiator, whether it is water soluble or monomer soluble. When the monomer soluble initiator such as azobisisobutylnitrile was used, the probable mechanism of nucleation is "monomer droplet nucleation" predicted based on the TEM and SEM images. Whereas, for the water soluble initiators such as ammonium persulphate leads to the "homogeneous coagulative nucleation" mechanism. Schematics of the two possible mechanisms are given in the Figure 11.12.

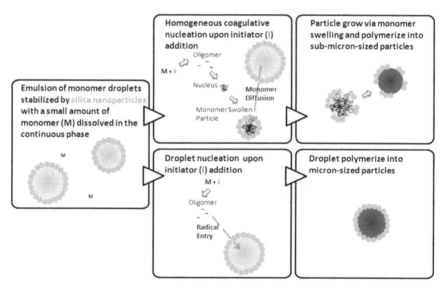

FIGURE 11.12 Schematics of the nucleation mechanism of the solid particle-stabilized polymer latexes [21] (Reprinted with permission from the author from Ma, H.; Luo, M.; Sanyal, S.; Rege, K.; Dai, L. L. The One-Step Pickering Emulsion Polymerization Route for Synthesizing Organic-Inorganic Nanocomposite Particles. *Materials.* **2010**, *3*, 1186–1202. https://creativecommons.org/licenses/by-nc-sa/3.0/)

The particle size distribution measured by dynamic light scattering technique for Laponite stabilized ST–BA copolymerization in presence of potassium persulphate, shows an increase in particle size during the polymerization. The particle size distribution, as shown in Figure 11.13, was recorded at regular intervals during the polymerization evidences, so that the polymerization might follow homogeneous coagulative mechanism and, hence, the particle size increases.

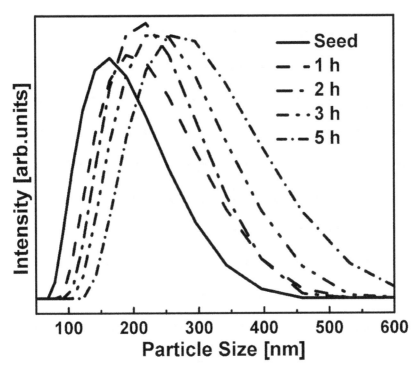

FIGURE 11.13 Variation in particle size during the emulsion polymerization of Laponite clay stabilized with 40% solid ST–BA system (unpublished).

During the surfactant-free emulsion polymerization of ST, direct evidence on the homogeneous coagulative mechanism was observed by the SEM images as shown in Figure 11.14.[38] SEM image captured 15 min after the initiation of polymerization, small particles of ~15 nm in diameter were observed. These particles are well separated by each other. The particle size slightly increases at ~45 min of polymerization; however, a tendency to coagulation of the particles was seen from Figure 11.14b. The particle size further increases with time, at 120 min of polymerization, the particle size reaches to ~190 nm (Fig. 11.14c). After 180 min of polymerization, the particle size reaches to ~235 nm and the surface of the latexes looks soften (Fig. 11.14d).

In summary, the small primary particles, which are generated in early stages of the polymerization, are colloidally unstable. These unstable particles start to aggregate into larger clusters; however, it undergoes a limited aggregation.[43] These clusters irreversibly aggregate, and subsequently grow by aggregation with polymerization time.

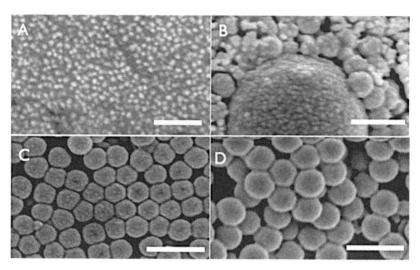

FIGURE 11.14 SEM micrographs of freeze-dried samples taken during a surfactant-free polymerization process at successive time lapses after initiation. (a) 15, (b) 45, (c) 120, and (d) 180 min. (a and b scale bars are 200 nm; c and d scale bars are 500 nm) [38] (Reprinted with permission from Dobrowolska, M. E.;Esch, J. H.; Koper, G. J. M. Direct Visualization of "Coagulative Nucleation" in Surfactant-Free Emulsion Polymerization. *Langmuir*. **2013**, *29*, 11724–11729. © 2013 American Chemical Society.)

11.5 SOME OF THE PROPERTIES OF THE PICKERING EMULSION POLYMERS

Pickering emulsion polymers are believed to be more stable than their counter part surfactant-stabilized emulsion polymers. Due to the unique properties of the solid particles (Pickering agents), the Pickering emulsion polymers show significantly improved thermal and mechanical properties compared with the surfactant-stabilized emulsion polymers. Generally, a higher glass transition temperature will be obtained for the solid-stabilized latexes compared with the surfactant-stabilized latexes, for example, the polystyrene latexes stabilized by silica particles show higher Tg than the polystyrene latexes stabilized by surfactants as shown in the differential scanning calorimetric profiles (Fig. 11.15).

It was well established that, the solid-stabilized latex polymer films exhibit better mechanical properties. The dynamic mechanical analysis (DMA) technique is used to find the variation in storage modulus and tan δ with applied temperature. Generally, the storage modulus of the solid-stabilized latex films will be on the higher side compared with the surfactant stabilized latex films as shown in Figure 11.16.

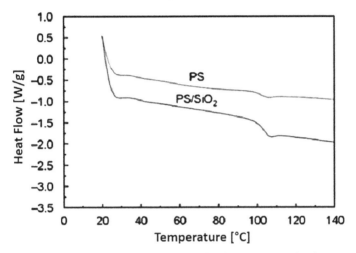

FIGURE 11.15 Differential scanning calorimetric thermograms of polystyrene latexes prepared by surfactant stabilization and silica particles stabilization [22] (Reprinted with permission from Zhang, W. H.; Fan, X. D.; Tian, W.; Fan, W. W. Polystyrene/Nano-SiO$_2$ Composite Microspheres Fabricated by Pickering Emulsion Polymerization: Preparation, Mechanisms and Thermal Properties. *eXPRESS Polym. Lett.* **2012**, *6*, 532–542.)

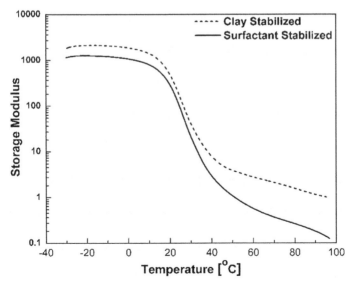

FIGURE 11.16 Dynamic mechanical analysis profiles of Laponite clay stabilized and surfactant stabilized emulsion films (unpublished).

Here, storage modulus curves for 40% solid ST–BA copolymer stabilized by Laponite clay, and surfactants are shown. As mentioned above a higher storage modulus was obtained for clay-stabilized latex films.

11.6 SOME OF THE APPLICATION/ADVANTAGES OF PICKERING EMULSION POLYMERS

The conventional surfactant-stabilized emulsion polymers are being used for paints, coatings, adhesives, textiles, and non-wovens, paper, and paperboard, and other applications. A recent survey report the market size of emulsion polymers will be $32.2 billion by 2020, at a significant growth rate of 7.7% between 2015 and 2020.[44] There are many industries which produce large quantity of conventional emulsion polymer, BASF SE, The Dow chemical Company, Wacker Chemie, Celenese, Arkema, etc. are just few to mention. However, there are many disadvantages of surfactant-stabilized emulsion polymers as mentioned in the beginning of this chapter. The Pickering emulsion polymers can be a potential alternate to overcome those issues and have great scope at industrial scale. It is already well known that solid particle-stabilized polymer latexes are more stable than the surfactant-stabilized latexes with respect to stability against coalescence, since removal of solid particles from the surface of the latex particles demands very high energy compared with the surfactant-stabilized latexes. Surfactants are the essential evil of the emulsion polymer, the major disadvantage of surfactant-stabilized emulsion is the water absorption of the film, particularly in the coating applications. Such disadvantages can be overcome by the use of Pickering emulsion polymers; also these emulsion polymers provide better thermal and mechanical properties due to the presence of solid particles. Apart from these, due to the enhanced properties of the Pickering emulsion polymers they are of great help in many other industrial applications. Polyaniline stabilized by graphene oxide has shown property of very good capacitor over polyaniline alone. They showed enhanced specific capacitance and better cycling stability than the pure polyaniline.[45] Some of the Pickering emulsions are used as templates for the interfacial atom transfer radical polymerization (ATRP) reactions.[46] Use of solid particles as stabilization will be helpful in high internal phase emulsions (HIPEs) which are highly porous polymers. The drawback of conventional HIPEs would be the removal of the surfactant from the poly HIPEs which will affect the properties of the resulting polymer. In such cases, Pickering emulsions will be of great help where there is no worry of surfactant removal and hence the instability in the polymer.[47]

ACKNOWLEDGMENTS

Authors would like to thank to Dr. Subrahmanya Shreepathi, Dr. V Mohan Rao, and Dr. B P Mallik for their suggestions and comments. We also thank the modern instrument lab—Asian Paints Ltd., who did the characterizations in time and also to the Asian Paints management for their kind support.

KEYWORDS

- polymerization
- emulsion
- latex
- Pickering
- interface

REFERENCES

1. Chern, C-S. *Principles and Applications of Emulsion Polymerization;* John Wiley & Sons, Inc.: Hoboken, New Jersey, 2008.
2. McDonald, C. J.; Devon, M. J. Hollow Latex Particles: Synthesis and Applications. *Adv. Colloid Interface Sci.* **2002,** *99,* 181–213.
3. Okubo, M.; Kanaida, K.; Matsumoto, T. Production of Anomalously Shaped Carboxylated Polymer Particles by Seeded Emulsion Polymerization. *Colloid Polym. Sci.* **1987,** *265,* 876–881.
4. Mock, E. B.; De, B. H.; Hawkett, B. S.; Gilbert, R. G.; Zukoski, C. F. Synthesis of Anisotropic Nanoparticles by Seeded Emulsion Polymerization. *Langmuir.* **2006,** *25,* 4037–43.
5. Flory, P. J. *Principles of Polymer Chemistry;* Cornell University Press: USA, 1995.
6. Zhao, C. L.; Holl, Y.; Pith, T.; Lambla, M.; FTIR-ATR Spectroscopic Determination of the Distribution of Surfactants in Latex Films. *Colloid Polym. Sci.* **1987,** *265,* 823–829.
7. Rawsden, W. Separation of Solids in the Surface-Layers of Solution and Suspensions. *Proc. Roy. Soc.* **1903,** *72,* 156.
8. Pickering, S. U. Emulsions. *J. Chem. Soc.* **1907,** *91,* 2001–2021.
9. Binks, B. P. Particles as Surfactants_Similarities and Differences, *Curr. Opin. Colloid In.* **2002,** *7,* 21–41.
10. Negrete-Herrera, N.; Putaux, J-L.; David, L.; Haas, F. D.; Lami, E. B. Polymer/Laponite Composite Latexes: Particle Morphology, Film Microstructure, and Properties. *Macromol. Rapid Commun.* **2007,** *28,* 1567–1573.
11. Wua, Y.; Hua, D.; Sub, Y-H.; Hsiaob, Y-L.; Youa, B.; Wu, L. Synthesis and Film Performances of SiO2/P(MMA-BA) Core–Shellstructural Latex. *Prog. Org. Coat.* **2014,** *77,* 1015–1022.

12. Negrete-Herrera, N.; Putaux, J-L.; Bourgeat-Lami, E. Synthesis of Polymer/Laponite Nanocomposite Latex Particles via Emulsion Polymerization Using Silylated and Cation-Exchanged Laponite Clay Platelets. *Prog. Solid State Ch.* **2006,** *34,* 121–137.

13. Bourgeat-Lami, E.; Herrera, N. N.; Putaux, J-L.; Perro, A.; Reculusa, S.; Ravaine, S.; Duguet, E. Designing Organic/Inorganic Colloids by Heterophase Polymerization. *Macromol. Symp.* **2007,** *248,* 213–226.

14. Reger, M.; Sekine, T.; Hoffmann, H. Pickering Emulsions Stabilized by Amphiphile Covered Clays. *Colloids Surf. A.* **2012,** *413,* 25–32.

15. Yilmaz, O.; Cheaburu, C. N.; Durraccio, D.; Gulumser, G.; Vasile, C.; Preparation of Stable Acrylate/Montmorillonite Nanocomposite Latex via in Situ Batch Emulsion Polymerization: Effect of Clay Types. *Appl. Clay Sci.* **2010,** *49,* 288–297.

16. Lee, G. J.; Son, H. A.; Cho, J. W.; Choi, S. K.; Kim, H. T.; Kim, J. W. Stabilization of Pickering Emulsions by Generating Complex Colloidal Layers at Liquid-Liquid Interfaces. *J. Colloid. Inter. Sci.* **2014,** *413,* 100–105.

17. Zhang, J.; Chen, K.; Zhao, H. PMMA Colloid Particles Armored by Clay Layers with PDMAEMA Polymer Brushes. *J. Polym. Sci. Pol. Chem.* **2008,** *46,* 2632–2639.

18. Chau, J. L. H.; Hsieh, C-C.; Lin, Y-M.; Li, A-K. Preparation of Transparent Silica-PMMA Nanocomposite Hard Coatings. *Prog. Org. Coat.* **2008,** *62,* 436–439.

19. Zhang, K.; Wu, W.; Meng, H.; Guo, K.; Chen, J. F. Pickering Emulsion Polymerization: Preparation of Polystyrene/Nano Sio2 Composite Microspheres with Core-Shell Structure. *Powder Technol.* **2009,** *190,* 393–400.

20. Colver, P. J.; Colard, C. A. L.; Bon, S. A. F. Multilayered Nanocomposite Polymer Colloids Using Emulsion Polymerization Stabilized by Solid Particles. *J. Am. Chem. Soc.* **2008,** *130,* 16850–16851.

21. Ma, H.; Luo, M.; Sanyal, S.; Rege, K.; Dai, L. L. The One-Step Pickering Emulsion Polymerization Route for Synthesizing Organic-Inorganic Nanocomposite Particles. *Materials.* **2010,** *3,* 1186–1202.

22. Zhang, W. H.; Fan, X. D.; Tian, W.; Fan, W. W. Polystyrene/Nano-Sio2 Composite Microspheres Fabricated by Pickering Emulsion Polymerization: Preparation, Mechanisms and Thermal Properties. *eXPRESS Polym. Lett.* **2012,** *6,* 532–542.

23. Lee, J.; Hong, C. K.; Choe, S.; Shim, S. E.; Synthesis Of Polystyrene/Silica Composite Particles By Soap-Free Emulsion Polymerization Using Positively Charged Colloidal Silica. *J. Colloid. Inter. Sci.* **2007,** *310,* 112–120.

24. Bachinger, A.; Kickelbick, G. Pickering Emulsions Stabilized by Anatase Nanoparticles. *Monatsh. Chem.* **2010,** *141,* 685–690.

25. Kim, Y. J.; Liu, Y. D.; Seo, Y.; Choi, H. J. Pickering Emulsion Polymerized Polystyrene/Fe2O3 Composite Particles and their Magnetoresponsive Characteristics. *Langmuir.* **2013,** *29,* 4959–4965.

26. Zopp, J. O.; Venditti, R. A.; Rojas, O. J.; Pickering Emulsions Stabilized by Cellulose Nanocrystals Grafted with Thermo-Responsive Polymer Brushes. *J. Coll. Inter. Sci.* **2012,** *369,* 202–209.

27. Bon, S. A. F.; Colver, P. J. Pickering Miniemulsion Polymerization Using Laponite Clay as a Stabilizer. *Langmuir.* **2007,** *23,* 8316–8322.

28. Ianchis, R.; Donescu, D.; Petcu, C.; Ghiurea, M.; Anghel, D. F.; Stanga, G.; Marcu, A. Surfactant-Free Emulsion Polymerization of Styrene in the Presence of Silylated Montmorillonite. *Appl. Clay Sci.* **2009,** *45,* 164–170.

29. Bourgeat-Lami, E.; Guimaraes, T. R.; Pereira, A. M. C.; Alves, G. M.; Moreira, J. C.; Putaux, J-L.; Santos, A. M. High Solids Content, Soap-Free, Film-Forming Latexes Stabilized by Laponite Clay Platelets. *Macromol. Rapid Commun.* **2010,** *31,* 1874–1880.

30. Kang, J-S.; Yu, C-l.; Zhang, F-A. Effect of Silane Modified Sio2 Particles on Poly(MMA-HEMA) Soap-Free Emulsion Polymerization. *Iran. Polym. J.* **2009,** *18,* 927–935.

31. Fuji, M.; Takei, T.; Watanabe, T.; Chikazawa, M.; Wettability of Fine Silica Powder Surfaces Modified with Several Normal Alcohols. *Colloid Surf. A.* **1999,** *154,* 13–24.

32. Schmid, A.; Scherl, P.;Armes, S. P.; Leite, C. A. P.; Galembeck, F. Synthesis and Characterization of Film-Forming Colloidal Nanocomposite Particles Prepared via Surfactant-Free Aqueous Emulsion Copolymerization. *Macromolecules.* **2009,** *42,* 3721–3728.

33. Blumstein, A. Polymerization of Adsorbed Monolayers. II. Thermal Degradation of the Inserted Polymer. *J. Polym. Sci. Pol. Chem.* **1965,** *3,* 2665–2672.

34. Kojima, Y.; Usuki, A.; Kawasumi, M.; Okada, A.; Kurauchi T.; Kamigaito, O. One-Pot Synthesis of Nylon 6–Clay Hybrid. *J. Polym. Sci. Pol. Chem.* **1993,** *31,* 1755–1758.

35. Teixeira, R. F. A.; McKenzie, H. S.; Boyd, A. A.; Bon, S. A. F. Pickering Emulsion Polymerization Using Laponite Clay as Stabilizer to Prepare Armored "Soft" Polymer Latexes. *Macromolecules.* **2011,** *44,* 7415–7422.

36. Greesh, N.; Sanderson R.; Hartmann, P. Preparation of Polystyrene–Clay Nanocomposites via Dispersion Polymerization Using Oligomeric Styrene-Montmorillonite as Stab. *Polym. Int.* **2012,** *61,* 834–843.

37. Sheibat-Othman, N.; Cenacchi-Pereira, A-M.; Santos, A. M. D.; Bourgeat-Lami, E. A. Kinetic Investigation of Surfactant-Free Emulsion Polymerization of Styrene Using Laponite Clay Platelets as Stabilizers. *J. Polym. Sci. Pol. Chem.* **2011,** *49,* 4771–4784.

38. Dobrowolska, M. E.;Esch, J. H.; Koper, G. J. M. Direct Visualization of "Coagulative Nucleation" in Surfactant-Free Emulsion Polymerization. *Langmuir.* **2013,** *29,* 11724−11729.

39. Tanrisever, T.; Okay, O.; Sonmezoclu, I. C. Kinetics of Emulsifier-Free Emulsion Polymerization of Methyl Methacrylate. *J. Appl. Polym. Sci.* **1996,** *61,* 485–493.

40. Sacanna, S.; Kegel, W. K.; Philipse, A. P.; Thermodynamically Stable Pickering Emulsions. *Phys. Rev. Lett.* **2007,** *98,* 158301–158304.

41. Kraft, D. J.; Folter, J. W. J.; Luigjes, B.; Castillo, S. I. R.; Sacanna, S.; Philipse, A. P.; Kegel, W. K. Conditions for Equilibrium Solid-Stabilized Emulsions. *J. Phys. Chem. B.* **2010,** *114,* 10347–10356.

42. Kralchevsky, P. A.; Ivanov, I. B.; Ananthapadmanabhan, K. P.; Lips, A. On the Thermodynamics of Particle-Stabilized Emulsions: Curvature Effects and Catastrophic Phase Inversion, *Langmuir,* **2005,** *21,* 50–63.

43. Gilbert, R. G. *Emulsion Polymerization: A Mechanistic Approach;* Academic Press: London, 1995.

44. http://www.marketsandmarkets.com/Market-Reports/emulsion-polymers-market-1269.html

45. Sun, J.; Bi, H. Pickering Emulsion Fabrication and Enhanced Super capacity of Grapheme Oxide-Covered Polyaniline Nanoparticles. *Mater. Lett.* 2012, *81,* 48–51.

46. Li, J.; Hitchcock, A. P.; Stover, H. D. H. Pickering Emulsion Templated Interfacial Atom Transfer Radical Polymerization For Microencapsulation. *Langmuir.* **2010,** *26,* 17926–17935.

47. Gurevitch, I.; Silverstein, M. S. Polymerized Pickering HIPEs: Effects of Synthesis Parameters on Porous Structure. *J. Polym. Sci. Pol. Chem.* **2010,** *48,* 1516–1525.

CHAPTER 12

CHEMICAL MODIFICATION OF MAJOR MILK PROTEIN (CASEIN): SYNTHESIS, CHARACTERIZATION, AND APPLICATION

SWETA SINHA*, GAUTAM SEN, and SUMIT MISHRA

Department of Applied Chemistry, Birla Institute of Technology, Mesra, Ranchi, Jharkhand, India

Corresponding author. E-mail: sweta.sinha2203@gmail.com

CONTENTS

ABSTRACT

In this study, we have synthesized poly (acrylamide)-grafted casein (CAS-g-PAM) without any radical initiator or catalyst using microwave (MW) irradiation. CAS-g-PAM was synthesized with 287% grafting at 700 W MW power for 4 min using poly (acrylamide) 15 g and casein 1 g/40 mL. MW-synthesized CAS-g-PAM copolymer was characterized by Fourier transform infrared spectroscopy, thermo gravimetric analysis, scanning electron microscopy, intrinsic viscosity measurement, elemental analysis, and number average molecular weight determination through osmometry; taking the starting material (casein) as reference. Maximum grafting was optimized by varying the MW irradiation time and monomer concentration. Flocculation efficacy of the graft copolymer was optimized in kaolin suspension and then studied in textile industry wastewater by *jar test* experiment. The effect of dosage and settling time on reduction of suspended/dissolved particles, organic waste, turbidity, and color of textile wastewater was studied. The results obtained found that grafted copolymers of casein were very effective in reduction of chemical oxygen demand, turbidity, total solids, total dissolved solids, heavy metals, and color. The flocculation efficacy was found to be at a maximum at 0.75 ppm dosage.

12.1   INTRODUCTION

Gradually increasing demand for a new material with friendly environmental impact has led to a great interest in the natural materials for the production of conventional and advanced polymer materials. The properties of these materials are limited by their natural state, so polymer modification can enhance various specific properties such as compatibility, impact response, thermal stability, multiphase physical responses, flexibility, and rigidity to the modified polymeric materials. Modifications make an insoluble polymer from a soluble one or vice versa. Therefore, surface modification of polymeric material improves the processibility of the polymers.

The major techniques for polymer modifications are grafting, cross linking, blending, and composite formation, which are all multicomponent polymer systems. Such materials have attracted considerable attention in the industrial field as they combine a variety of functional components in a single material.

Graft copolymerization is the most effective synthesis route of modifying natural polymers (e.g., polysaccharides).[18–20] The process of graft copolymer synthesis starts with a preformed polymer (protein in this case). An external agent is used to create free radical sites on this preformed polymer, once the free radical sites are formed on the polymer backbone (i.e., preformed polymer), the monomer (i.e., vinyl or acrylic compound) can get added up through the chain propagation step, leading to the formation of grafted chains. The various methods of graft copolymer synthesis actually differ in the ways of generation of the free radical sites on the preformed polymer. The most contemporary technique in graft copolymer synthesis involves the use of microwave (MW) radiation to initiate the grafting reactions. Superiority of this technique over others has been well discussed in earlier studies.[1–11]

12.1.1   TEXTILE WASTEWATER AND ITS CHALLENGES

A lot of industries such as textile, paper, plastics, and dyestuffs consume extensive volume of water and at the same time use chemicals during manufacturing and dyes to color their products. Because of these activities, a considerable amount of polluted wastewater was generated.[12] The unused materials from the processes are discharged as wastewater that is high in color, biological oxygen demand, chemical oxygen demand (COD), pH, temperature, turbidity, and toxic chemicals. Thus, water becomes limited resource because of the industrialization and urbanization that caused the pollution in the water environment. Furthermore, if the discharge water untreated, the toxic effluents from the industries become the major source of aquatic pollution and will cause considerable damage to the receiving waters.[13]

The main challenge with the textile wastewater is to eliminate the color, which is due to the remaining dyes. Effluent from textile industries contains different types of dyes having high molecular weight and complex structures, showing very low biodegradability.[14–15] In order to cope with these new restrictions and due to ineffectiveness of conventional biological treatment method in decolonization and degradation of textile wastewater, the research interest to find the other alternative which is simple and low-cost technologies for the on-site treatment of wastewaters. Therefore, coagulation and flocculation is a promising technique in order to reduce COD and color removal. Some researchers have reported that coagulation–flocculation is widely used for dyes removal.[15,16]

The objective of the study is to reduce COD and color removal by using biodegradable flocculants which is casein. Casein was extracted in order to obtain proline component that was used as biodegradable flocculant. Proline and histidine (amino acids) consist of cyclic nitrogen and carboxylic acid which is a functional group responsible for metal biosorption. This chapter reported that casein resulted in the highest removal of color as compared to synthetic polymers such as polyacrylamide (PAM), and chemical coagulant, such as alum. In this study, textiles wastewater will be treated by using CAS and CAS-g-PAM as the biodegradable flocculants in order to analyze the efficiency of the flocculation process toward the textile wastewater. Color of textile wastewater can be removed either by adsorption onto solid-state graft copolymer of casein or by coagulation–flocculation using dissolved-state poly (acrylamide) grafted casein (CAS-g-PAM).

12.2 EXPERIMENTAL

12.2.1 MATERIALS

Casein was procured from Loba Chemie Pvt Ltd, Mumbai, India. Acryl-amide was procured from Sisco research laboratory, Mumbai, India. Acetone was purchased from Rankem, New Delhi, India. All the chemicals were used as received; without further modification.

12.2.2 SYNTHESIS OF GRAFT COPOLYMER

12.2.2.1 MICROWAVE-INITIATED SYNTHESIS OF POLY (ACRYLAMIDE)-GRAFTED CASEIN (CAS-g-PAM)

A suspension of 1 g of casein in 40 mL of distilled water was prepared. Desired amount of acrylamide diluted in 10 mL water was added to the casein suspension. The reaction mixture was transferred to the MW reactor (Catalyst™ system CATA 4 R), stirred well and irradiated with MW radiation (700 W) for the intended amount of time (3–5 min), keeping the irradiation cut-off temperature at 70°C. The irradiation time was strictly maintained, compensating for the cut off time. Once the MW irradiation procedure got complete, the gel like mass left in the reaction vessel was cooled and poured into excess of acetone. The precipitated grafted polymer was collected and was dried in hot air oven. Subsequently, it was pulverized and sieved. The synthesis details of various grades of the graft copolymer have been shown in Table 12.1.

TABLE 12.1   Synthesis Details of CAS-g-PAM.

Polymer grades	Wt. of casein (g)	Wt. of acrylamide (g)	Irradiation time (min)	% grafting	Number average molecular weight (kDa)	Intrinsic viscosity
Casein	–	–	–	–	30	0.89
CAS-g-PAM 1	1	10	4	70.8	460	1.60
CAS-g-PAM 2	1	20	4	198	850	5.083
CAS-g-PAM 3	1	**15**	**4**	**287**	**1072**	**10.96**
CAS-g-PAM 4	1	15	3	153	800	4.52
CAS-g-PAM 5	1	15	5	218	890	7.99

$$\% \text{ Grafting} = \frac{\text{weight of graft copolymer} - \text{weight of protein}}{\text{weight of protein}} \times 100$$

12.2.2.2   PURIFICATION OF THE GRAFT COPOLYMER BY SOLVENT EXTRACTION METHOD

Any occluded PAM formed due to competing homopolymer formation reaction was removed by extraction with acetone for 24 h. The synthesis procedure of the graft copolymer has been summarized in Scheme 12.1.

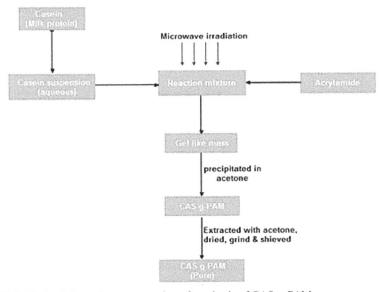

SCHEME 12.1   Schematic representation of synthesis of CAS-g-PAM.

12.2.3 CHARACTERIZATION

12.2.3.1 INTRINSIC VISCOSITY MEASUREMENT

Viscosity measurements of the polymer solutions were carried out with an Ubbelohde viscometer (constant: 0.003899) at 25°C. The viscosities were measured in aqueous solutions at neutral pH. The time of flow for solutions was measured at four different concentrations (0.1%, 0.05%, 0.025%, and 0.0125%). From the time of flow of polymer solutions (t) and that of the solvent (t_0, for distilled water), relative viscosity (ηrel $= t/t_0$) was obtained. Specific viscosity (η_{sp}), reduced viscosity (η_{red}), and inherent viscosity (η_{inh}) was mathematically calculated as:

$$\eta_{sp} = \eta_{rel} - 1$$
$$\eta_{red} = \eta_{sp}/C$$
$$\eta_{inh} = \ln \eta_{rel}/C$$

where C represents polymer concentration in g/dL.

The reduced viscosity (η_{red}) and the inherent viscosity (η_{inh}) were simultaneously plotted against concentration. The intrinsic viscosity was obtained from the point of intersection after extrapolation of two plots (i.e., η_{red} versus C and η_{inh} versus C) to zero concentration. [21]. The intrinsic viscosity thus evaluated for various grades of the graft copolymer has been reported in Table 12.1.

12.2.3.2 ELEMENTAL ANALYSIS

The elemental analysis of casein and that of CAS-g-PAM 3 (best grade) was carried out by an Elemental Analyzer (make: M/s Elementar, Germany; model: Vario EL III). The estimation of three elements, that is, carbon, hydrogen, and nitrogen were undertaken. The results have been summarized in Table 12.2.

12.2.3.3 DETERMINATION OF NUMBER AVERAGE MOLECULAR WEIGHT

The number average molecular weights of casein and various grades of CAS-g-PAMs were determined in aqueous medium by Osmometry (A+ Advanced Instruments, Inc. Model 3320, Osmometer).

12.2.3.4 FTIR SPECTROSCOPY

The FTIR spectrums of casein and of CAS-g-PAM 3 (Fig. 12.1) were recorded in solid state, by KBr pellet method, using FTIR spectrophotometer (Model IR-Prestige 21, Shimadzu Corporation, Japan) in the fingerprinting region of IR (400 and 4000 cm^{-1}).

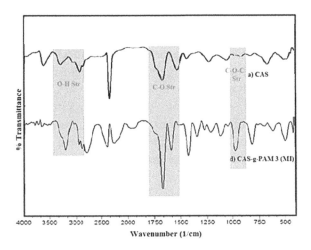

FIGURE 12.1 FTIR spectra.

12.2.3.5 SCANNING ELECTRON MICROSCOPY

Surface morphology of casein (Fig. 12.2a) and CAS-g-PAM 3 (Fig. 12.2b) were analyzed by scanning electron microscopy (SEM) in powdered form (Model: JSM-6390LV, Jeol, Japan).

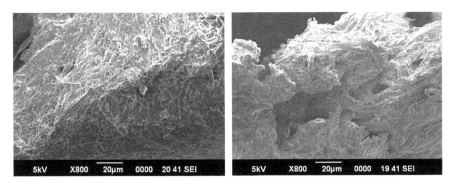

FIGURE 12.2 (a) SEM of CAS, and (b) SEM of CAS-g-PAM 3.

12.2.4 FLOCCULATION STUDIES

12.2.4.1 FLOCCULATION STUDY IN KAOLIN SUSPENSION

Flocculation efficacy of CAS, various grades of synthesized CAS-g-PAM, coagulant (alum), and poly (acrylamide) which is used as commercial flocculant were evaluated in 0.25% kaolin suspension by standard *jar test* procedure as done in earlier studies.[17–18]

The procedure involved addition of the polymeric flocculant under examination in highly concentrated form, to affect flocculant dosage ranging between 0 and 1.25 ppm, in increments of 0.25 ppm; in six identical jars. The test protocol involved rotation at 150 rpm for 30 s, 60 rpm for 5 min, followed by 15 min of settling time. Afterward, supernatant was collected from each jar and turbidity was measured in calibrated naphaloturbidity meter (Digital Nephelo-Turbidity Meter 132, Systronics, India). Flocculation profile was plotted in each case (*turbidity* vs. *flocculant concentration*) as in Figure 12.3.

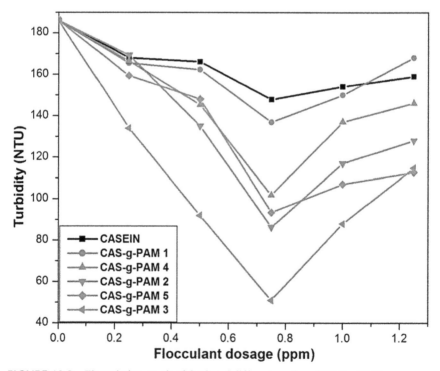

FIGURE 12.3 Flocculation graph of CAS and different grades of CAS-g-PAM.

Settling test employs a 100 mL stoppered graduated cylinder containing 0.25% kaolin suspension. The polymeric flocculent to be investigated is added in concentration solution form to effect the optimized dosage as decided by *jar test* experiment as above. The cylinder was turned upside down 20 times and then kept undisturbed. The height of the clear *water suspension* was tracked with respect to time. The experiment was performed for CAS, alum, PAM, and all the grades of CAS-g-PAM and the result was expressed as *height of interfacial layer* vs. *time* plot (Fig. 12.4).

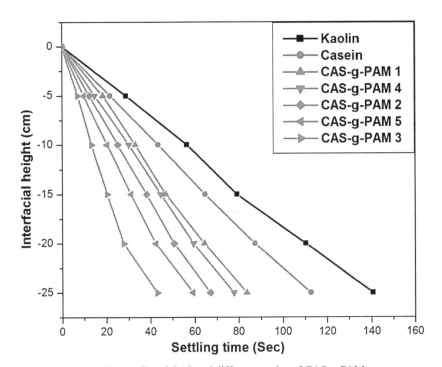

FIGURE 12.4   Settling profile of CAS and different grades of CAS-g-PAM.

12.2.4.2   STUDY OF FLOCCULATION EFFICACY OF CAS-g-PAM IN TEXTILE INDUSTRY WASTEWATER

The flocculation efficacy of CAS-g-PAM 3 (optimized grade) at optimized dosage (0.75 ppm as determined by *jar test* procedure in kaolin suspension) in textile wastewater was determined and compared to that in case of raw material (CAS) and that with wastewater without flocculant (BLANK); all three subjected to identical protocol of *jar test* procedure as above (i.e., 150 rpm stirring for 30 s, 60 rpm stirring for 5 min, and then 15 min settling; followed by drawing of supernatant). The three sets of *jar test* batches were as:

SET A: Textile water without flocculant (BLANK set)
SET B: Textile water with 0.75 ppm (optimized dose) of casein
SET C: Textile water with 0.75 ppm (optimized dose) of CAS-g-PAM 3

The supernatants drawn were analyzed by standard procedures[19] and reported in Table 12.3. Color measurement was performed with each of the three supernatants by spectrophotometer (DR/2400, Hach®) and expressed as *Absorbance* vs. *wavelength* plot as in Figure 12.5.

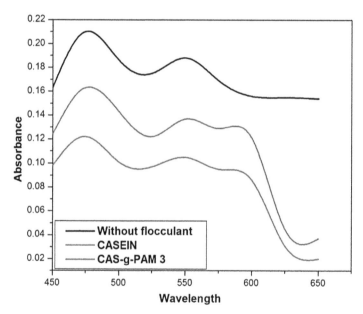

FIGURE 12.5   Color profile of supernatants.

12.3   RESULTS AND DISCUSSION

12.3.1   SYNTHESIS OF CAS-g-PAM BY MICROWAVE-INITIATED METHOD

CAS-g-PAM has been synthesized by MW-initiated method. Various grades of the graft copolymer were synthesized by varying MW irradiation time and acrylamide (monomer) concentration. The synthesis details have been tabulated in Table 12.1.The synthesis was optimized first in terms of monomer concentration, keeping the MW *irradiation time* constant (CAS-g-PAM 1, 2, and 3), then optimized in terms of *irradiation time*, keeping the monomer concentration as optimized earlier (CAS-g-PAM 3, 4, and 5). The optimized grade has been determined through its highest percentage grafting and intrinsic viscosity (which is proportional to molecular weight). From Table 12.1, it is obvious that the grafting is optimized at acrylamide concentration of 15 g in the reaction mixture and MW irradiation time of 4 min, maintaining the MW power at 700W (maximum power of the reactor). The grafting follows a mechanism similar to the mechanism proposed in studies on MW-initiated grafting of polysaccharides.[17] The proposed mechanism of the MW-initiated synthesis of CAS-g-PAM has been depicted in Scheme 12.2.

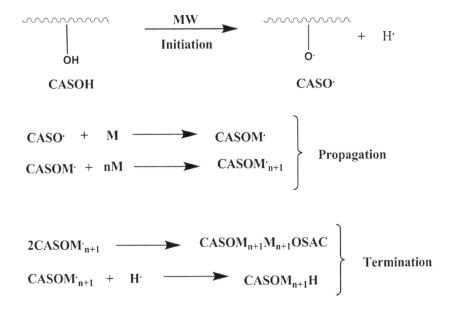

CASOH : Casein

**M : Monomer
(Acrylamide)**

MW : Microwave radiation

SCHEME 12.2   Proposed mechanism for synthesis of CAS-g-PAM.

12.3.2 CHARACTERIZATION

12.3.2.1 ESTIMATION AND INTERPRETATION OF INTRINSIC VISCOSITY

The intrinsic viscosity was evaluated for casein and the various grades of CAS-g-PAM, as shown in Table 12.1. Intrinsic viscosity is practically the hydrodynamic volume of the macromolecule in solution (aqueous solution in this case). It is obvious from Table 12.1 that the intrinsic viscosities of all the grades of CAS-g-PAM are greater than that of casein. This can be explained by the increase in hydrodynamic volume due to the grafting of the PAM chains on the main polymer backbone, that is, casein. The grafted PAM chains increase hydrodynamic volume by two ways:

1. By uncoiling of the protein chain through steric hindrance to intra-molecular bonding;
2. By contributing their own hydrodynamic volume.

Further, this is in good agreement with Mark–Houwink–Sakurada relationship (intrinsic viscosity $\eta = KM^a$ where K and α are constants, both related to stiffness of the polymer chains), applying which we can explain the increase in intrinsic viscosity as a result of increase in molecular weight due to the grafted PAM chains.

As evident from supplementary Figure 12.2, intrinsic viscosity grows exponentially with percentage grafting.

12.3.2.2 ELEMENTAL ANALYSIS

The results of elemental analysis for casein and that of the best grade of polyacrylamide-grafted casein (i.e., CAS-g-PAM 3) are given in Table 12.2. Casein which is a starting material has low percentages of nitrogen (i.e., 12.77%) than CAS-g-PAM 3. The higher percentage of nitrogen is

TABLE 12.2 Elemental Analysis Result.

Polymer Grade	% C	% H	% N
Casein (CAS)	47.66	7.778	12.77
PAM	50.08	7.69	19.76
CAS-g-PAM 3	48.74	7.435	17.17

contributed by the grafted PAM chains (which have higher nitrogen content than the casein).

12.3.2.3 NUMBER AVERAGE MOLECULAR WEIGHT OF THE POLYMERS

As evident from Table 12.1, higher the percentage grafting, higher is the number average molecular weight. Thus, there is predictable correlation between the three parameters, that is, percentage grafting (%G), intrinsic viscosity (η), and number average molecular weight. The empirical relations are as below:

Intrinsic viscosity (η) = $A_1 + A_2 \times G + A_3 \times G^2$

Number average molecular weight = $B_1 + B_2 \times G + B_3 \times G^2$,

where $A_1, A_2, A_3, B_1, B_2,$ and B_3 are constants and G is % grafting.

12.3.2.4 FTIR Spectroscopy

Figures 12.1a and 12.2b show the FTIR spectra of starting material casein and best grade of CAS-g-PAM 3, respectively. Major regions of the FTIR spectra of casein (Fig. 12.1a) showed absorption peaks at 3626 cm⁻¹ (O–H stretching vibrations), 2924 cm⁻¹ (C–H stretching vibrations), 1674 cm⁻¹ (C=O stretching), 1531 cm⁻¹ (N–H bending), 1442 cm⁻¹ (aromatic C=C stretching), and 1230 cm⁻¹ (saturated C–C stretching).

In the case of CAS-g-PAM 3 (Fig. 12.1b), the additional peak of CAS-g-PAM 3 at 1120 cm⁻¹ is assigned to C–O–C asymmetric stretching vibrations. C–O–C bond is present in grafted casein and is absent in both casein as well as PAM and thus is the confirmation of grafting. The peak at 3190 cm⁻¹ is due to the overlapping of O–H stretching band of hydroxyl group of casein and N–H stretching band of amide group of PAM with each other.

12.3.2.5 SCANNING ELECTRON MICROSCOPY ANALYSIS

It is evident from the SEM micrographs of CAS (Fig. 12.2a) and that of the best grade of CAS-g-PAM 3 (Fig. 12.2b) that profound morphological changes have been taken place in form of transition from granular structure (of CAS) to flaky structure (of CAS-g-PAM 3), because of grafting of PAM chains onto the protein backbone.

12.3.3 FLOCCULATION STUDY

12.3.3.1 FLOCCULATION STUDY AND DOSAGE OPTIMIZATION IN KAOLIN SUSPENSION

Flocculation efficacy of CAS, alum, PAM, and all synthesized grades of CAS-g-PAM were analyzed by standard *jar test* procedure, in 0.25% kaolin suspension. All the grades of CAS-g-PAM showed better flocculation efficacy than both CAS and alum due to the higher intrinsic viscosity (Fig. 12.3). Higher the percentage grafting of the graft copolymer, higher is the intrinsic viscosity and consequently, higher is the flocculation efficacy. Thus, the grade of CAS-g-PAM with highest percentage grafting showed highest flocculation efficacy (i.e., CAS-g-PAM 3). This is in fine agreement with *Extended Singh's easy approachability model*[20] as well as Brostow, Singh, and Pal's model of flocculation.[21]

Figure 12.4 shows the settling characteristics in kaolin suspension for CAS and different grades of CAS-g-PAM at optimized dosage (0.75 ppm). In synergism with *jar test* above, CAS-g-PAM 3 has the most rapid settling rate. Further, higher the percentage grafting, higher is the settling rate. The settling rate is another reliable indicator of flocculation efficacy, having direct relation with radius of gyration.[21] The higher the settling rate of the floc, superior will be its flocculation performance.

12.3.3.2 FLOCCULATION STUDY IN TEXTILE WASTEWATER

As consistent with the flocculation study in kaolin suspension, the flocculation efficacy of CAS-g-PAM 3 in textile wastewater is much higher than that of CAS (raw material). The reduction in pollutant load in the supernatant collected is much lower in case of CAS-g-PAM 3 (SET C) than in case of raw material as flocculant (SET B) as well as the CONTROL SET (SET A); as evident in Table 12.3.

TABLE 12.3 Comparative Study Performance of Best Grade Of CAS-g-PAM 3 for the Treatment of Textile Wastewater.

Parameter	Supernatant liquid SET A	Supernatant liquid SET B	Supernatant liquid SET C
Turbidity (NTU)	233	162	79
Total solid (ppm)	12,830	7920	4560
TDS (ppm)	7200	5110	3726

TABLE 12.3 *(Continued)*

Parameter	Supernatant liquid SET A	Supernatant liquid SET B	Supernatant liquid SET C
TSS (ppm)	5630	2810	834
COD (ppm)	1069	671	288
Total Iron (ppm)	7.01	5.25	1.53
Cr VI (ppm)	5.64	1.7	0.48

As evident from the color profiles (*absorbance* vs. *wavelength*) of supernatants (Fig. 12.5), CAS-g-PAM 3 is highly effective in color reduction of textile wastewater; color being the most pressing issue in this case.

12.4 CONCLUSION

A novel graft copolymer of casein (CAS-g-PAM) was synthesized by *microwave-initiated* method. The synthesized grades of the graft copolymer were characterized through various physicochemical techniques (intrinsic viscosity measurement, FTIR spectroscopy, elemental analysis, osmometry, and SEM). Flocculation efficacy of the synthesized grades was studied in kaolin suspension. It was evident that higher the percentage grafting, higher is the intrinsic viscosity as well as flocculation efficacy. Further, the best grade of the graft copolymer (CAS-g-PAM 3) was found to be highly effective as flocculant for the treatment of textile industry wastewater.

KEYWORDS

- casein
- flocculation
- turbidity
- waste water
- polyacrylamide

REFERENCES

1. Rani, P.; Sen, G.; Mishra, S.; Jha, U. *Carbohydr. Polym.* **2012,** *89,* 275–281.
2. Mishra, S.; Sen, G.; Rani, U.; Sinha, S. *Int. J. Biol. Macromol.* **2011,** *49,* 591–598.
3. Mishra, S.; Mukul, A.; Sen, G.; Jha, U. *Int. J. Biol. Macromol.* **2011,** *48,* 106–111.
4. Mishra, S.; Sen, G. *Int. J. Biol. Macromol.* **2011,** *48,* 688–694.
5. Mishra, S.; Rani, U.; Sen, G. *Carbohydr. Polym.* **2012,** *87,* 2255–2262.
6. Pal, S.; Ghorai, S.; Dash, M. K.; Ghosh, S.; Udayabhanu, G. *J. Hazard. Mater.* **2011,** *192,* 1580–1588.
7. Pal, S.; Sen, G.; Ghosh, S.; Singh, R. P. *Carbohydr. Polym.* **2012,** *87,* 336–342.
8. Sen, G.; Pal, S. *Int. J. Biol. Macromol.* **2009,** *45,* 48–55.
9. Sen, G.; Kumar, R.; Ghosh, S.; Pal, S. *Carbohydr. Polym.* **2009,** *77,* 822–831.
10. Sen, G.; Mishra, S.; Jha, U.; Pal, S. *Int. J. Biol. Macromol.* **2010,** *47,* 164–170.
11. Sen, G.; Ghosh, S.; Jha, U.; Pal, S. *Chem. Eng. J.* **2011,** *171,* 495–501.
12. Hassan, M. A. A.; Hui, L. S.; Noor, Z. Z. Removal of Boron from Industrial Wastewater by Chitosan via Chemical Precipitation. *J. Chem. Nat. Resour. Eng.* **2009,** *4*(1), 1–11.
13. Al-Momani, F.; Touraud, E., Degorce-Dumas, J. R.; Roussy, J.; Thomas O. Biodegradability Enhancement of Textile Dyes and Textile Wastewater by VUV Photolysis. *J. Photochem. Photobiol. A: Chem.* **2002,** *153*(1–3), 191–197.
14. Pala, A.; Tokat, E. Color Removal from Cotton Textile Industry Wastewater in an Activated Sludge System with Various Additives. *Water Res.* **2002,** *36*(11), 2920–2925.
15. Kim, T.H.; Lee, J.K.; Lee, M.J. Biodegradability Enhancement of Textile Wastewater by Electron Beam Irradiation. *Radiat. Phys. Chem.* **2007,** *76*(6), 1037–1041.
16. Zheng, H.; Zhu, G.; Jiang, S.; Tshukudu, T.; Xiang, X.; Zhang, P.; He, Q. Investigations of Coagulation–flocculation Process by Performance Optimization, Model Prediction and Fractal Structure of Flocs. *Desalination* **2011,** *269*(1–3), 148–156.
17. Mishra, S.; Sen, G. Microwave Initiated Synthesis of Polymethylmethacrylate Grafted Guar (GG-g-PMMA), Characterizations and Applications. *Int. J. Biol. Macromol.* **2011,** *48,* 688–694.
18. Mishra, S.; Sen, G.; Rani, G.U.; Sinha, S. Microwave Assisted Synthesis of Polyacrylamide Grafted Agar (Ag-G-PAM) and its Application as Flocculant for Wastewater Treatment. *Int. J. Biol. Macromol.* **2011,** *49,* 591–598.
19. *Standard Method of Examination of Water and Wastewater, 20th Edition*; Greenberg, Ed.; American Association of Public Health, 1999; pp 1–1220.
20. Singh, R. P. Advanced Drag Reducing and Flocculating Materials Based on Polysaccharides. In *Polymers and Other Advanced Materials: Emerging Technologies and Business Opportunities;* Prasad, N., Mark, J. E., Fai, T. J., Eds.; Plenum Press: New York, pp 227–249.
21. Brostow, W.; Pal, S.; Singh, R. P. A Model of Flocculation. *Mat. Lett.* **2007,** *61,* 4381–4384.

CHAPTER 13

ELECTRON BEAM DEPOSITION OF HYDROXYAPATITE COATING ON POLYAMIDE SUBSTRATE FOR BIOMEDICAL APPLICATION

HARIHARAN K.* and ARUMAIKKANNU G.

Department of Manufacturing Engineering, Anna University, Chennai, India

*Corresponding author. E-mail: hariharancim28@gmail.com, arumai@annauniv.edu

CONTENTS

ABSTRACT

Polyamide (PA) has a wide range of applications in the medical field such as scaffolds for tissue engineering, prostheses, bone repair, dentistry, medical devices, etc. There are a number of ways to fabricate this kind of medical devices and prostheses using PA, but a notable technique among these is additive manufacturing (AM). AM is a modern manufacturing technique which constructs the parts by adding the material in layer-by-layer manner directly from the 3D CAD (Computer Aided Design) model without any tooling process. This work aims to deposit a layer of hydroxyapatite (HA) over the PA substrate which will augment the biocompatibility of prostheses or medical devices. PA substrate was fabricated using selective laser sintering, which is one of the AM techniques. Wet chemical process was used to synthesize HA powder and the coating was done by physical vapor deposition with electron beam (EB-PVD) as a source. The surface morphology was investigated by means of scanning electron microscopy, elemental composition of coated layer was analyzed using energy-dispersive X-ray spectroscopy, and the crystalline phase of coated layer was investigated by X-ray diffraction (XRD). The adhesion strength of layer was estimated using Rockwell adhesion test. The 3T3 fibroblast cells were seeded on the coated sample and its viability was studied. The results revealed that HA layer was deposited successfully over the PA substrate using EB-PVD and the cell possesses 92.3% of viability, which may significantly improve the biocompatibility of prostheses and medical devices.

13.1 INTRODUCTION

A material which is intended to interface with the living tissue to evaluate, treat, or replace any tissue, organ, or function of the body is known as biomaterial.[1] The most important aspect of the biomaterial is biocompatibility, which has to perform with an appropriate host response (a structural and biological bond between the material and host tissue) in a specific application.[2] There are a number of biomaterials which are grouped as metals (stainless steel, cobalt chromium, and titanium alloys), ceramics (hydroxyapatite (HA), alumina, titania, and zirconia), polymers (polyethylene, polyurethane, polyamide (PA), polytetrafluoroethylene, polymethyl methacrylate, etc.), and composites (made from various metals, ceramics, and polymers). The application of these biomaterials include the fabrication of 3D scaffolds for tissue engineering, orthopedic joints, soft contact lens, drug delivery, and fixation devices, etc.[3] Nowadays these devices are fabricated very precisely using additive manufacturing (AM) technology. AM is a fast-growing technology in which parts are fabricated directly from the 3D CAD model without any tooling or process planning. Using AM, complex structured parts can be fabricated in considerable time. These models can be used in various applications from aerospace, automobile, teaching aids, jewellery, and medical modeling for design verification, form, and fit.[4] Modeling of medical devices such as implants, prosthesis, and biomaterials encompassed with AM is termed as bio-AM (BAM).[5] BAM approach produces the physical models of patient with extremely detailed features that can serve as excellent templates for the creation of custom implants. To produce these implants, various techniques of AM were used such as selective laser sintering (SLS), fused deposition modeling, three-dimensional printing, and stereolithography apparatus. All other techniques lack behind the SLS due to their construction with complex internal and external geometries. Virtually, any powdered material that will fuse but not decompose under laser beam can be used for fabrication. Metallic and metallic alloy implants were very popular in replacing the bone. However, it was well known that these materials were prone to corrode in the body fluid environment and lead to fatigue failure which ultimately causes to loosening of the implants.[6] Currently, implant industries are marching toward the newer biomaterial which overcomes the difficulties faced by metallic implants and opt for biopolymers. Among various biopolymers, PA was one of the preferred candidate to produce medical devices, dental screws sutures, and cancellous bone implants by virtue of its biostability, good mechanical property, and broad temperature tolerance as other biomaterials.[7] To enhance the osseointegration, bioactive materials such as HA, zirconia, and alumina can be deposited over the implants.[8] HA powder was a commonly used bioactive material since its chemical composition and properties are very similar to the human bone. The HA powder coatings on implant are widely used in orthopedic applications, and quantitative understanding of HA powder coating deformation might enable us to elucidate the mechanical stability of such implants in clinical applications. Laboratory research also supports the conclusion that the early bone growth is accelerated by implants coated with HA powder. This work aims to deposit HA powder over the laser sintered PA substrate and its cell viability was investigated.

13.2 METHODS AND MATERIALS

13.2.1 FABRICATION OF PA SUBSTRATE

The substrate was fabricated using SLS. A 3D CAD model was generated and the data were sliced into layers. The model was loaded on to SLS
machine (Fig. 13.1) and a computer-directed CO_2 laser sintered layers of PA
powder together. After each solidification layer, another layer of powder was
deposited, and again sintering took place until the part was completed.[9,10]

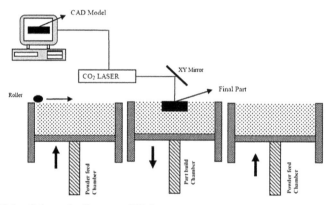

FIGURE 13.1   Schematic diagram of SLS.

13.2.2 SYNTHESIS OF HA TARGET

HA powder was prepared by wet chemical precipitation method. In this
method, calcium hydroxide and ammonium phosphate (dibasic) powder was
utilized as a precursor in the ratio of 10:6. These precursors were made into
solution and precipitate was obtained. Then it was calcined at a temperature
of 500°C to obtain HA powder. The prepared HA powder was made into
pellets with high-pressure hydraulic press with 400-MPa load and sintered
at 800°C for 5 h.[11]

13.2.3 ELECTRON BEAM DEPOSITION OF HA

The substrate to be coated was ultrasonically cleaned using acetone and ethyl
alcohol. The prepared HA powder was made into pellets and sintered at 300°C
for 8 h. The pellets were made free from the moisture and contaminates. For

coating, physical vapor deposition with electron beam source (EB-PVD) (Hind High Vaccum, India) was used. At the early stage of deposition, the substrate was preheated to a temperature of 50°C by means of e-beam with a 40 mA current. Ar^+ ion beam was used to clean the substrate for the period of 10 min; after cleaning, the chamber was evacuated of a base pressure of 5×10^{-4} torr. The high vacuum was attained at the range of 10^{-7} to 10^{-6} torr. During the deposition process the substrate was allowed to rotate at a speed of 30 rpm, where an e-beam was generated at operating voltage of 5 kV.[12]

13.2.4   IN VITRO ANALYSIS

13.2.4.1   CELL CULTURING

The cellular response to the coatings was assessed in terms of cell viability and proliferation to the surface. 3T3 fibroblast cells were purchased from National Center for Cell Science (NCCS, Pune). This kind of cells is suitable for screening large number of samples for cytotoxic compound and also used in the rapid evaluation of the biomaterial surface qualities. Cells were grown in 75-cm^2 flask containing Dulbecco's modified Eagle's medium (DMEM; Sigma). The mediums were supplemented with 10% fetal bovine serum (FBS; Invitrogen), 1.5 g/L sodium bicarbonate, 10,000 units/mL penicillin, 10 mg/mL streptomycin, and 25 µg/mL amphotericin B. Cells were cultured as monolayers in culture flasks at 37°C under a humidified atmosphere of 5% CO_2 in air.

13.2.4.2   CELL SEEDING ON HA-COATED SUBSTRATE

The coated substrate was sterilized with 70% ethanol for 30 min and then it was autoclaved for 30 min, followed by drying at room temperature for 2 h. The cells were seeded approximately 1×10^5 cells/sample. The cell proliferation and viability were assessed and observed at specific time period of 1st, 7th, and 15th day, respectively.

13.2.5   CELL VIABILITY

The viability of cells was assessed by standard MTT (3-(4,5-dimethylthiazol-2-yl)-2,5-diphenyltetrazolium bromide) assay.[13] This assay is based on the reduction of soluble yellow tetrazolium salt to insoluble purple formazan crystals by metabolically active cells. Only live cells are able to take up the tetrazolium salt. The enzyme (mitochondrial dehydrogenase) present in the mitochondria of the live cells is able to convert internalized tetrazolium salt to formazan crystals, which are purple in color. Then the cells were dissolved in DMSO (Dimethyl Sulfoxide) solution. The color developed is then determined in an ELISA (Enzyme Linked Immunosorbent Assay) reader at 570 nm of UV (Ultra violet)-absorbance wavelength.

13.2.6 MICROSTRUCTURAL ANALYSIS

The microstructure of coated surface was examined using scanning electron microscopy (SEM) (Philips XL30 FEG) and energy-dispersive X-ray spectroscopy (EDX) (oxford instrument). Analysis was carried out to find the stoichiometry of coated layer for 60 s and electron beam energy of 15 keV. The XRD measurement was performed to observe the crystalline phases in the coated layer. The adhesion of layer was examined by Rockwell adhesion test according to VDI 3198 standard. A standard Rockwell hardness tester was used to create an indentation over the surface of the coated layer of thickness 200 nm and this will create plastic deformation on the substrate and form bulging over the region. The indentation was examined by using microscope with magnification of 100×. The standard indentation marks (Fig. 13.2) were classified into six levels (HF1 to HF6), in which first four levels are acceptable (HF1 to HF4) which will cause small amount of delamination. The remaining two standards (HF5 and HF6) are unacceptable since they cause large-area delamination.

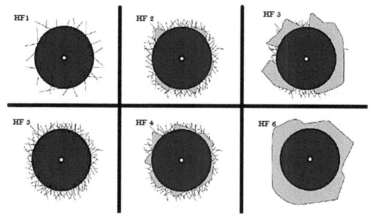

FIGURE 13.2   Standard indentation mark.

13.3 RESULTS AND DISCUSSION

13.3.1 SURFACE MORPHOLOGY ANALYSIS

The surface of HA coating was characterized using SEM. Figure 13.3(a) shows that a dense and uniform layer was formed over the surface; further examination revealed that the surface of the coating was free from defects such as cracks, pores, and large voids. Some isolated particles were identified on the layer (Figure 13.3(b)); this may be due to agglomeration of smaller particles. This kind of particle agglomerations will increase the surface roughness and help in improving osseointegration. Figure 13.3(c) shows the cross-sectional image of HA layer with a thickness of 200 nm. The average surface roughness (R_a) of deposited layer was 3.41 µm measured using stylus probe technique. It has been demonstrated that rough surface will stimulate better osseointegration.[14,15]

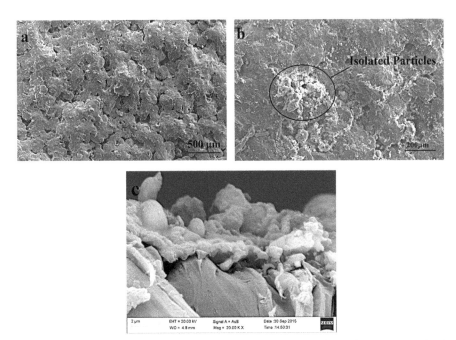

FIGURE 13.3 SEM micrograph: (a) surface, (b) isolated particle, and (c) cross-sectional image.

13.3.2 ELEMENTAL ANALYSIS

The elemental composition of a coated layer was analyzed using EDX and the spectrum was shown in Figure 13.4. It shows that the characteristic peaks of calcium, phosphate, and oxygen elements confirming that the layer was HA. These elements are main constitutes of hard tissue, teeth, and tendons. The ratio of calcium and phosphate was approximately 1.68, which is close to that of standard HA.

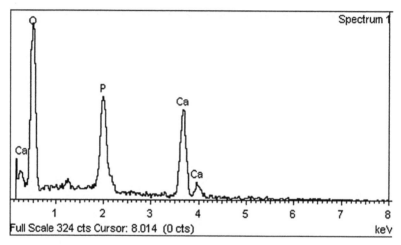

FIGURE 13.4   EDX spectrum of coated layer.

13.3.3 XRD ANALYSIS

The XRD diffraction pattern of HA layer was shown in Figure 13.5. It confirms that the formation of HA with sharp diffraction at 2θ value of 25.9° and 32.62°. Some amorphous peaks were also absorbed at 46.58° and 49.34° corresponding to the relative intensity of diffraction of standard data (ICSD 087727). The major peaks and some shoulder peaks of the layer indicate a mixture of amorphous and crystalline structure of HA, since the coating was carried out at low substrate temperature. Better bone–tissue integration and osseointegration may be achieved due to this phase mixture in layer.[16]

13.3.4 ROCKWELL ADHESION TEST

The standard Rockwell C indentation procedure was followed. A conical diamond indenter penetrates into the coated surface inducing massive plastic deformation to the substrate thereby fracturing the coating. The indentation was examined using video measuring system and the indentation on the coated layer was shown in Figure 13.6. The obtained image was compared with the standard grades and the adhesion strength was in a range of HF1 or HF2, since it had a very small amount of delamination.

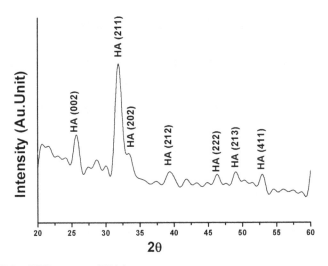

FIGURE 13.5 XRD pattern of HA layer.

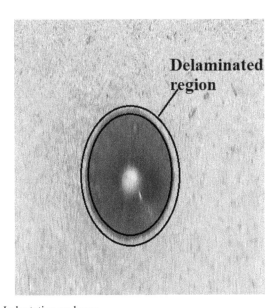

FIGURE 13.6 Indentation on layer.

13.3.5 *IN VITRO CELL RESPONSE*

The in vitro analysis of HA-coated substrates was carried out by measuring the viable cells seeded on the substrates after particular time interval of 1st, 7th and 15th day. Figure 13.7 shows that the cells can grow more effectively on the coated substrates. From the first day observation, the cell growth on coated substrate was slow due to low cell–substrate interaction. The amount of cells on the coated substrate was increased significantly for longer incubation time. This increment cell viability was due to release of Ca^{2+} ions, surface roughness, and crystallinity of HA layer and also it was believed that the cell–substrate interaction was directly related to that of organic and inorganic interface. Fluorescent microscopic image of cells seeded on coated sample is shown in Figure 13.8; it was noted that the viable cells were green in color.

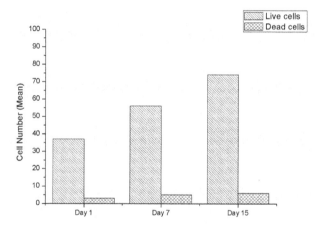

FIGURE 13.7   Cell viability of live and dead cells.

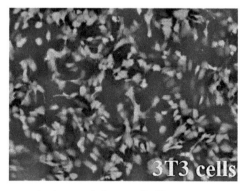

FIGURE 13.8   Fluorescent microscopic image of cells.

13.4 CONCLUSION

In this work, wet chemical synthesized HA powder was coated over the PA substrates using physical vapor deposition with electron beam source. The surface morphology, elemental composition, crystallinity, and adhesion strength were studied using SEM, EDX, XRD, and Rockwell adhesion tester. Fibroblast cells were seeded on the HA-coated substrate and the cell viability was studied. The results show that a dense layer without any cracks or pores of HA was deposited along with some isolated particles. It was believed that this isolated particle will increase surface roughness of the layer and promote osseointegration. Also, the Ca/P ratio of the layer was 1.68 which was close to that of standard HA. Due to low deposition temperature, the coating consists of mixture of amorphous and crystalline phases. The HA layer shows very good adhesion strength toward the surface. In vitro results reveal that the layer has better cell viability. Finally, it was concluded that EB-PVD was a promising technique to deposit HA powder over PA substrate. It is more effective in improving the cell viability and has implication for enhancing the quality of biomaterials including dental and other bone-related implants.

KEYWORDS

- additive manufacturing
- electron beam deposition
- hydroxyapatite
- polyamide
- characterization
- in vitro analysis

REFERENCES

1. William, D. F. *The William Dictionary of Biomaterials*; Liverpool University Press: Liverpool, 1999.
2. Black, J. *Biological Performance of Materials: Fundamentals of Biocompatibility*; CRC Press, Boca Raton, USA, 2005.

3. *ASM Hand Book, Materials for Medical Devices;* Vol. 23, ASM International, The Materials Information Society, 2012.

4. Hosni, Y. A.; Harrysson, O. L. A. *Design and Manufacturing of Customized Implants*, IERC 2002, Orlando, Florida, USA, May 19–21, 2002.

5. Bourell, D. L.; Leu, M. C.; Rosen, D. W. *Road Map of Additive Manufacturing*; University of Texas: Austin, 2009.

6. Antunes, R. A.; Rodas, A. C. D.; Lima, N. B.; Higa, O. Z.; Costa, I. Study of the Corrosion Resistance and In Vitro Biocompatibility of PVD TiCN-coated AISI 316L Austenitic Stainless Steel for Orthopedic Applications. *Surf. Coat. Technol.* **2010**, *205*(7), 2074–2081.

7. Jagur-Grodzinski, J. Biomedical Application of Functional Polymers. *React. Funct. Polym.* **1999**, *39*(2), 99–138.

8. Radin, S. R.; Ducheyne, P. The Effect of Calcium Phosphate Ceramic Composition and Structure on In Vitro Behavior. II. Precipitation. *J. Biomed. Mater. Res.* **1993**, *27*(1), 35–45.

9. Salmoria, G. V., Leite, J. L.; Paggi, R. A. The Microstructural Characterization of PA6/PA12 Blend Specimens Fabricated by Selective Laser Sintering. *Polym. Test.* **2009**, *28*(7), 746–751.

10. Van Hooreweder, B.; Moens, D.; Boonen, R.; Kruth, J.-P.; Sas, P. On the Difference in Material Structure and Fatigue Properties of Nylon Specimens Produced by Injection Molding and Selective Laser Sintering. *Polym. Test.* **2013**, *32*(5), 972–981.

11. Monmaturapoj, N. Nano-size Hydroxyapatite Powders Preparation by Wet-chemical Precipitation Route. *J. Metals Mater. Miner.* **2008**, *18*(1), 15–20.

12. Hamdi, M.; Ektessabi, A.-I. Calcium Phosphate Coatings: A Comparative Study Between Simultaneous Vapor Deposition and Electron Beam Deposition Techniques. *Surf. Coat. Technol.* **2006**, *201*(6), 3123–3128.

13. Mosmann, T. Rapid Colorimetric Assay for Cellular Growth and Survival: Application to Proliferation and Cytotoxicity Assays. *J. Immunol. Meth.* **1983**, *65*(1), 55–63.

14. Costa-Rodrigues, J.; Fernandes, A.; Lopes, M. A.; Fernandes, M. H. Hydroxyapatite Surface Roughness: Complex Modulation of the Osteoclastogenesis of Human Precursor Cells. *Acta Biomater.* **2012**, *8*(3), 1137–1145.

15. Gittens, R. A.; McLachlan, T.; Olivares-Navarrete, R.; Cai, Y.; Berner, S.; Tannenbaum, R.; Schwartz, Z.; Sandhage, K. H.; Boyan, B. D. The Effects of Combined Micron-/submicron-scale Surface Roughness and Nanoscale Features on Cell Proliferation and Differentiation. *Biomaterials* **2011**, *32*(13), 3395–3403.

16. Wang, G.; Zreiqat, H. Functional Coatings or Films for Hard-tissue Applications. *Materials* **2010**, *3*(7), 3994–4050.

CHAPTER 14

SYNTHESIS AND CHARACTERIZATION OF TEMPLATED POLYANILINES: A NEW CLASS OF POLYMERIC MATERIALS

JYOTHI LAKSHMI AVUSULA, SHANTANU KALLAKURI, and SUBBALAKSHMI JAYANTY*

Department of Chemistry, Birla Institute of Technology and Science, Pilani-Hyderabad Campus, Jawaharnagar, Shameerpet Mandal, Ranga Reddy, Hyderabad 500078, Telangana, India

*Corresponding author. E-mail: jslakshmi@hyderabad.bits-pilani.ac.in

CONTENTS

ABSTRACT

Polyaniline (PANI) is one of the most extensively studied conducting polymers. The advantages, coupled with its chemical stability and high conductivity in doped state, make its commercial application quite attractive. Polyethylenedioxythiophene–polystyrenesulfonate, trisodium citrate, and polyethylene glycol-templated PANIs were prepared through chemical synthesis. Stable aqueous colloidal suspensions obtained were evaporated to obtain powders. Infrared spectra showed corresponding characteristic peaks for each of the templated PANI. Interestingly, scanning electron microscopy revealed blocklike features with different particle sizes. Powder X-ray diffraction revealed the presence of nanosized particles and semicrystalline nature in all these compounds. Four-probe room temperature conductivity measurements on powders exhibited semiconducting behavior.

14.1 INTRODUCTION

Conducting polymers have attracted a great deal of attention for possible applications in electrochromic devices, redox batteries, and so on. Polyaniline (PANI), in particular, is one of the most extensively studied materials among the conducting polymers due to its simple and facile synthesis by chemical and electrochemical means,[1–5] having probably the best combination of stability and low cost[1,2] and for possessing good environmental stability too. The advantages, coupled with its chemical stability and high conductivity in doped state, make its commercial application quite attractive. In recent years, PANI and its composites have been shown to exhibit several submicro and nano architectures[6–10] with the expectation that such materials would possess advantages of both low-dimensional and organic conducting systems. PANI nanostructures could be made by template-guided polymerization where the templates could be either hard[6,11–13] or soft.[14,15] Studies on the effect of polyelectrolyte template molecular weight on PANI interestingly showed that morphology and conductivity of the PANI films depend on the template molecular weight.[9]

Recently, great progress has been made in the process of PANI.[13,16,17] Due to the increasing concern on environmental, safety, and processing methodologies, conducting PANI process in aqueous solution is becoming increasingly important. PANI doped with bifunctional water-soluble organic acids showed strong tendency to create water suspensions and solutions.[18] Investigations have been made to improve the poor mechanical properties and the processability of PANI[19] by blending it with other polymers[20–22] or by substitution and copolymerization. Blending of PANI with other components helps in achieving good stabilities, mechanical properties, and ease of fabrication. Synthesis and studies of directionally conductive PANI hydrogel by vapor deposition have also been reported.[23] Sodium citrate is often used as a stabilizing agent for the preparation of gold nanoparticles. It has been observed that addition of Au–citrate nanoparticles to the salt and base form of PANI[24] resulted in different morphologies due to electrostatic attractions. However, detailed characterization and studies have not been reported. Currently, PEDOT-PSS is the most widely spread conducting polymer. Dispersions of PEDOT-PSS were applied for the inkjet printing.[25,26] Recently, new-generation PEDOT:PSS/polyprrole and PEDOT:PSS/PANI bilayered organic nano films with enhanced thermoelectric performance were successfully fabricated by electrochemical polymerization.[27] PEDOT-PSS also finds application in touch screens due to its transparent nature. The intrinsic polymers such as PANI, PEDOT-PSS, etc. are identified for their use in mobile phones to replace the current conducting metal–carbon material technologies. Both PEDOT-PSS and PANI are electrochromic materials which have been studied over the past decade. Synthesis of various conducting polymers/ polyanion composites[28] and their application in fuel cells and supercapacitors have been reported by Pickup et al. Polyethylene glycol (PEG) has wide application ranging from industries to medicine. PEG-grafted PANI copolymers were synthesized and characterized, which also exhibited effectiveness in preventing platelet adhesion.[29] PANI-PEG has been synthesized by modification of poly (ethylene) glycol at both the ends with aniline. The achiral rod–coil copolymer was proposed to form helical superstructures due to the π–π stacking of the rigid segments of copolymers.[30] PANI-PEG electroactive and ion-conductive copolymer was utilized to modify carbon–LiFePO$_4$ by facile in situ chemical copolymerization method as a cathode material for high-performance lithium ion batteries.[31] Hybrid materials based on PANI-PEG-CdS have been prepared and characterized. It has been reported that the in situ polymerization of aniline on anionic CdS resulted in hybrids and utilization of PEG helped in diffusion-limited growth of PANI and CdS, resulting in nanosized hybrid materials.[32]

However, study on the effect of small molecules such as trisodium citrate (TSC) and polymers such as PEG or PEDOT-PSS as templates through simple chemical oxidative polymerization has not been reported so far. Powder X-ray diffraction (XRD) study revealed glassy/crystalline nature in the samples. PEDOT-PSS-templated PANI (PANI-PEDOT:PSS) exhibited higher conductivity. Differential scanning calorimetry suggests crystalline transformation in TSC-templated PANI (PANI-TSC), softening in poly(ethylene glycol)-templated PANI (PANI-PEG), and cross-linking in PANI-PEDOT:PSS. Scanning electron microscopy (SEM) showed blocklike morphologies. Detailed characterization of all the samples suggests better organization and symbiotic interactions between the π-donor molecular ions and the polyionic polymers leading to an improved polymer chain alignment. Hence, it is always of challenge and considerable interest to examine morphological changes and material properties obtained therein by studying the effect of small molecules or polyions on PANI to evolve new design strategies.

14.2 EXPERIMENTAL

14.2.1 MATERIALS AND SYNTHESIS

Analytical grade aniline, ammonium persulfate (AP), TSC, and HCl (12 N) were purchased from SD Fine chemicals. Poly(3,4-ethylenedioxythiophene)/poly(4-styrenesulfonate, PEDOT:PSS), PEG, and methanol were obtained from Sigma Aldrich. All solutions were prepared in Milli-Q water. To 0.42 mL (4.6 mmol) of aniline 15 mL of 3 N HCl was added and this mixture was stirred for 15 min at temperature 0–5°C. Low temperature is more preferred for the synthesis of PANI since the oxidation polymerization of aniline is exothermic. To this solution, TSC (3.9 g), PEDOT:PSS (4.42 mL), and PEG (0.697 mL) each of 13.5 mmol were added. AP (1.02 g, 4.4 mmol) was added dropwise with continuous stirring over 20–30 min to undergo oxidative polymerization. The reaction mixture was stirred for 3 h more to get green colloidal suspension, centrifuged for three times, and supernatant was drained out. Methanol was added and kept aside overnight to remove oligomers or any other impurities; the supernatant was drained out again and the colloidal suspension obtained was kept for sonication for 20 min. Acetone was added to get precipitate through evaporation. Finally, green-colored product was collected and dried. Possible reaction and molecular interactions in all samples are shown in Scheme 14.1. Hydrogen bonding and intermolecular interactions seem to play a dominant role.

14.2.2 FOUR-PROBE CONDUCTIVITY MEASUREMENT

DC conductivity for powders of PANI-PEG, PANI-PEDOT:PSS, and PANI-TSC was measured at room temperature using four-probe method. Constant current source of 0.1–1 µA was passed using Keithley Model 224 constant current source and voltage drop was measured using a Keithley Model 175 Multimeter.

14.2.3 DIFFERENTIAL SCANNING CALORIMETRY

All samples were characterized using Shimadzu differential scanning calorimeter (DSC)-60 in the range of temperature 35–300°C with heating and cooling rate of 10°C min⁻¹. DSC curves were obtained using alumina pan containing 4 mg of sample under nitrogen atmosphere at the rate of 100 mL min⁻¹. An empty alumina pan was used as reference.

14.2.4 ELEMENTAL ANALYSIS

Compositional analysis was carried out on vario MICRO V3.1.1 of Elementar Analysensysteme GmbH make. Samples were considered to be of approximately equal weights.

14.2.5 SCANNING ELECTRON MICROSCOPY

Morphology of powders on glass substrates was studied under Philips XL 30 ESEM SEM. Glass plates were cleaned thoroughly with soap solution, pure water, and acetone prior coating and were dried in hot air oven at 110°C.

14.3 RESULTS AND DISCUSSION

14.3.1 CHARACTERIZATION

FT-IR spectra were recorded on Jasco FTIR-4200 spectrophotometer. Samples in the form of KBr pellets were made from vacuum drying. Our synthesis produced PANI in the emeraldine salt form. Backbone of emeraldine salt form is positively charged. Absorption at 1569, 1468, 2919, and 2855 cm⁻¹ for PANI-PEG; 1543, 1475, 3019, and 2902 cm⁻¹ for PANI-PEDOT:PSS, and 1510, 1402, 2924, and 2907 cm⁻¹ observed for PANI-TSC correspond to C=C stretch of the quinonoid and benzenoid rings and to the asymmetric and symmetric aliphatic C–H stretches, respectively (Fig. 14.1). PANI-PEDOT:PSS and PANI-TSC form stable colloidal suspension through electrostatic repulsions as PEDOT:PSS and TSC are ionic. PEG behaves more like a surfactant. Hence, PANI-PEG could more probably form via steric repulsions. All samples showed absorption of O–H (H-bonded) stretching frequency with strong and broad feature; also overlapping with N–H and aromatic C–H stretch, extending from 3000 to 3650 cm^{-1} which was not intense in PANI-PEDOT:PSS, due to weak dipole moment of OH bonded to sulfur in PEDOT:PSS.

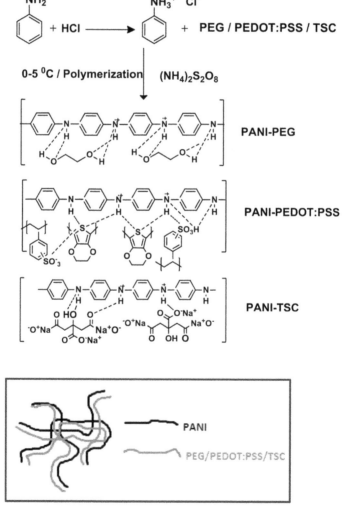

SCHEME 14.1 Proposed preparation and possible interactions in templated polyanilines obtained as PANI-PEG, PANI-PEDOT:PSS, and PANI-TSC.

Free OH (3685 cm^{-1}) was found to exist in all the three samples. C–O was observed at 1100–1130 cm^{-1}; also absorption at 1126, 1124, and 1116 cm^{-1} in particular correspond to the C–H bending of protonated PANI (Fig. 14.1). C=O (1720 cm^{-1}) was observed only in PANI-TSC. C=N and C–N peaks were found at 1635 cm^{-1} and 1225 cm^{-1}, respectively. Absorption of C–H and N–H bending frequencies (1298 and 747 cm^{-1}) was found in all the three samples. Characteristic peaks of S=O, S–O, and S–C at 1056 cm^{-1}, 666 cm^{-1}, and 593 cm^{-1} were observed in PANI-PEDOT:PSS. Weak bands associated with major characteristic peaks could be attributed to the doping effect, in our case it can be correlated to templated PANIs. N–H stretch (2355–2358 cm^{-1})[33] was observed in all samples as shown in Figure 14.1. These typical vibrational frequencies show that templates have strong interaction with PANI. Our earlier studies showed interesting dependency of conductivity on molecular weight of the template polyelectrolyte.[9]

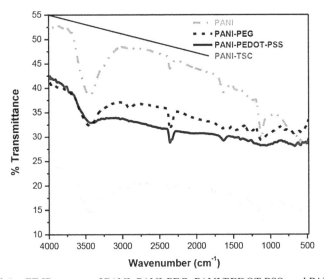

FIGURE 14.1 FT-IR spectra of PANI, PANI-PEG, PANI/PEDOT-PSS, and PANI-TSC.

Powder XRD (Fig. 14.2) showed PANI-PEG to be of more crystalline nature when compared to PANI-TSC and PANI-PEDOT:PSS. Powder XRD was employed to characterize the structure of the chemically synthesized powders. Pure PANI shows primary characteristic peaks at 20.0°C and 25.0°C attributed to the alternating distance between layers of polymer chains and periodicity perpendicular to PANI chains.[34,35] XRD pattern for our compounds were very much different from the individual PANI powder and PEDOT:PSS film's XRD spectra.[36,37] Figure 14.2 shows the XRD pattern with broad and sharp features indicating semicrystalline nature. It is noticed that the full width at half maximum are slightly broadened, typical feature of nanostructured materials. Interestingly, we found several additional peaks, confirming the formation of the templated PANIs. The broad region in the range of 2(θ) from 35 to 50 indicates glassy nature of PANI-PEDOT:PSS, this is also in agreement with the SEM image showing platelike structure with smaller thickness when compared to the blocks in PANI-TSC and PANI-PEG. Thus, PANI-PEDOT:PSS could form better films by various coating techniques. PANI-PEG shows intense and more number of sharp peaks at higher 2θ values indicating periodicity perpendicular to the polymer chain. Probably this leads to a low conductivity value. PANI-TSC have broad intense peaks at low 2 values and less intense sharp peaks riding over hump at higher 2θ's, possessing conductivity in between PANI-PEDOT:PSS and PANI-PEG. Though, all the three samples have slightly broad humps, this feature was observed to be more prominent in PANI-PEDOT:PSS,

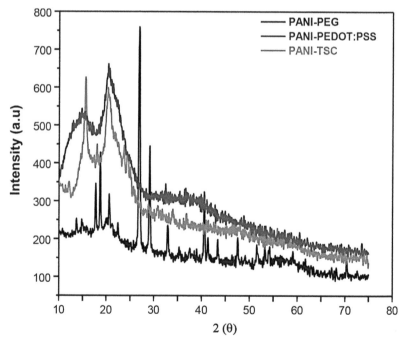

FIGURE 14.2 Powder X-ray diffraction plots for PANI-PEG, PANI/PEDOT-PSS, and PANI-TSC.

indicating that the chain ordering among the PANI and the counterion molecules is most likely due to short range[4] and periodicity parallel to the polymer chain.[36] Thus, PANI-PEDOT:PSS shows improved conduction. The average crystalline size could be calculated using the Scherrer formula.[38,39] Thermal properties and molecular interactions in templated PANIs were analyzed by DSC study. DSC is most widely used in order to determine the transition temperatures such as the glass transition temperature, decompositions, crystallization temperature, and melting cross-linking reactions. However, it deals with the measurement of the total heat flow and the total amount of entire thermal transitions within the sample. The representative DSC curves for the templated PANIs considered in this study are shown in Figure 14.3. Glass transition temperature (T_g) of PANI-PEG was observed at 105°C (Fig. 14.3), broad exothermic peak observed in the range from 225°C to around 255°C correspond to the softening of the polymer chain. The rise at 300°C may be due to the degradation of the polymer backbone.

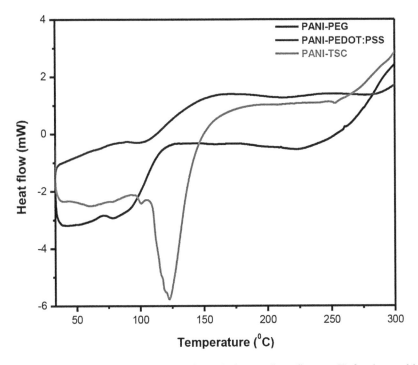

FIGURE 14.3 Plot of differential scanning calorimetry (heat flow vs. T) for the resulting polyanilines considered in this study.

T_g of PANI-PEDOT:PSS was estimated at 125°C. Exothermic peaks were not clear. However, a rise in exothermic range (Fig. 14.3) could be observed at 270°C, probably due to cross-linking, a strong rise at 300°C indicates decomposition. Glass transition temperature of PANI-PEG was found to be less than that of PANI-PEDOT:PSS, probably indicating plasticization effect of PEG as the template, also such effect was observed and reported in PANI–sulfonic acid blends.[40] PANI-TSC exhibited sharp endothermic peak at 125°C leading significantly to crystallization phenomenon. Again, as noted earlier, the sharp rise at 300°C indicates decomposition of the polymer. Endothermic peaks were not prominently observed in PANI-PEG or PANI-PEDOT:PSS. Similar phenomenon of the T_g, softening, cross-linking, and decomposition properties were also reported in PANI–polyvinylacetate blends[41] and in the study of thermal properties of HCl-doped PANIs and poly (o-toluidines).[42] The conductivity in these compounds is attributed to the π-conjugated chain in the polymer backbone.

14.3.2   CONDUCTIVITY AND MORPHOLOGY

Electrical conductivity is facilitated by the delocalization of the π-electrons along the polymer chain and transport of the electronic charges among the PANI and templates. Table 14.1 shows average particle sizes and conductivities of templated PANIs. These samples possess strong interaction between the quinonoid ring of PANI and templates (PEDOT:PSS, PEG, and TSC) leading to the origin of charge transport between the PANI domains.

TABLE 14.1   Powder Conductivity and Average Particle Sizes for PANI-PEDOT:PSS, PANI-TSC, and PANI-PEG.

Compounds	Particle size (nm)	Conductivity ($\times 10^{-2}$ S cm^{-1})
PANI-PEDOT:PSS	9.9	21.0
PANI-TSC	19.3	10.6
PANI-PEG	42.9	1.4

High conductivity in PANI-PEDOT:PSS could be due to better packing density with the presence of very small- to large-sized plates/block and also probably, due to the higher molecular weight of PEDOT-PSS. PANI-TSC also showed effective packing density with better uniform thickness observed among the blocks. However, due to the small molecule and lower molecular weight, conductivity in PANI-PEG is found to be slightly lower than PANI-PEDOT:PSS. Low conductivity in PANI-PEG could be due to the increase in the average molecular weight of the particles and decrease in packing density. An increase in particle size leads to an increase in the surface area which further leads to the decrease in conductivity (Fig. 14.4), also in agreement with earlier reports.[6] This observation also suggests that small molecules or polymers act as nanoreactor templates which control the polymerization process.[43] We have investigated the conductivities of various composite samples of PANI for comparison with our samples. Conductivities in powders of PANIs doped with inorganic acids were reported to be of 0.1–1 S cm^{-1} where surfactants were added as stabilizers.[8] PANI-CMC composites possessed conductivities of 10^{-1} to 10^{-5} S cm^{-1}.[44] PEDOT:PSS films are known to exhibit conductivity of ~10^{-1} S cm^{-1}. Electrical conductivity of aqueous solution of TSC is ~10^{-3} S cm^{-1}. Conductivity of PEG measured by Hampton Research Corporation suggests ~10^{-6} S cm^{-1}. In our case, small molecule (TSC) and polymers (PEDOT:PSS, PEG) assisted synthesis of PANI-enhanced conductivity by several orders of magnitude. Samples considered in this study overall possessed conductivities of about

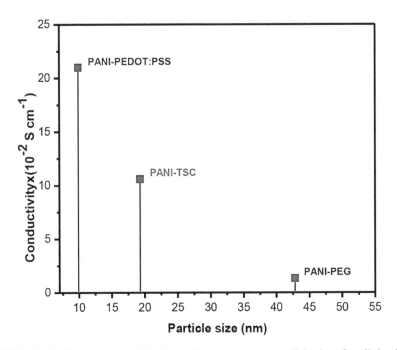

FIGURE 14.4 Plot of conductivity dependence on average particle sizes for all the three samples considered in this study.

10^{-1} to 10^{-2} S cm^{-1} (Table 14.1), prepared by direct chemical method without any further addition of stabilizing agents. Compositional analysis for all the three samples considered in our study is shown in Table 14.2. The amounts of carbon, hydrogen, nitrogen, and sulfur obtained indicate proficient protonation in PANI-PEDOT:PSS, also consistent with the observed high conductivity value among the three samples considered in this study.

TABLE 14.2   Compositional Analysis of PANI-PEDOT: PSS, PANI-TSC, and PANI-PEG.

Sr. No.	Compounds	C (%)	H (%)	N (%)	S (%)
1	PANI-PEDOT: PSS	44.38	5.29	5.34	12.64
2	PANI-PEG	44.71	6.20	8.32	4.31
3	PANI-TSC	36.49	5.48	6.41	2.58

We have also examined morphology of powders. SEM images (Figs. 14.5 and 14.6) showed block-like features not observed among the powdered PANIs that have been reported so far. Prominent difference in the overall morphological features when compared among PANI-PEDOT:PSS, PANI-PEG, and PANI-TSC has not been observed. However, these possess appreciable packing density.

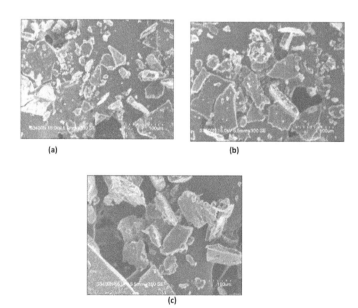

(a)   (b)

(c)

FIGURE 14.5   Scanning electron micrographs of (a) PANI/PEDOT:PSS, (b) PANI-PEG, and (c) PANI-TSC at high magnification (100 m).

FIGURE 14.6 Scanning electron micrographs of (a) PANI/PEDOT: PSS, (b) PANI-PEG, and (c) PANI-TSC under low magnification (500 m).

Thickness of the blocks showed certain variations. The aspect ratio of PANI-PEDOT:PSS (70:1) was found to be higher than PANI-PEG (50:1) and PANI-TSC (5:1), suggesting blocklike features in PANI-TSC. High aspect ratio in PANI-PEDOT:PSS indicated platelike features. PANI-PEG possessed both plate and block feature with aspect ratio in between PANI-PEDOT:PSS and PANI-TSC. Thus, though films could be of direct relevance to applications, we could compare and correlate the impact of the template on PANI on powder compactions forming microstructures with blocklike semicrystalline features, not reported so far on PANIs.

14.4 CONCLUSIONS

Effect of polymer (PEDOT:PSS and PEG) and small molecule (TSC) as counterions on PANI has been successfully investigated through simple facile chemical synthesis on powder compactions as composite materials. Powder XRD characterization indicated glassy/crystalline nature in the samples. PANI-PEDOT:PSS exhibited higher conductivity. DSC suggests crystalline transformation in PANI-PEG and PANI-PEDOT:PSS. SEM reveals blocklike morphologies not reported among the PANIs that have been studied so far. Detailed characterization of all the samples suggests better organization and symbiotic interactions between the π-donor molecular ions and the polyionic polymers leading to an improved polymer chain alignment. All samples formed stable colloidal suspensions. Investigation on spin-coated films obtained from the stable colloidal suspensions could lead to their direct applicability in electronic devices. Preliminary investigations on the mechanical properties of these blends reveal better tensile strength in PANI-TSC.

ACKNOWLEDGMENT

We thank financial support from the Department of Science and Technology, New Delhi, India, under Fast Track Scheme. We gratefully acknowledge the support of Prof. T. P. Radhakrishnan, School of Chemistry, University of Hyderabad, for providing his laboratory facility for conductivity measurements.

KEYWORDS

- composites
- conducting polymers
- differential scanning calorimetry
- morphology
- nanoparticles

REFERENCES

1. Trivedi, D. C.; Nalwa, H. S. *Handbook of Organic Conductive Molecules and Polymers;* John Wiley: London, 1997; Vol. 2, pp.509–511.
2. Metzger, R. M.; Day. P.; Papavassilou, G. C. *Low Dimensional Metals and Molecular Electronics;* NATO, ASI Series, Plenum Press: New York, 1990.
3. Huang, J.; Virji, S.; Weiller, B. H.; Kaner, R. B. *J. Am. Chem. Soc.* **2003,** *125,* 314–315.
4. Abdulla, H. S.; Abbo, A. I. *Int. J. Electrochem. Sci.* **2012,** *7,* 10666–10678.
5. Michaelson, J. C.; Mc Evoy, A. J.; Kuramoto, N. *React. Polym.* **1992,** *17,* 197–206.
6. Parthasarathy, R. V.; Martin, C. R. *Chem. Mater.* **1994,** *6,* 1627–1632.
7. Planèes, J.; Cheguettine, Y.; Samson, Y. *Syn. Metals* **1999,** *101,* 789–790.

8. Zhang, Z.; Wei, Z.; Wan, M. *Macromolecules* **2002,** *35,* 5937–5942.

9. Jayanty, S.; Prasad, G. K.; Sridhar, B.; Radhakrishnan, T. P. *Polymer* **2003,** *44,* 7265–7270.

10. Li, G.; Zhang, Z. *Macromolecules* **2004,** *37,* 2683–2685.

11. Wu, C. G.; Bein, T. *Science* **1994,** *264,* 1757–1758.

12. Martin, C. R. *Chem. Mater.* **1996,** *8,* 1739–1746.

13. Wang, C. W.; Wang, Z.; Li, M. K.; Li, H. L. *Chem. Phys. Lett.* **2001,** *341,* 431–434.

14. Wei, Z.; Zhang, Z.; Wan, M. *Langmuir.* **2002,** *18,* 917–921.

15. Huang, L.; Wang, Z.; Wang, H.; Cheng, X.; Mitra, A.; Yan, Y. *J. Mater. Chem.* **2002,** *12,* 388–391.

16. Karami, H.; Mousavi, M. F.; Shamsipur, M. *J. Power Sourc.* **2003,** *124,* 303–308.

17. Qui, H.; Wan, M. *J. Polym. Sci. Part A: Polym. Chem.* **2001,** *39,* 3485–3497.

18. Laska, J.; Widlarz, J. *Syn. Metals* **2003,** *135,* 261–262.

19. Li, D.; Kaner, R. B. *Chem. Commun.* **2005,** 3286–3288.

20. Laska, J. *J. Mol. Struc.* **2004,** *701,* 13–18.

21. Bae, W. J.; Jo, W. H.; Park, Y. H. *Synth. Met.* **2003,** *132,* 239–244.

22. Chen, C.; LaRue, J. C.; Nelson, R. D.; Kulinsky, L.; Madou, M. J. *J. Appl. Polym. Sci.* **2012,** *125,* 3134–3141.

23. Reed, E. W. Thesis submitted for the partial fulfillment of Master of Engineering University of Louisville, 2009

24. Sanches, E. A.; Lost, R. M.; Soares, J. C.; Zucolotto, V.; Mascarenhas, Y. P. *Reunião Annual da SociedadeBrasileria de Quimica.* 34.

25. Hrehorova, E.; Rebros, M.; Pekarovikova, A.; Fleming, P. D.; Bliznyuk, V. N. *Taga J.* **2008,** *4,* 219–231.

26. Perinka, N.; Kim, C. H.; Kaplanova, M.; Bonnassieux, Y. *Phy. Procedia.* **2013,** *44,* 120–129.

27. Shi, H.; Liu, C.; Xu, J.; Song, H.; Jiang, Q.; Lu, B.; Zhou, W.; Jiang, F. *Int. J. Electrochem. Sci.* **2014,** *9,* 7629–7643.

28. Pickup, P. G.; Kean, C. L.; Lefebvre, M. C.; Li, G. C.; Qi, Z. Q.; Shan, J. N. *J. Newmat. Electrochem. Sys.* **2000,** *3,* 21–26.

29. Wang, P.; Tan, K. L. *Chem. Mater.* **2001,** *13,* 581–587.

30. Yan, L.; Tao, W. *J. Polym. Sci. Part A: Polym. Chem.* **2008,** *46,* 12–20.

31. Gong, C.; Deng, F.; Tsui, C.P.; Xue, Z.; Ye, Y. S.; Tang, C. Y.; Zhou, X.; Xie, X. *J. Mater. Chem. A,* **2014,** *2,* 19315–19323.

32. Singh, A.; Singh, N. P.; Singh, R. A. *Bull. Maert. Sci.* **2011,** *34,* 1017–1026.

33. Saxena, R.; Shaktawat, V.; Jain, N.; Saxena, N. S.; Sharma, K.; Sharma, T. P. *Ind. J. Pure App. Phys.* **2008,** *46,* 414–416.

34. Pouget, J. P.; Jozefowick, M. E.; Epstein, A. J; Tang, X.; MacDiarmid, A. G. *Macromolecules.* **1991,** *24,* 779–789.

35. Leff, D. V.; Brandt, L.; Heath, J. R. *Langmuir.* **1996,** *12*(20), 4723–4730.

36. Peng, H.; Ma, G.; Ying, W.; Wang, A.; Huang, H.; Lei, Z. *J. Power Source.* **2012,** *211,* 40–45.

37. Yoo, D.; Kim, J.; Kim, J. H. *Nano Res.* **2014,** *7*(5), 717–730.

38. Bagheri-Mohagheghi, M.; Shokooh-Saremi, M. *Semicond. Sci. Technol.* **2004,** *19*(6), 764–769.

39. Li, D.; Fang, X.; Deng, Z.; Zhou, S.; Tao, R.; Dong, W.; Wang, T.; Zhao, Y.; Meng, G. *J. Phys. D: App. Phys.* **2007,** *40,* 4910–4915.

40. Cardoso, M. J. R.; Lima, M. F. S.; Lenz, D. M. *Mater. Res.* **2007,** *10*(4), 425–429.
41. Hosseini, H. S.; Entezami, A. A. *Iran. Polym. J.* **2005,** *14*(3), 201–209.
42. Kumar, D.; Chandra, R. Indian. *J. Eng. Mater. Sci.* **2001,** *8*, 209–214.
43. Liu, W.; Cholli, A.L.; Nagarajan, R.; Kumar, J.; Tripathy, S.; Bruno, F. F.; Samuelson, L. *J. Am. Chem. Soc.* **1999,** *121*, 11345–11355.
44. Basavaraja, C.; Kim, J. K.; Thinh, P. X.; Huh, D. S. *Polym. Compos.* **2012,** *33*, 1541–1548.

CHAPTER 15

EFFECT OF PROCESSING PARAMETERS ON PHYSICOMECHANICAL PROPERTIES OF VISIBLE LIGHT CURE COMPOSITES

P. P. LIZYMOL* and C. VIBHA

Biomedical Technology Wing, Sree Chitra Tirunal Institute for Medical Sciences and Technology, Poojappura, Thiruvananthapuram 695012, Kerala, India

Corresponding author. E-mail: lizymol@rediffmail.com

CONTENTS

ABSTRACT

Polymeric composites are usually used as tooth-colored restorative materials. Composite restorative materials represent one of the many successes of modern biomaterials research, since they replace biological tissue in both appearance and function. The current state of the art of dental composites includes a wide variety of materials with a broad range of mechanical properties, handling characteristics, and aesthetic possibilities. Although the commercially available dimethacrylate-based composite materials have become vital for dental restorations due to their superior aesthetic quality, they experience a considerable mechanical challenge during function. The polymerization rate can regulate the cross-linking density, which is very important because a reduction in the effective cross-linking density of a cured resin will lead to a decrease in its mechanical strength, solvent resistance, and glass transition temperature. Thus, improving the mechanical properties is one of the most important research tasks in this field. For restorations that are subjected to large masticatory stresses, high flexural strength is desired. The main focus of the present work is to synthesize novel inorganic–organic hybrid resins containing alkoxides of zinc with polymerizable methacrylate groups. The effect of inorganic content present in the resin on the physicomechanical properties of the visible light cured composite was evaluated.

15.1 INTRODUCTION

With increasing aesthetic demands in modern dentistry, the use of dental composites has increased tremendously. The ideal restorative material should be identical to the natural tooth structure, both in appearance and strength. The expanded use of these materials in a wide range of applications puts great demands on their properties and performance. Dental composites habitually consist of organic phase, inorganic phase, and a coupling agent. Coupling agent binds the dispersed glass or silica filler with the resin-based restorative material. Commercially available composites are based on dimethacrylate organic resin matrix. The organic matrix is formed by free radical polymerization of dimethacrylates, which are nontoxic and capable of rapid polymerization in the presence of oxygen and water, because the restorations are polymerized in situ in a tooth cavity. Several investigations were undertaken to optimize the polymerization rate, mechanical properties, etc. of these monomers.[1,2] Photopolymerization increases molecular weight by polymer conversion and cross-linking of existing macromolecules.

Increasing the conversion of double bonds during the photopolymerization is critical for the optimization of mechanical properties, biocompatibility, and color stability of light-activated dental resins. The mechanical strength of dental resin-based composites is reliant upon the complex intraoral forces such as compressive, tensile, and shear introduced during mastication and it has a significant influence on the performance of dental restorations. Various static-load-to-failure strength testing techniques, that is, compressive, diametral tensile, and flexure (bending) have been employed for the determination of the mechanical strength of resin based composites (RBCs).[3] The appropriate polymerization of resin composites is a crucial factor in the clinical performance of restorations. Several problems can be associated with inadequate polymerization, such as reduced physical properties, solubility in the oral environment, microleakage, and pulp irritation. The process of composite resin polymerization occurs through the conversion of the monomer molecules into a polymer network, accompanied by the narrowing of the space between the molecules which causes polymerization shrinkage in the composite. If inadequate levels of conversion are achieved in the polymerization, mechanical properties and wear performance can be compromised; amount of leachable residual monomer increase and color stability may decline. However, if conversion is maximized to reduce these difficulties, then problems inherent polymerization shrinkage of the composite becomes more critical. Recent works in dental biomaterial research targets on the development of bioactive restorations such as development of novel dental cements, innovative light-curable therapeutic resin-based restorative materials to promote mineral precipitation in mineral-depleted dental hard tissues at the bonding interface, etc. by compromising the mechanical properties.[4] However, to date, there have been no published studies of dental restorative composites containing bioactive inorganic organic hybrid resin as the matrix. Our previous studies reported that use of inorganic–organic hybrid resins containing alkoxides of calcium in dental restoratives can diminish polymerization shrinkage and improve wear resistance.[5] This inspired us to synthesize novel inorganic–organic hybrid resins containing alkoxides of zinc with polymerizable methacrylate groups. Alkoxides of zinc was selected due to its reported ability to reduce enamel solubility with other metal ions and it can modify crystal growth of the calcium phosphates which results in remineralization.[6,7] The inorganic content in the resin has a direct influence on the structural variation of these resins. In the present study, we investigated the effect of inorganic content on physicomechanical properties of visible light cure dental composites based on these synthesized resins.

15.2   MATERIALS AND METHODS

15.2.1   MATERIALS

The materials used in this study for the synthesis of resin were 3-trime-
thoxysilyl propyl methacrylate, triethylene glycol dimethacrylate, (Aldrich
Chem. Co. Milwaukee), Laboratory Rasayan (LR) grade zinc acetate, AR
grade sodium hydroxide, LR grade diethyl ether (S.D. Fine Chemicals,
Mumbai, India).

15.2.2   METHODS

15.2.2.1   SYNTHESIS OF INORGANIC–ORGANIC HYBRID RESINS

Inorganic–organic hybrid resins with polymerizable methacrylate group
were synthesized through modified sol-gel technique,[8] using 3-trimethoxy
silyl propyl methacrylate (TSPM) as the starting material. The inorganic part
consists of alkoxide of zinc.

15.2.2.2   CHARACTERIZATION OF INORGANIC–ORGANIC HYBRID RESINS

Synthesized resins were characterized using Fourier transform infrared spectroscopy (FTIR). FTIR spectra of TSPM, hydrolyzed silane (bare resin), and
zinc-incorporated hydrolyzed silane were recorded using a JASCO FTIR
spectrophotometer (Model 6300, Japan; resolution 0.07 cm 21) by applying
a thin layer on NaCl cell.

15.2.2.3   PREPARATION OF DENTAL COMPOSITES

Dental composites were prepared using the modified procedure which was
previously patented.[9] The resin matrix, filler, diluent, initiator, and accel-
erator were masticated in an agate mortar to get a uniform paste was packed
in to the mold and exposed to visible light for duration of 60 s on both sides
using Prolite (Caulk/ Dentsply), which is used to polymerize visible-light-
activated materials.

15.2.2.4   EVALUATION OF DENTAL COMPOSITES

15.2.2.4.1   Polymerization Shrinkage

The internal diameter of the mold was calculated accurately by a digital venire caliper (Mitutoyo, Japan). Six measurements were taken in all directions and the mean value was calculated. Samples were preparation is same as that for diametral tensile strength (DTS) as per American Dental Association (ADA) specification.[10] The sample was then taken from the mold and the diameter of the cured sample was measured at six points and the mean value was calculated. The percentage shrinkage can be calculated from the equation given below:

Polymerization shrinkage = (diameter of ring − diameter of composite/ diameter of ring) × 100.

15.2.2.4.2   Diametral Tensile Strength (DTS)

After measuring Vickers hardness (VHN), the same samples were used for the measurement of DTS because hardness measurement is a nondestructible property. Samples were prepared as per ADA specification. The DTS was evaluated using the Universal Testing Machine (Instron, Model 3363, UK). The load at which break occurs was noted, and DTS was calculated using the following equation.

$$DTS \text{ (MPa) } 2P \mathbin{/} \pi DL,$$

where P—the load in newtons, D—the diameter, and L—the thickness of the specimen in mm. Mean and standard deviation of six samples were calculated.

15.2.2.4.3   Flexural Strength

Flexural strength (FS) test specimens were prepared as per ISO (International Organization for Standardization) specification No. 4049–2000 (E).[11] The FS was evaluated using Universal Testing Machine (Instron, Model 3363, UK). Load at break was noted and FS was calculated using the following equation:

$$FS \text{ (MPa) } FL \mathbin{/} 2bd^2,$$

where F—the load at break in newtons, L—length of the specimen between two metal rods at the base plate in mm, b—width of the specimen in mm, and d—depth of the specimen in mm.

15.2.2.4.5 Depth of Cure

Depth of cure of the specimens was measured as per ISO specification No. 4049—2000 (E). Samples for depth of cure were prepared using a brass molds with 3 mm diameter and 6 mm depth. Cure one side of the sample with light source at 60 min. and scrap out the uncured potion. After that the composite was removed from the mold and measured the length of cured composite using vernier calipers.

15.3 RESULTS AND DISCUSSION

TSPM exhibited peaks at 1165 cm^{-1} and 1719 cm^{-1} that indicates Si–O–CH$_3$ bond and C=O group, respectively (Fig 15.1). Characteristic peaks of O–Si–O bond, and aliphatic C–H group were seen at the range of 1000–1200 cm^{-1} and 2800–3000 cm^{-1}. A peak observed at 1636 cm^{-1} indicates the presence of C = CH$_2$ group. After hydrolysis, broad spectrum bands were observed around 3479 cm^{-1} which indicates the presence of Si–OH peak for bare and zinc-containing resin. Two spectra showed the characteristics peaks of silane groups around 1000–1200 cm^{-1}, which indicates the presence of O–Si–O bond and typical carboxyl group peak (1716–1725 cm^{-1}) and methacrylate group peak (1636 cm^{-1}).

Composite materials based on Bis-GMA experienced a substantial mechanical encounter during function despite the fact they have superior aesthetic quality. So improving the mechanical properties is one among the essential research tasks in dentistry. FS of the composite designates the quantity of flaws within the material which are prospective to cause catastrophic failure when subjected to loading. Another material property for characterizing dental composite is DTS because low tensile strength leads to early let-down of the performance of the materials. DTS of the composites was comparable and it does not show any significant change as moved from Zn0 to Zn0.4. The visible light cure dental composites exhibited gradual increment in FS as the concentration of inorganic content was increased from 0 to 0.4 (Fig. 15.2). It implies that the inorganic content has a substantial impact in imparting better properties to the composites. The effective bonding between the resin matrix and the inorganic filler may be the reason for better mechanical properties.

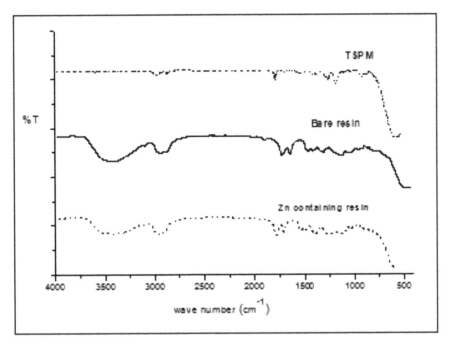

FIGURE 15.1   Overlay of FTIR spectra of TSPM, bare resin, and Zn containing resin.

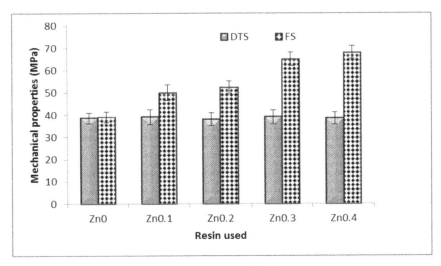

FIGURE 15.2   Effect of inorganic content on mechanical properties.

Polymerization shrinkage, marginal leakage, and nonbioactivity beset with Bis-GMA limit its application and switch the dental research to look forward for other monomers. Possible approaches to increasing the longevity of restorations is to reduce polymerization shrinkage and to promote remineralization of tooth structure.[12]

Polymerization shrinkage of Zn0.4 resin-based visible light cure dental composites was lower with highest depth of cure it also implies that the effective bonding of inorganic filler with resin matrix and between the organic and inorganic parts within the resin (Fig. 15.3).

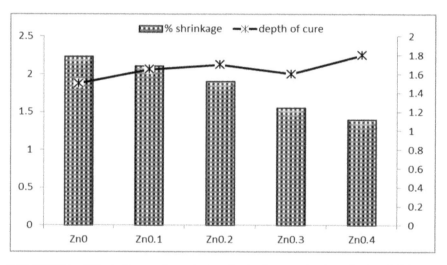

FIGURE 15.3   Effect of inorganic content on depth of cure and polymerization shrinkage.

15.4   CONCLUSION

The selection of alkoxide and specific organic monomers having functional groups responsible for effective chemical bond formation and enhanced properties of the inorganic–organic hybrid materials is important. The better properties may be due to the effective bonding between the organic and inorganic parts within the resin. We propose that the presence of inorganic content in the resin induces structural and compositional changes within the resin, which enable the dental composites to have better mechanical properties as it is subjected to many masticatory forces.

ACKNOWLEDGMENT

Financial support as a project grant from Kerala State Council for Science, Technology and Environment, Government of Kerala, India is gratefully acknowledged. We thank Dr. V. Kalliyana Krishnan, SIC, Dental Products Laboratory, Head, Biomedical Technology Wing, and Director, SCTIMST for extending the facilities in BMT Wing.

KEYWORDS

- resins
- dental
- composites
- restorative
- visible light curing

REFERENCES

1. Ilie, N.; Hickel, R.; Valceanu, A. S. Huth, K. C. Fracture Toughness of Dental Restorative Materials. *Clin. Oral. Invest.* **2012**, *16*, 489–498.
2. Manhart, J.; Kunzelmann, K. H.; Chen, H. Y.; Hickel, R. Mechanical Properties of New Composite Restorative Materials. *J. Biomed. Mater. Res. (Appl. Biomater.).* **2000**, *53*, 353–361
3. Chung, S. M.; Yap, A. U. J.; Chandra, S. P.; Lim, C. T. Flexural Strength of Dental Composite Restoratives: Comparison of Biaxial and Three-Point Bending Test. *J. Biomed. Mater. Res. Part B: Appl. Biomater.* **2004**, *71* B, 278–283.
4. Sauro, S.; Osorio, E.; Watson, T. F.; Toledano, M. Therapeutic Effects of Novel Resin Bonding Systems Containing Bioactive Glasses on Mineral-Depleted Areas within the Bonded-Dentine Interface. *J. Mater. Sci. Mater. Med.* **2012**, *23*, 1521–1532.
5. Lizymol, P. P. Studies on New Organically Modified Ceramics Based Dental Restorative Resins. *J. Appl. Polym. Sci.* **2010**, *116*, 509–517.
6. Santiago, B. M.; Ventin, D. A.; Primo, L. G.; Barcelos, R. Microhardness of Dentine Underlying ART Restorations In Primary Molars: An In Vivo Pilot Study. *Br. Den. J.* **2005**, *2*, 199.
7. Banerjee, A.; Sherriff, M.; Kidd, E. A. M.; Watson, T. F. A Confocal Microscopic Study Relating the Autofluorescence of Carious Dentine to its Microhardness. *Br. Den. J.* **1999**, *12*, 187.
8. Pampadykandathil, L. P.; Chandrababu, V. A Process for the Synthesis of Inorganic-Organic Hybrid Resins Comprising of Alkoxides or Mixture of Alkoxides of Calcium,

Magnesium, Zinc, Strontium, Barium and Manganese with Polymerisable (Di/Tetra) Methacrylate Groups. Patent Application Number, 4027/CHE/2014.

9. Pampadykandathil, L. P.; Chandrababu, V. A Visible Light Cure Dental Restorative Composite with Excellent Remineralization ability with Good Physico- Mechanical Properties and Low Shrinkage Based on a Novel Calcium Containing Inorganic Organic Hybrid Resin with Polymerizable Methacrylate Groups. Patent Application Number, 4996/CHE/2014.

10. American National Standard. American Dental Association Specification n. 27 for Resin-based filling materials; 1993.

11. International Organization for Standardization, ISO 4049 Dentistry—Polymer-Based Filling, Restorative and Luting Materials, 4th ed., 2009.

12. Sauro, R.; Osorio, S.; Fulgencio, R.; Watson, T. F.; Cama, G.; Thompsona, I.; Toledanob, M. Remineralisation Properties of Innovative Light-Curable Resin-Based Dental Materials Containing Bioactive Micro-fillers. *J. Mater. Chem. B.* **2013,** *20*, 2624–2638.

CHAPTER 16

EFFECT OF CARBON BLACK ON MICROSTRUCTURE AND MACROMOLECULAR BEHAVIOR OF NITRILE BUTADIENE RUBBER/ THERMOPLASTIC POLYURETHANE COMPOSITES

MOHAMMED SAHAL[1*] and G. UNNIKRISHNAN[2]

[1]*Department of Mechanical Engineering, NIT, Calicut 673601, Kerala, India*

[2]*Polymer Science and Technology Laboratory, NIT, Calicut 673601, Kerala, India*

Corresponding author. E-mail: mohammedsahal160@gmail.com

CONTENTS

ABSTRACT

Polyurethane (PU) was blended with nitrile butadiene rubber (NBR) (up to 100 phr) and the blends were cured with dicumyl peroxide. The blends are cured at a suitable temperature according to their blend ratio. These blends are considered to be cross-linked networks (interpenetrating networks— IPN) since all the studies were done on the mold of the blends that had been cured at a suitable temperature. The curing and mechanical characteristics of the NBR were investigated as a function of PU loadin in the presence and absence of carbon black (CB). The addition of PU to NBR increased the cure rate of both unfilled and CB-filled NBR. The mechanical properties of the blends were substantially enhanced in the presence of CB. There was a fivefold increase in the tensile and tear properties of the NBR/PU blends in the presence of CB and a notable increase in the Shore A hardness. Thermogravimetric analysis (TGA) of the blends was also performed. The hold temperature was up to 600°C. The results indicated an increase in the thermal stability of the NBR/PU blends in the presence of CB. An investigation of the SEM micrographs of NBR containing PU indicated phase separation at the micro level. The addition of CB to the NBR/PU blends homogenized the blend morphology to a certain extent and reduced the phase separation, which could be cited as a primary reason for the enhancement of the mechanical properties and thermal stability of the NBR/PU blends.

16.1   INTRODUCTION

Developing nitrile butadiene rubber (NBR) blends with sufficient thermal stability and mechanical strength is of industrial importance, especially for the application of O-rings. NBR is the most widely used rubber in industry as a sealant and especially for the lip seals of ball bearings due to its moderate cost, excellent resistance to oils, fuels and greases, processability, and very good resistance to swelling by aliphatic hydrocarbons.[1-6] Polyurethane (PU) exhibits high stiffness and strength.[7-9] Several attempts have been made previously to enhance the mechanical properties of the NBR. The blending together of natural rubber (NR) and NBR was intended to produce a vulcanizate with the best properties from each component, i.e., NBR's high resistance to swelling by oils and NR's good strength properties but it did not yield the expected results.[10-12] Carbon black (CB) is the most important reinforcing filler for rubber, and its reinforcement plays an important role in enhancing the mechanical properties.[13–18] Studies indicated the effect of CB on rubber in increasing the mechanical properties. Fu et al. studied the reinforcing properties of CB on NR. Degrange et al. suggested the use of NBR with CB for oil seal applications, especially lip seals. This study is devoted to understanding the effect of PU loading on the curing and mechanical and thermal properties of NBR in the presence and absence of CB.

16.2   EXPERIMENTAL

16.2.1   MATERIALS

Polyether polyurethane rubber (E34 millathane) was purchased from West Coast Polychem Ltd, Mumbai. NBR (DN202) was supplied by Ashok Rubber Industries Kottayam, India. The vulcanizing agent used, dicumyl peroxide (DCP), was of commercial grade. The CB (ISAF) was supplied by Intermix Plant, Industrial Estate, Kottayam.

16.2.2   PREPARATION OF SAMPLE

The mixing was performed in a two-roll mixing mill (150 mm × 300 mm) with a nip gap of 1.3 mm and at a friction ratio of 1:1.4 at room temperature. The temperature was maintained constant by circulating water through the rolls. The vulcanizing agent, DCP, was incorporated as per ASTM D15-627. The blend compositions were named P0, P40, P80, P100, P120, P160, and P200 for compositions without carbon and from PC0 to PC200 for compositions with carbon, where P represents polyurethane and C represents carbon black. The number represents the weight amount of PU in the blend, varying from pure NBR (P0) to pure PU (P200). In addition, 5 g of DCP was added to all of the formulations. The formulations are listed in Tables 16.1 and 16.2.

TABLE 16.1   Formulations of Blend without CB.

Recipe	Formulation used for vulcanization (phr)
NBR	100, 80, 60, 50, 40, 20, 0
PU	0, 20, 40, 50, 60, 80, 100
DCP	2.5

TABLE 16.2 Formulations of Blend with CB.

Recipe	Formulation used for vulcanization (phr)
NBR	100, 80, 60, 50, 40, 20, 0
PU	0, 20, 40, 50, 60, 80, 100
CB	30
DCP	2.5

16.2.3   CURE CHARACTERISTICS

The cure characteristics of the blends were determined using a Monsanto R-100 rheometer at a rotational frequency of 100 rpm. The mixes were cured at 160°C on a hydraulic press along the mill grain direction under a pressure of 6.7 MPa (mold dimensions: 150 mm × 150 mm × 2 mm).

16.2.4   MORPHOLOGY

The blend samples for scanning electron microscopy (SEM) were prepared by cryogenically fracturing the samples in liquid nitrogen. The samples were sputter coated with gold, and images were taken using a field-emission scanning electron microscope (FESEM) (variable pressure Hitachi SU6600 FESEM).

16.2.5   MECHANICAL PROPERTIES

Mechanical properties, such as the tensile strength, elongation at break, modulus of elasticity, and tear strength, were examined on a universal testing machine (series IX Automated Materials Testing System 1.38, model-441, Instron Corporation, USA) at a cross head speed of 500 mm/min and at 252°C. The tensile properties of the blends were examined according to ASTM D-412. The experimental conditions and equipment for the tear measurements were the same as those for the tensile testing. The tear testing was conducted as per ASTM D-624 using 90° angle test pieces.

16.2.6   THERMAL ANALYSIS

Thermogravimetric analysis (TGA) was performed to assess the thermal stability (weight loss with respect to temperature change) of the blends with a TGA-50 thermogravimetric analyzer (Shimadzu, Japan). All of the samples were kept in a TGA furnace at 600°C hold temperature in an air atmosphere for 30 min to remove water content in the samples before the TGA characterization. The typical heating rate was 10°C/min in a platinum cell.

16.2.7   MATHEMATICAL MODELING OF TENSILE STRENGTH

Mathematical equations can be used for the prediction of technological information from experimental data. There are two common methods for obtaining an equation to fit experimental data. The first method is the derivation of an equation from the experimental data, and the second method is fitting an established equation with experimental data. The latter approach is adopted in this study.

16.2.7.1   EINSTEIN EQUATION

For the tensile property of matrices with perfect interaction between the phases, Einstein proposed the following equation:

$$Mc = Mm \, (1 + 1.25 Vf), \tag{16.1}$$

where Mc is the tensile strength of the concerned matrix, Mm is the tensile strength inclusion of the second phase, and Vf is the volume fraction of the inclusion.

16.2.7.2   MOONEY EQUATION

The Mooney equation considers the interaction of the strain field around the two phases. The Einstein equation is applicable at low volume fraction of the compatibilizer, while the Mooney equation is able to represent the data at higher volume fractions and is given by the following equation:

$$Mc = Mm \left(\frac{\exp 2.5(Vf)}{1 - SVf} \right), \tag{16.2}$$

where S is the crowning factor.

16.2.7.3   BRODNYAN EQUATION

Brodnyan modified the Mooney equation by adding an aspect ratio using the following equation:

$$Mc = Mm \, \exp(2.5Vf + 0.407 \, (p-1)^{1.508} \, Vf), \hspace{1cm} (16.3)$$

where p is the aspect ratio $1 < p < 15$.

16.2.7.4   GUTH EQUATION

Guth modified the Einstein equation for the perfect bonding between phases as

$$Mc = Mm(1 + 2.5Vf + 14.1(Vf^2)) \hspace{1cm} (16.4)$$

16.2.7.5   KERNER EQUATION

The Kerner equation is given as

$$Mc = Mm\left(1 + \frac{Vf15(1-Vm)}{Vm(8-10Vm)}\right) \hspace{1cm} (16.5)$$

16.3   RESULTS AND DISCUSSION

16.3.1   CURING BEHAVIOR

16.3.1.1   EFFECT OF PU LOADING ON CURE TIME

All of the formulations were cured in the presence of DCP. phr indicates parts per hundred rubber. The addition of PU to NBR gradually reduces the cure time; however, for the blend composition of 50:50 NBR/PU (P100), the cure time increases and then decreases except for pure PU. The rheographs of the cured samples without CB are presented in Figure 16.1. Insight into the cross-linking within a phase and the cross-linking between the two phases can be obtained from the cure characteristics and mechanical properties of the rubbers. Insufficient cross-linking between the phases occurs

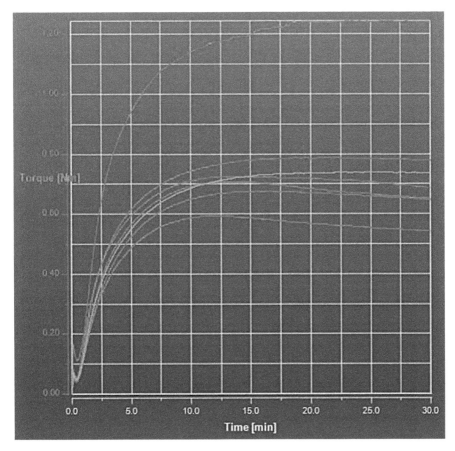

FIGURE 16.1   Rheographs of NBR/PU blends without CB.

if the two rubbers do not mix properly, which will be reflected in the poor mechanical properties. The cure temperature was 160°C for all the blends. All the blends prepared were a mixture of synthetic rubbers. There was no substantial difference in the cure time when PU was loaded with NBR. An important finding was that the cure time decreases when PU was loaded with NBR up to 50 phr, which indicates that the NBR/PU blends containing less than 50 phr were more processable than pure NBR cured with DCP. The cure time increased when more than 50 phr PU was incorporated into the NBR matrix. The presence of more than 50 phr PU in the NBR matrix causes difficulty in the formation of cross-links. The cure results of the blends with and without CB are presented in Table 16.3.

TABLE 16.3 Cure Characteristics of Blends with and without CB.

Cure characteristics	NBR/PU blends without CB						
Sample name	P0	P40	P80	P100	P120	P160	P 200
Cure time, (t90) (min)	8.95	8.20	6.66	7.77	7.09	6.87	9.56
Scorch time (min)	4.99	3.92	6.7	5.46	4.52	3.93	2.36
Alpha parameter	–	6.76	−20.27	−8.10	−4.05	−1.35	67.57
CRI (s^−1)	13.59	14.74	20.12	16.42	17.89	17.86	12.16
Maximum torque (dNm)	0.74	0.79	0.59	0.68	0.71	0.73	1.24
Cure characteristics	**NBR/PU blends with CB**						
Sample name	PC0	PC40	PC80	PC100	PC120	PC160	PC200
Cure time, (t90) (min)	9.22	9.21	7.09	8.63	8.74	8.91	7.62
Scorch time (min)	1.78	2.8	6.61	9.16	2.91	4.87	4.76
Alpha parameter	–	−18.31	−53.67	−52.20	−26.47	−40.44	−63.01
CRI (s^−1)	12.04	12.57	17.89	15.36	13.42	14.2	16.77
Maximum torque (dNm)	1.36	1.09	0.63	0.65	1	0.81	0.73

16.3.1.2 EFFECT OF PU LOADING ON CURE TIME OF NBR WITH CB

The addition of CB to the blends slightly increased the cure time. However, there was a visible reduction in the cure time for NBR with 40 phr PU. The scorch time was maximum for the blend with 50 phr PU, indicating a better reinforcing effect for this blend. However, there was only small variation in the cure time with the addition of CB. The cure results with the addition of CB in the blends are presented in Table 16.3. The rheographs of the blends are presented in Figure 16.2. The slight increase in cure time is attributed to the presence of CB. CB induces a more pronounced reinforcement of PU on NBR. It should be noted that the CB does not affect the processability of the blend negatively but produces a better reinforcing effect. In practical sense, this slight increase is not significant when the material is processed for engineering applications. However, providing a better reinforcement enhances the mechanical properties.

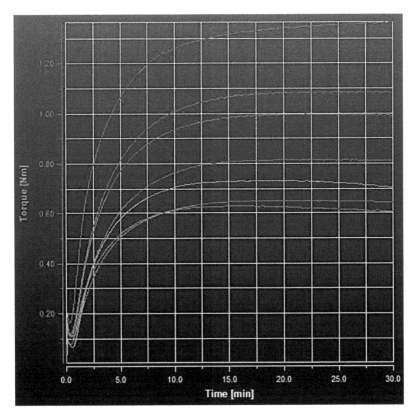

FIGURE 16.2   Rheographs of NBR/PU blends with CB.

16.3.1.3   EFFECT OF CB ON CRI, SCORCH TIME, AND ALPHA PARAMETER

The cure rate index (CRI) is the measure of activation of the cure reaction during the curing process. The CRI was reduced for the pure NBR blend containing CB but increased for PU loading up to 40 phr. Loading of PU alone up to 40 phr makes the NBR more processable. The CB addition slightly reduced the CRI of the blends; however, the CRI was still greater than that of pure NBR cured with DCP. The reinforcement potential and scorch time are measures of the reinforcement of a foreign particle in a base matrix; an increased scorch time indicates good reinforcement properties. The CB addition yields enhanced reinforcement up to 40 phr, as indicated by the scorch time graph. The PU loading in NBR provides good reinforcement without CB. The CRI is given by

$$CRI = \frac{100}{T90 - Ts2},\tag{16.6}$$

where $T90$ is the cure time and $Ts2$ is the induction time.

The alpha parameter is the reinforcement potential and is given by

$$\alpha = \frac{D\max - D^0 \max}{D\max},\tag{16.7}$$

where $D\max$ is the maximum torque of blended material and $D^0\max$ is the maximum torque of pure material without additive. The cure time, CRI, scorch time, and alpha parameter (reinforcement potential) of the blends with and without CB are listed in Table 16.3. Alpha parameter cannot be calculated for the initial values as maximum and minimum values are same. A comparison of the cure time, CRI, scorch time, and alpha parameter of the blends with and without CB is presented in Figures 16.3–16.6.

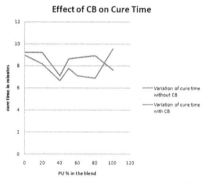

FIGURE 16.3   Comparison of cure time of blends with the addition of CB.

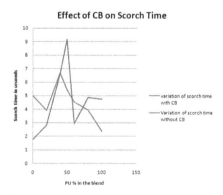

FIGURE 16.4   Comparison of scorch time of blends with the addition of CB.

16.3.2   MECHANICAL BEHAVIOR

16.3.2.1   EFFECT OF PU LOADING ON TENSILE PROPERTIES

The mechanical properties of NBR increased for up to 50 phr loading of PU with the maximum increase being obtained at 40 phr loading. The properties do not further improve with further PU loading. The tensile property increase may be attributed to the reinforcing effect of the PU, as is evident from the increase in the scorch time up to 40 phr. However, the reinforcing abilities do not improve with further addition of PU. The percent elongation also exhibits the same trend.

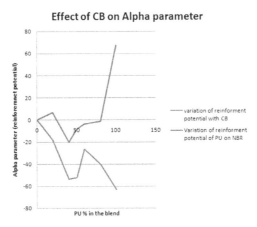

FIGURE 16.5   Comparison of alpha parameter of blends with the addition of CB.

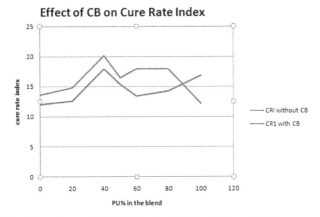

FIGURE 16.6   Comparison of CRI of blends with the addition of CB.

The percent elongation increased up to 40 phr and decreased with further loading. The modulus at 100% elongation also follows the same trend. The tensile properties of the blends are presented in Table 16.4.

TABLE 16.4 Tensile Properties of Blends with and without CB.

Sample name	Tensile strength	% Elongation	Modulus at 100% elongation
P0	0.45	53.4	–
P40	1.2	115.87	1.070
P80	1.39	124.83	1.239
P100	1.23	101.41	1.070
P120	1.11	91.44	0.713
P160	0.83	60.19	0.5
P200	–	–	–
PC0	2.42	21.21	–
PC40	3.97	45.19	–
PC80	4.8	132.19	3.085
PC100	6.44	198.9	1.977
PC120	5.62	113.84	4.508
PC160	2.16	115.02	1.768
PC200	1.14	58.14	–

16.3.2.2 EFFECT OF PU LOADING ON TEAR PROPERTIES

The tear strength increased initially for the PU loading of 20 phr, which was the maximum and then gradually decreased up to 50 phr. The sudden increase at 20 phr may be attributed to the piling up of the PU particles in the NBR matrix. However, the tear strength of pure NBR vulcanizate was less than that of all of the compositions. An important finding was that the small inclusions impart more tear strength. The blend containing 20 phr NBR with 80 phr PU exhibited the maximum tear strength. The blend containing 20 phr PU in 80 phr NBR also exhibited very good tear strength compared with the other blends. The tear results are presented in Table 16.5.

TABLE 16.5 Tear Results of NRB/PU Blends with and without CB.

Sample name	Tear strength (N/mm)
P0	2.54
P40	6.09
P80	5.04
P100	5.86
P120	5.33
P160	9.06
P200	1.73
PC0	32.04
PC40	25.07
PC80	24.85
PC100	24.52
PC120	23.11
PC160	16.64
PC200	20.54

16.3.2.3 EFFECT OF PU LOADING ON HARDNESS

The hardness increased gradually and did not decrease up to 80 phr with an increase of phr of PU. The blends exhibited higher hardness than pure PU vulcanizate and pure NBR vulcanizate. The hardness results are presented in Table 16.6.

TABLE 16.6 Hardness Results of NBR/PU Blends with and without CB.

Sample name	Shore A hardness
P0	50
P40	52
P80	56
P100	53
P120	58
P160	58
P200	51
PC0	70
PC40	65
PC80	68
PC100	62
PC120	61
PC160	61
PC200	60

16.3.2.4　EFFECT OF PU LOADING ON TENSILE PROPERTIES OF NBR WITH CB

The tensile properties were considerably enhanced with the addition of CB to the blends. The addition of PU to NBR also increased the tensile strength of NBR. A cumulative effect of CB and PU was observed at 50%, yielding substantial tensile strength enhancement. The tensile strength was enhanced by 500% with the addition of CB at all compositions. The mechanism may be the development of a large polymer-filler interface, which was the most important factor for the degree of reinforcement provided by the filler. With the incorporation of CB, the number of rubber chains entangling with CB aggregates as well as that of the cross-links increases, which results in an enhancement of the tensile properties. The entanglement of polymer chains with CB increases the torque, making the CB serve as a physical cross-link in rubbers. The percent elongation increased substantially for the combination of 50:50:30 NBR:PU:CB. The tensile properties of the NBR/PU blends with CB are listed in Table 16.4.

16.3.2.5　EFFECT OF PU LOADING ON TEAR PROPERTIES OF NBR WITH CB

The tear properties were significantly enhanced with the incorporation of CB to the blends. A 600% increase in the 50:50:30 combination of CB was observed. The tear properties of NBR/PU blends with CB are presented in Table 16.5.

16.3.2.6　EFFECT OF PU LOADING ON HARDNESS OF NBR WITH CB

The hardness properties were substantially enhanced with the addition of CB. However, the maximum hardness was obtained for pure NBR with CB. The hardness results are presented in Table 16.6.

16.3.2.7　EFFECT OF CB ON MECHANICAL PROPERTIES

The enhancement of the tensile properties with the addition of CB is shown in Figures 16.7–16.9. The effect of CB on the tear properties is shown in

Figure 16.10. The effect of CB on the hardness is shown in Figure 16.11. It is evident that CB enhanced the tensile, tear, and hardness properties of the NBR/PU blends. The stress–strain curve presented in Figure 16.12 shows the increase in the area under the curve with the addition of CB. The enhancement of the tensile properties may be attributed to the effective transmission of the load in the presence of CB. The presence of CB creates a physical link between the polymer phases and, in turn, the load is distributed evenly throughout the two phases of NBR and PU.

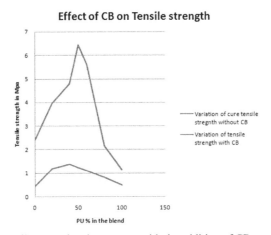

FIGURE 16.7   Tensile strength enhancement with the addition of CB.

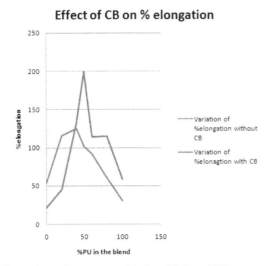

FIGURE 16.8   Elongation enhancement with the addition of CB.

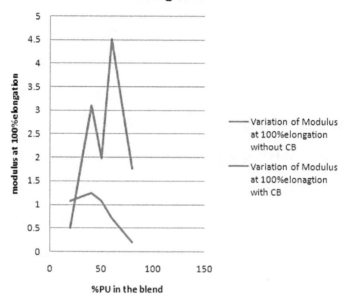

FIGURE 16.9　Enhancement of modulus at 100% elongation with the addition of CB.

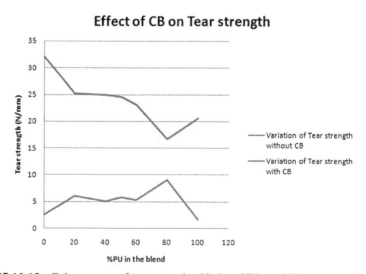

FIGURE 16.10　Enhancement of tear strength with the addition of CB.

Effect of CB on SHORE A hardness

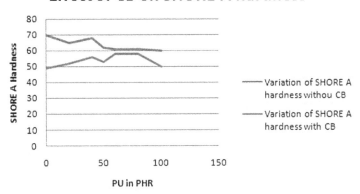

FIGURE 16.11   Enhancement of Shore A hardness with the addition of CB.

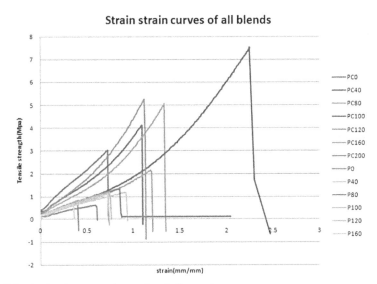

FIGURE 16.12   Stress–strain graphs of all samples.

16.3.3   THERMAL BEHAVIOR

16.3.3.1   EFFECT OF PU LOADING ON THERMAL BEHAVIOR

The thermal characteristics of NBR/PU blends were investigated. The thermal analysis was performed up to 600°C, and the TGA graphs were recorded using a TGA-50 analyzer. The heating rate was taken as 10°C/min. The TGA graph of pure NBR is presented in Figure 16.13, which indicates that rapid degradation starts at approximately 400°C and ends at 500°C; after this temperature, the residue does not degrade further. The amount of residue was approximately 2 mg. However, the material underwent combustion at some points after 500°C. The TGA graph of NBR with 20 phr PU degraded more steadily than that of pure NBR. The blends underwent combustion at some points after 500°C, indicating the vulnerability of PU above 500°C.

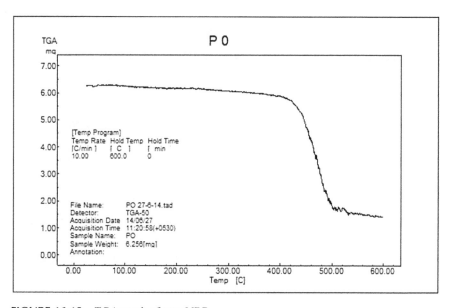

FIGURE 16.13   TGA graph of pure NBR.

The TGA graph of NBR with 40 phr PU underwent more degradation than pure NBR, leaving less residue. However, there were no signs of combustion. The TGA graph of NBR with 50 phr also followed a similar trend. The TGA graph of NBR with 60 phr, as shown in Figure 16.14, revealed a definite demarcation at approximately 460°C, indicating the initiation of the degradation of the PU phase. However, there was no definite demarcation in NBR with 80 phr PU. The residue of this blend is much less, indicating almost complete degradation. The TGA graph of pure PU, as presented in Figure 16.15, reveals rapid degradation at 400°C, indicating the importance of NBR in chemical resistance.

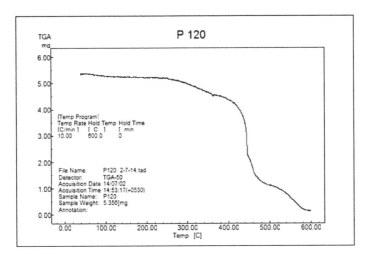

FIGURE 16.14   TGA graph of NBR with 60 phr PU.

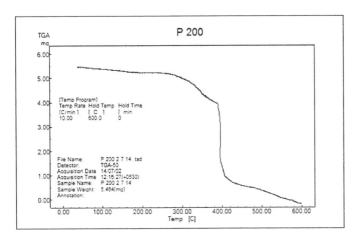

FIGURE 16.15   TGA graph of pure PU.

16.3.3.2   EFFECT OF PU LOADING ON THERMAL BEHAVIOR OF NBR WITH CB

The thermal characteristics of NBR/PU blends with CB were investigated. The thermal analysis was performed up to 600°C, and the TGA graph was recorded using a TGA-50 analyzer. The heating rate was taken as 10°C/min. The TGA graph of NBR with 20 phr PU with CB is presented in Figure 16.16. The volatility was observed to decrease with the addition of

CB. An important observation can be made in the TGA graph of NBR with 40 phr PU with CB; the degradation of PU was delayed due to the presence of CB. The TGA graph of NBR with 50 phr PU with CB, as presented in Figure 16.17, demonstrates that the degradation becomes smooth with the addition of CB, indicating a considerable reduction in vulnerability. The TGA graph of NBR with 60 phr PU with CB showed less degradation, indicating an enhancement in the chemical and thermal stability.

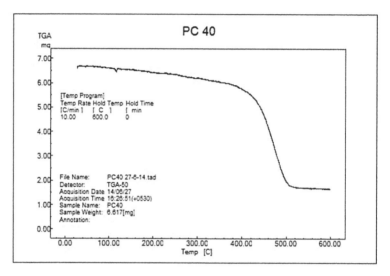

FIGURE 16.16   TGA graph of NBR with 20 phr PU and CB.

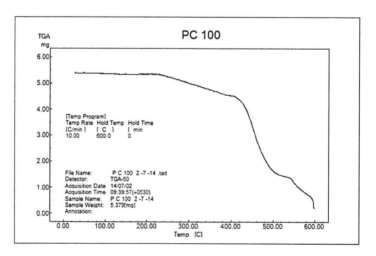

FIGURE 16.17   TGA graph of NBR with 50 phr PU and CB.

The TGA graph with 80 phr PU with CB showed a smoother but persistent degradation. The TGA graph of pure PU with CB, presented in Figure 16.18, shows less degradation than pure PU without CB, demonstrating the effect of CB in delaying the PU degradation.

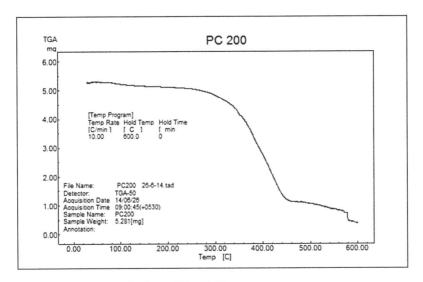

FIGURE 16.18   TGA graph of pure PU with CB.

16.3.3.3   EFFECT OF CB ON THERMAL BEHAVIOR OF NBR/PU BLENDS

TGA is a measure of thermal stability. The thermal analysis was performed in the presence of nitrogen. The sample exhibited significant improvement in the thermal stability when CB was incorporated in the blends. The increase in the thermal stability was due to the delayed degradation of PU in the presence of CB. The sample without CB exhibited less thermal stability because the sample degraded rapidly after 400°C. The blends with CB exhibited increased thermal stability due to the interaction between the polymer chains. In general, an increase in the thermal stability is compromised by a decrease in the physical properties; however, in this study, the reverse occurred. CB increased the interactions between the polymer chains. TGA graphs of all the blends without CB are presented in Figure 16.19. TGA graphs of all the blends with CB are presented in Figure 16.20. The graphs become smoother with less degradation with the addition of CB.

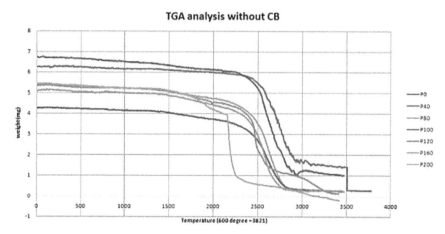

FIGURE 16.19 TGA graph of blends without CB.

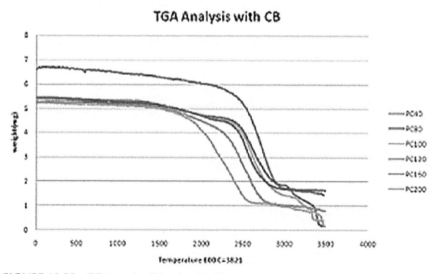

FIGURE 16.20 TGA graph of blends with CB.

16.3.4 SEM CHARACTERIZATION

SEM micrographs were acquired to characterize the blends. Figure 16.21 presents an SEM micrograph of vulcanized NBR without PU that shows the formation of cross-linking within NBR. Figure 16.22 presents an SEM image of pure PU with no distinct features. Figure 16.23 presents an SEM image of 50:50 NBR/PU at the 10 micron level, revealing distinct phase separation. Figure 16.24 shows the 50:50 NBR/PU with CB at the 10 micron level. This figure reveals reduced phase separation; however, the phase separation is not fully overcome. The minimized phase separation leads to improved mechanical properties. However, the NBR/PU blend is not fully homogenized.

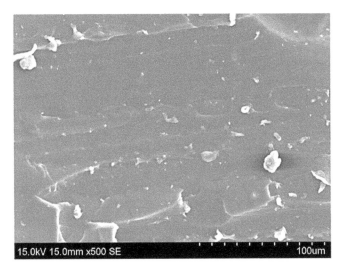

FIGURE 16.21   SEM micrograph of pure NBR.

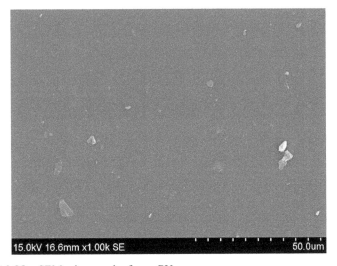

FIGURE 16.22   SEM micrograph of pure PU.

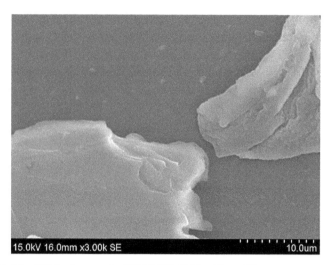

FIGURE 16.23   SEM micrograph of NBR/PU 50:50 without CB.

FIGURE 16.24   SEM micrograph of NBR/PU 50:50 with CB.

16.3.5   MATHEMATICAL MODELING OF TENSILE STRENGTH

Mathematical equations can be used for the prediction of technological prop-erties from experimental data. There are two common methods for obtaining an equation to fit the experimental data. The first method is the derivation of an equation from the experimental data, and the second method involves fitting an established equation with experimental data. The latter approach was adopted in this study. The equations are presented in Section 16.2 of this chapter. Figure 16.25 shows the mathematical modeling of tensile strength using the abovementioned equations. The Kerner equation fits the best with the experimental data.

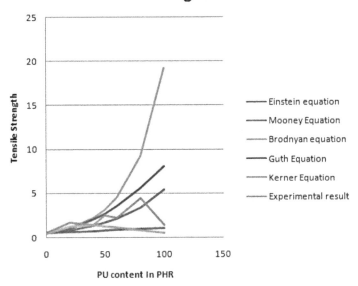

FIGURE 16.25   Mathematical modeling of tensile strength.

16.3.6   VOLUME SWELL OF BLENDS

Volume swells of some the blends were calculated and compared with the pure NBR vulcanizate. The volume swell of some of the blends was calculated and compared with the pure NBR vulcanizate. The pure NBR possessed the minimum volume swell, and the volume swell increased with PU addition but decreased when CB was added. The volume swell of pure NBR was not attained by the blends; however, the volume swells of the blends were comparable to that of pure NBR. A further reduced volume swell of blends is desired for engineering applications. A volume swell experiment was performed by immersing the blends into petrol for one day and measuring the weight difference. The volume swell of the blends is given in Table 16.7.

TABLE 16.7   Volume Swell of Blends.

Sample name	Weight before swell (mg)	Weight after swell (mg)	Volume swell for one day immersed in petrol (%)
P0	1.1781	1.7938	39.82
P80	1.48	2.354	86.37
P160	1.266	2.073	106.78
PC80	1.4054	2.0461	66.79
PC100	1.5592	2.39	75.66

16.3.7   COMPARISON OF PROPERTIES OF PURE NBR, NBR/ PU, AND OIL SEAL MATERIAL REQUIREMENTS

The O-ring is the most common type of fluid seal. Different types of rubber compounds exist in the market today because of the different temperatures, chemical exposures, and environments to which O-rings are subjected. The simple design of an O-ring lends itself to a multitude of sealing uses, including static, reciprocal, oscillating, and dynamic applications at low speed and pressure. In elevated operating pressure environments, the addition of a backup ring will greatly increase the extrusion resistance of an O-ring. The most basic shape of an O-ring is a simple circular sealing torus, usually made from an elastomeric material. In Table 16.8, the material requirements of an O-ring and a comparison between the properties of pure NBR and the NBR/PU blends are provided. NBR/PU is observed to exhibit enhanced properties; however, a decrease in volume swell is desired.

TABLE 16.8   Oil Seal Material Requirements and Comparison of Properties of Pure NBR and NBR/PU Blend.

Property	Oil seal material requirements	NBR/PU with CB (50:50:30)	Pure NBR
Tensile strength (MPa)	<8	5.42	0.45
Elongation after fracture (%)	<300	152.48	53.4
Tear strength (N/mm)	14	17.05	2.54
Shore A hardness	60	65	40
Cure time (s)	Optimum	13.35	15.15
Volume swell (%)	<15	75	39.42

16.4 CONCLUSION

NBR was blended with PU. The incorporation of PU in the NBR matrix reduced the cure time up to 40 phr of PU inclusion without CB. The inclusion of PU in the NBR matrix increased the CRI. However, the inclusion of CB increased the cure time only slightly. CB enhanced the reinforcing effect of PU in NBR, as observed in the SEM image. The addition of CB increases the processability of pure PU. All of the blend combinations exhibited significant increases of more than 500% in their tensile, tear, and hardness properties. SEM characterization of the NBR/PU/CB blend revealed the improved homogenization of the blends with the addition of CB. The phase separation of the NBR/PU blends was reduced, resulting in superior mechanical properties. The thermal stability of the blends increased in the presence of CB without compromising the mechanical properties. The improved interaction between the polymer chains in the presence of CB enhanced the thermal stability. Mathematical modeling of the blend properties was performed for the tensile strength; the Kerner equation was observed to best fit the data. A potential material for oil seal applications was suggested.

KEYWORDS

- polymer
- interface
- stress transfer
- scanning electron microscopy
- compression molding

REFERENCES

1. Radhesh Kumar, C.; Karger-Kocsis, J. *Eur. Polym. J.* **2002**, *38*, 2231–2237.
2. Biswas, T.; Das, A.; Debnath, S. C.; Naskar, N.; Das, A. R.; Basu, D. K. *Eur. Polym. J.* **2004**, *40*, 847–854.
3. Degrange, J.-M.; Thomine, M.; Kapsa, P.; et al. *Wear* **2005**, *259*, 684–692.
4. Moua, H.; Shen, F.; Shi, Q.; Liu, Y.; Wua, C.; Guo, W. *Eur. Polym. J.* **2012**, *48*, 857–865.
5. Wang, Q.; Yang, F.; Yang, Q.; Chen, J.; Guan, H. *Mater. Des.* **2010**, *31*, 1023–1028.
6. Wang, L. L.; Zhang, L. Q.; Tian, M. *Mater. Des.* **2012**, *39*, 450–457.

7. Findik, F.; Yilmaz, R.; Kksal, T. Investigation of Mechanical and Physical Properties of Several Industrial Rubbers. *Mater. Des.* **2004,** *25,* 269–276.
8. Minnath, M. A.; Unnikrishnan, G.; Purushothaman, E. *J. Membr. Sci.* **2011,** *379,* 361–369.
9. Huang, S.-L.; Lai, J.-Y. *Eur. Polym. J.* **1997,** *33* (10-2), 1563–1567.
10. Oertel, G.; Abele, L. *Polyurethane Handbook: Chemistry, Raw Materials, Processing, Application, Properties,* Hanser Publishers. Distributed in USA by Scientific and Technical Books, Macmillan; 1985.
11. Van Beek, L. K. H.; Van Pul, B. I. C. F. *J Appl. Polym. Sci.* **1962,** *6* (24), 651–655.
12. Payne, A. R. J. Appl. Polym. Sci. **1962,** *6* (19), 57–63.
13. Tinker, A. J.; Jones, K. P. *Blends of Natural Rubber*, 1st ed., Chapman & Hall: London,; 1998.
14. Degrange, J.-M.; et al. *Wear* **2005,** *259,* 684–692.
15. Liua, N.; Liua, X.; Zhanga, X.; Zhub; L. *Mater. Charact.* **2008,** *59,* 1440–1446.
16. Li, Z. H.; Zhang, J.; Chen, S. J. *eXPRESS Poly. Lett.* **2008,** *2* (10), 695–704.
17. Medalia, A. I. *Rubber Chem. Technol.* **1978,** *51* (3), 437–523.
18. Sahal, M. Characterization of NiP Coating on AZ91D Magnesium Alloy with Surfactants and Nanoadditives. *J. Magnes. Alloy.* **2014,** *2* (4), 293–298.

CHAPTER 17

SYNTHESIS OF "SEA-URCHIN"-LIKE MICROSTRUCTURE h-MoO$_3$ BY CHEMICAL BATH DEPOSITION

NEHA D. DESAI*, VIJAY V. KONDALKAR, PALLAVI. B. PATIL, RAHUL M. MANE, ROHINI R. KHARADE, and POPATRAO N. BHOSALE

Materials Research Laboratory, Department of Chemistry, Shivaji University, Kolhapur 416004, India

Corresponding author. E-mail: nehadesai323@gmail.com

CONTENTS

ABSTRACT

In the present investigation, sea-urchin-like microstructured h-MoO$_3$ was successfully synthesized by chemical bath deposition technique. The thermal stability, structural details, morphology, and composition of MoO$_3$ were analyzed using thermogravimetry, X-ray diffraction (XRD), scanning electron microscopy, and X-ray photoelectron spectroscopy techniques, respectively. The surface area measurements and pore volume determination were carried out by BET (Brunauer–Emmett–Teller) analysis. The TGA (thermogravimetric analysis)–DTA (differential thermal analysis) analysis showed a sharp, exothermic peak indicating change in crystallinity around 409°C. Change in crystallinity was confirmed by X-ray diffraction study of as-deposited and annealed sample. X-ray diffraction pattern showed hexagonal–to-orthorhombic phase transition after annealing the sample up to 450°C. Scanning electron microscopy images of as-synthesized sample revealed well-oriented nanorods with hexagonal cross-section assembled together to form a sea-urchin-like architecture while that of annealed MoO$_3$ sample revealed a 2D layer-by-layer growth. The oxidation state and chemical environment of elements were confirmed by X-ray photoelectron spectroscopy (XPS). XPS study confirms the Mo^{6+} and O^{2-} in single phase without any impurities. The surface area and pore volume of as-synthesized and annealed sample were recorded by BET under N$_2$ adsorption–desorption isotherm. In this chapter, we report synthesis of hexagonal MoO$_3$ and its characteristic properties suitable in fabrication of electrochromic device.

17.1 INTRODUCTION

In the present era, transition metal oxides (TMOs) have been paid more attention in the field of material science due to their various crystal phases and specific properties. The TMOs are synthesized by variety of ways in one-dimensional and two-dimensional structures. These TMOs exhibit quite different properties than others.

Molybdenum trioxide has various polymorphs such as α-MoO$_3$, β-MoO$_3$, and h-MoO$_3$. α-MoO$_3$ has a particular two-dimensional layered structure.[1] It has higher thermodynamic stability with orthorhombic crystal structure.[2] h-MoO$_3$ is a metastable form with hexagonal crystal structure. However, h-MoO$_3$ allows interesting intercalation chemistry with unique chemical, electrochemical, electronic, and catalytic properties. The potential applications of MoO_3 are photocatalysis, electrochromism, lithium-ion batteries, solar cells, and lasers.[3,4]

MoO_3 exhibits a variety of nanostructures such as nanowires, nanobelts, nanoflakes, and nanoplates, etc. They were fabricated mainly by solvothermal reaction, thermal evaporation, and modified hot-plate method. Metastable h-MoO_3 exhibits novel or enhanced properties; but it is difficult to synthesize.[9]

Considering all these facts, we are reporting here a simple and cost-effective chemical bath deposition technique which was used to synthesize h-MoO_3. The XRD study confirmed formation of h-MoO_3 with space group (P6₃/m). The TGA-DTA data showed an irreversible phase change from h-MoO_3 to α-MoO_3. Further, the samples were investigated by SEM, XPS, and BET analysis. SEM image shows "sea-urchin"-like morphology and two-dimensional layer-by-layer growth for h-MoO_3 and α-MoO_3, respectively. XPS study confirms Mo^{6+} and O^{2-} in single phase without any impurities.

17.2 GROWTH AND REACTION MECHANISM

Hexagonal MoO_3 is synthesized by chemical bath deposition method using ammonium heptamolybdate (AHM) as reactant precursor and concentrated HNO_3 as oxidizing agent. The preparative parameters such as pH, concentration, temperature, time of the reaction are optimized to obtain desired morphology. In a typical synthesis, 0.05 M aqueous solution of AHM was prepared. It was stirred for 15 min using magnetic stirrer with dropwise addition of concentrated HNO_3. This solution was placed in preheated water bath kept at 70°C for half an hour. In the formation of MoO_3, overall reaction takes place as follows:

$$(NH_4)_6 \, Mo_7O_{24}{\cdot}4H_2O_{(aq)} + 6HNO_3 \rightarrow 7MoO_{3(s)} + 6NH_4NO_{3(aq)} + 7H_2O.$$

MoO_6 is considered as initial seed nucleus for the formation of MoO_3.[5] Once these initial seed nuclei attain critical size, they act as nucleation centers for the further growth of crystal. Hence, growth of larger crystallites takes place at the expense of smaller ones according to the Ostwald ripening law.[6]

17.3 STRUCTURE ANALYSIS

The phase identification of MoO_3 samples was confirmed on the basis of XRD pattern (Fig. 17.1). The sharp intense diffraction peaks in the XRD pattern[5] of the as-synthesized MoO_3 powder were well matched with JCPDS

Card No.21-0569. This confirmed the formation of single phase, pure, and crystalline h-MoO$_3$. After annealing at 450° phase, transformation from hexagonal MoO$_3$ to orthorhombic MoO$_3$ was observed and XRD peaks of annealed MoO$_3$ were well matched with JCPDS Card No.05-0508. The thermal analysis of as-synthesized MoO$_3$ shows a sharp exothermic peak at 409°C indicating change in crystallinity.

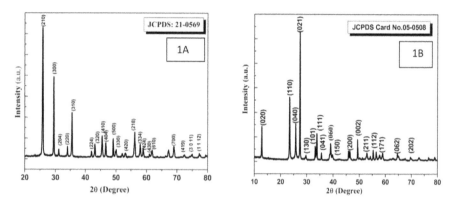

FIGURE 17.1   (A) XRD pattern of as-synthesized MoO$_3$ and (B) XRD pattern of annealed MoO$_3$.

17.4   MORPHOLOGICAL STUDY

The SEM image of as-synthesized sample shows well-oriented rods with hexagonal cross section.[7] These rods assembled together to form "Sea-urchin"-like architecture. After annealing, 1D rods were transformed into 2D microplate-like structure.

FIGURE 17.2 (A) SEM image of h-MoO$_3$ and (B) SEM image of α-MoO$_3$.

17.5 XPS STUDY

The X-ray photoelectron spectroscopy was used to determine the chemical bonding and surface composition of the samples. The Mo3d core-level spectrum of as-synthesized h-MoO$_3$ shows spin-orbit doublet at Mo3d 3/2 = 235.12 eV and Mo3d 5/2 = 232.05 eV.[8] It corresponds to Mo in (+6) oxidation state.

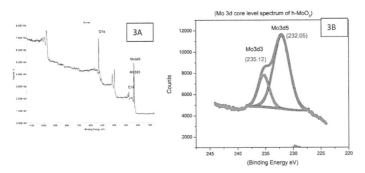

FIGURE 17.3 (A) XPS survey spectrum and (B) XPS core-level spectrum of Mo3d.

17.6 CONCLUSIONS

In the present research work, we have successfully synthesized h-MoO$_3$ by chemical bath deposition method. After annealing at 450°C, phase transition was observed from hexagonal MoO$_3$ to orthorhombic MoO$_3$. The SEM images of as-synthesized h-MoO$_3$ show the presence of one-dimensional hexagonal rods with sea-urchin-like architecture, while that of annealed MoO$_3$ show the presence of microplate-like structure. The XPS analysis confirms (+6) oxidation state for molybdenum. All these results reveal that h-MoO$_3$ is a better candidate in order to study electrochromic properties.

KEYWORDS

- sea-urchin-like MoO$_3$
- chemical bath deposition
- hexagonal-to-orthorhombic phase conversion
- XPS of MoO$_3$

REFERENCES

1. Badica, P. *Cryst. Growth Des.* **2007,** *7*(4), 794–801.
2. Irmowati, R.; Shafizah, M. *Int. J. Basic Appl. Sci.* **2009,** *9*(9), 34–36.
3. Huang, L.; Xu, H.; Zhang, R.; Cheng, X.; Xia, J.; Xu, Y.; Li, H. *Appl. Surf. Sci.* **2013,** *283*, 25–32.
4. Rao, M. C.; Ravindranadh, K.; Kasturi, A.; Shekhawat, M. S. *Res. J. Recent Sci.* **2013,** *2*(4), 67–73.
5. Chithambararaj, A.; Bose, A. *Beilstein J. Nanotechnol.* **2011,** *2,* 585–592.
6. Kharade, S. D.; Pawar, N. B.; Mali, S.; Hong, C. K.; Patil, P. S.; Bhosale, P. N. *J. Mater. Sci.* **2013,** *48*, 7300–7311.
7. Pan, W.; Tian, R.; Jin, H.; Chu, W. *Chem. Mater.* **2010,** *22*, 6202–6208.
8. Zou, J.; Schraderb, G. L. *J. Catal.* **1996,** *161*, 667.
9. Zheng, L.; Xu, Y.; Jin, D.; Xie, Y. *Chem. Mater.* **2009,** *21*, 5681–5690.

CHAPTER 18

DEFECT-ASSEMBLED NANOPARTICLES IN LIQUID CRYSTALLINE MATRICES

SAMO KRALJ[1,2], DALIJA JESENEK[2], MARTA LAVRIČ[2], MAJA TRČEK[2], GEORGE CORDOYIANNIS[2,3,4], JAN THOEN[4], GEORGE NOUNESIS[5], and ZDRAVKO KUTNJAK[2,6]

[1]*Faculty of Natural Sciences and Mathematics, University of Maribor, Koroška 160, 2000 Maribor, Slovenia*
[2]*Jožef Stefan Institute, Jamova 39, 1000 Ljubljana, Slovenia*
[3]*Department of Physics, University of Athens, 15784 Athens, Greece*
[4]*Department of Physics and Astronomy, KU Leuven, 3001 Leuven, Belgium*
[5]*Biomolecular Physics Laboratory, National Centre for Scientific Research "Demokritos", 15310 Aghia Paraskevi, Greece*
[6]*Jožef Stefan International Postgraduate School, Jamova 39, 1000 Ljubljana, Slovenia*
Corresponding author. E-mail: samo.kralj@ijs.si

CONTENTS

ABSTRACT

We studied mechanisms stabilizing topological defects (TDs) and their inter-actions with nanoparticles (NPs) in liquid crystalline (LC) matrices. In LC shells, we analyzed within the Landau–de Gennes approach the impact of Gaussian curvature on position and number of TDs. We demonstrated that electrostatic analogy could be exploited to predict depinning transitions in which pairs {defect, antidefect} are formed. We also considered interactions between NPs and lattices of TDs in representative blue phases (BPs) and twist grain boundary (TGB) phases using the Landau–de Gennes–Ginzburg approach. We estimated influence of NPs on temperature stability range of representative BP and TGB structures. Theoretical predictions are supported by high-precision calorimetry measurements in mixtures of LCs and CdSe NPs.

18.1 INTRODUCTION

Topological defects (TDs) comprise an unavoidable consequence of contin-uous symmetry-breaking which is ubiquitous in nature.[1] Physics of TDs is of interest for all branches of physics, spanning fields of particle physics, condensed matter, and cosmology.[1,2] In general, TDs correspond to topologi-cally stable singularities in the corresponding order parameter field, which describes a symmetry broken phase hosting defects. TDs are locally stabi-lized due to topological reasons, which are independent of systems details. A single TD cannot be removed by local continuous transformations for fixed boundary conditions. Consequently, their behavior often exhibits several universalities and the theory of TDs is strongly interdisciplinary. This is clearly demonstrated by the fact that the first comprehensive theory of coars-ening of TDs was developed in cosmology in order to explain coarsening dynamics of the Higgs field in the early universe.[3] Furthermore, several studies on defects have been carried out in relatively easily experimentally accessible condensed matter systems,[2] such as superconductors, superfluids, and liquid crystals (LCs). This was done in order to reveal fundamental features related to the structure of the universe, where controlled experi-ments are difficult or even impossible to be realized.

The key feature of TDs is represented by the topological charge, which is a discrete and conserved quantity.[1] In most systems, topological charges can be either positive or negative and often they behave like electric charges.[4,5] One commonly refers to TDs with positive and negative topological charge as *defects* and *antidefects*, respectively. In general, charges of opposite (the same) sign attract (repeal) each other. Furthermore, TDs of opposite charge can annihilate into a defectless state.

For example, in two-dimensional (2D) systems of magnetic spins, where the order is described by a vector field $\vec{m}$, the topological charge is equivalent to the winding number m.[1,4] The latter is determined by $\vec{m}$ surrounding a topological defect. It counts the total angular reorientation of $\vec{m}$ on encircling TD in a counterclockwise manner (using an arbitrary path), divided by 2π. For a vector field m can possess integer values. For a pseudospin, where orientations $\pm\vec{m}$ are equivalent, m can also possess half integers. Some typical examples of TDs are schematically shown in Figure 18.1. At the center of defect $\vec{m}$ is not uniquely defined. Consequently, the degree of order is there melted. In general, the presence of TDs is energetically costly and in such cases systems avoid them.

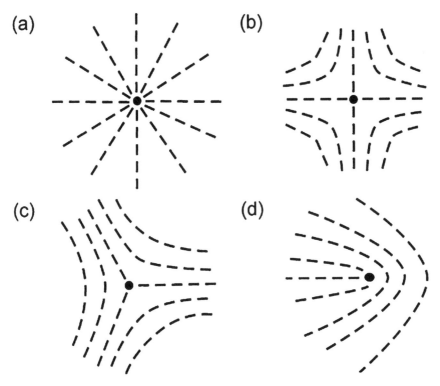

FIGURE 18.1 Schematic presentations of topological defects in orientational ordering in a two-dimensional plane. *Dashed curves* indicate the spatial distribution of the director field. With a *circle* we mark the origin of a defect. (a) $m = 1$, (b) $m = -1$, (c) $m = -1/2$, and (d) $m = 1/2$.

In order to illustrate the reasons why TDs are at least temporarily created let us consider a textbook example of a bulk temperature-driven paramagnetic–magnetic phase transition, where a ferromagnet is quenched into a ferromagnetic phase. The paramagnetic phase exhibits a continuous orientational symmetry, where all orientations are equivalent. On the other hand, in the ferromagnetic phase this symmetry is broken and a local magnetization $\vec{m} = |\vec{m}|\vec{e}$ aligns along a symmetry-breaking direction. The order parameter of the phase transition consists of two components,[6] which exhibit qualitatively different behavior if the degree of ordering is perturbed. If locally perturbed, the *hard* component $|\vec{m}|$ of the order parameter typically recovers a bulk-like order on the material-dependent order parameter correlation length ξ. Note that the system locally prefers a single value of $|\vec{m}|$. On the other hand, the unit vector $\vec{e}$ represents the *gauge* component of the order parameter field. In the absence of external (electric, magnetic, effective surface) fields and, if elastically isolated, all its orientations are equivalent, mirroring the lost symmetry of the higher symmetry unbroken (in this case paramagnetic) phase. Consequently, if frustrating values are imposed to it, the gauge field responds on the geometrically available length scale. In general, the hard and gauge component of an order provide information on the degree of ordering and on which symmetry-breaking option was selected in a symmetry-breaking phase transition, respectively.

Following the sudden quench (with respect to the characteristic order parameter relaxation time), a local direction $\vec{m}$ is selected by fluctuation imposed local preference.[2] Because different areas are causally disconnected and because of the finite speed that information propagates, in general, different directions of $\vec{m}$ are chosen within them. Consequently, a domain-type pattern is formed in $\vec{m}$ well characterized by a single average domain size length ξ_d. At the domain intersections, TDs almost unavoidably appear. Soon after the quench, the scaling regime is entered where the domain growth follows the scaling law $\xi_d \propto t^{\gamma}$ in order to minimize the domain wall volume. In bulk[7,8] the universal exponent equals $\gamma \sim 1/2$ or $\gamma \sim 1/3$ for nonconserved and conserved order parameters characterizing the forming phase, respectively. In this process, larger domains grow on the expense of smaller domains and this process is enabled by the annihilation of *defects* and *antidefects*. In the absence of impurities, the final system's structure consists of a single domain. However, if impurities are present they could pin the TDs and, thus, the domain structure might become thermodynamically stable or metastable. Note that a similar process follows the quench in any continuous broken phase whose bulk equilibrium is described by a spatially homogeneous value of at least the hard component of the order parameter.

Various LC[9] phases are especially adequate to study phenomena related with the physics of TDs. In general, they display a liquid-like behavior and at the same time some kind of orientational or/and translational (quasi) long-range order. In addition, their stability range is often close to room temperatures. Such a behavior is enabled by relatively weakly interacting anisotropic LC molecules. LCs are typical representatives of soft matter systems. Their softness is due to Goldstone modes in the gauge component of the order parameter,[6] which is a consequence of continuous symmetry-breaking. They play an important role in nature and also several technological applications because of their unique combination of liquid character, softness, optic anisotropy, and transparency. Furthermore, there exists a rich spectrum of different LC phases and structures that have potential to display various physical phenomena. Consequently, they are often exploited as an experimental testing ground for fundamental physics.[2,4]

For simplicity and illustration purposes, we henceforth restrict our focus on rodlike thermotropic LC molecules[9] that, on a mesoscopic scale, possess a uniaxial cylindrical symmetry. We also focus on particular LC phases of interest. In bulk, at relatively high temperatures the system exhibits the isotropic (I) phase characterized by isotropic orientational symmetry. Therefore, it is equivalent to an ordinary liquid phase. On decreasing temperature for nonchiral LC molecules, a nematic (N) phase is commonly entered via first-order phase transition exhibiting only long-range orientational order. At mesoscopic level local orientation LC ordering is commonly described by the nematic director field, where $\pm \vec{n}$ directions are physically equivalent and $|\vec{n}| = 1$. In bulk the thermodynamically stable nematic phase $\vec{n}$ is spatially homogeneously aligned along a single symmetry-breaking direction as illustrated in Figure 18.2. On further decreasing temperature a smectic A (SmA) phase is entered, possessing in addition also one-dimensional translational order. At this transition the continuous translational symmetry is broken. The N–SmA phase transition is either of first or second order. The simple SmA phase consists of a stack of parallel smectic layers, characterized by a layer distance d as it can be seen in Figure 18.2. Note that the smectic layers exhibit relatively strong fluctuations, giving rise to quasi long-range translational order.

In case of chiral LC molecules the cholesteric (N^*) phase replaces the nematic phase. It consists of 2D nematic plates forming a helical structure characterized by the wavevector q_{ch}, see Figure 18.2. Therefore, on decreasing temperature, one observes the phase sequence I–N^*–SmA. On the other hand, for strong enough chirality, different blue phases (BPs)[10] could intervene between the I and the N^* phases. An illustrative example is schematically depicted in the right side of the bottom panel of Figure 18.2. Three distinct thermodynamically stable BPs can exist, named blue phase I (BPI), blue phase II (BPII), and blue phase III (BPIII). All three phases are dominated by lattices of line TDs in the orientational order, referred to as disclinations. These lattices exhibit a simple cubic symmetry, a body-centered cubic symmetry, and an amorphous structure in BPI, BPII and BPIII, respectively. Note that all these structures exhibit a high potential for various applications, such as three-dimensional (3D) lasers and fast displays. However, their extensive use in the industry was until recently hampered by the narrow temperature range at which they usually appear.[11] Furthermore, in strongly chiral LCs twist grain boundary phases (TGBs)[12,13] can appear in a temperature interval between the N^* and the SmA phases. Our study is confined to its simplest representative, the so-called TGB$_A$ phase that can be seen in the left bottom panel of Figure 18.2. It consists of a one-dimensional lattice of line screw dislocations, corresponding to TDs in translational ordering. They reside in planar grain boundaries (GB) that separate blocks of smectic A-type order. Note that they represent the LC analogue of the Shubnikov phase exhibiting Abrikosov flux vortex configurations in type-II superconductors.[12,14] In summary, for strongly chiral LC molecules the I–BPs–N^*–TGB_A–SmA phase sequence could be observed on decreasing temperature.

The interest in studying the interactions of colloidal particles and nanoparticles (NPs) with TDs in fluid matrices has rapidly increased in the last years.[15–21] Several studies have revealed that the particles are often attracted to locally elastically distorted LC regions and, consequently, to TDs.[22–27] In the latter case, the key reason behind this attraction is the so-called defect core replacement (DCR) mechanism.[11] According to it, the elastic penalty of TDs is reduced by the partial replacement of the energetically expensive defect core volume by a nonsingular volume of trapped particles. This mechanism is effective only if it does not significantly perturb the corresponding gauge field of the phase hosting TDs. In addition, the so-called saddle splay elasticity might play an important role.[11,27,28] The latter is related to the Gaussian curvature of a hypothetical surface,[28] in which $\vec{n}$ plays the role of the surface normal. These mechanisms could be exploited toward the controlled targeting and assembling of appropriate NPs into diverse structures and superstructures of interest for future applications in nanotechnology. Therefore, it is of interest to understand and to master the underlying principles.

In this contribution, we focus on the fundament understanding of phenomena which could enable the controlled assembling of NPs in diverse configurations exhibiting different symmetries. Of particular interest is the formation of robust tunable configurations of NPs that could be sensitively controlled by external means.

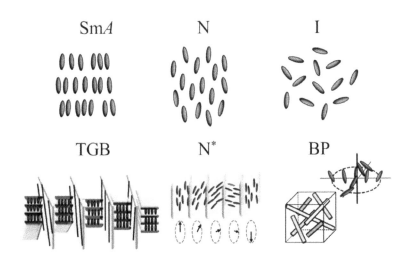

FIGURE 18.2   The liquid-crystalline phases of interest. On decreasing temperature one goes through I, N, and SmA phases, respectively. In chiral LCs, N^* appears instead of N. In some strongly chiral compounds, blue phase (BP) order may appear between I and N^* and TGB order between N^* and SmA. The BP order, schematically depicted in the bottom right part of the figure, is characterized by a lattice of disclinations. They appear due to double-twist deformations in nematic director field (shown just above the BP structure) which due to topological reasons enforce TDs.

The structure of this chapter is as follows: In Section 18.2, we present a Landau–de Gennes–Ginzburg model containing the key free energy contributions behind the mechanisms of interest. In Section 18.3, it is demonstrated how the Gaussian curvature could be used to form almost arbitrary patterns of TDs on effectively 2D nematic films on frozen curved surfaces. We show that the electrostatic analogy might be exploited to estimate the critical conditions for depinning TDs. This mechanism is of interest to increase the number of TDs. We illustrate the curvature-driven depinning and controlled positioning of TDs in the case of nematic shells exhibiting toroidal and spherical topology. In Section 18.4, we demonstrate that TDs efficiently trap proper types of NPs. First, the key mechanisms enabling the efficient trapping of NPs are analytically identified. Afterward, the corresponding experimental proofs are presented. In the last section, we summarize the results and present potential future research directions in the respective field.

18.2 MESOSCOPIC MODEL

In the following, we present a simple mesoscopic model which we use to describe the physics of our interest. The orientational LC order is described by the traceless and symmetric nematic tensor order parameter[9]

$$\underline{Q} = \sum_{i=1}^{d} s_i \vec{e}_i \otimes \vec{e}_i. \tag{18.1}$$

Here s_i and $\vec{e}_i$ are the i-th eigenvalue and the i-th eigenvector of Q, respectively. The quantity d stands for the space dimensionality, which is in our treatment either set to $d = 2$ (2D) or $d = 3$ (3D) and $\otimes$ stands for the tensorial product. In a 3D uniaxial ordering, it is commonly parametrized as

$$\underline{Q} = s(\vec{n} \otimes \vec{n} - \underline{I}/3), \tag{18.2}$$

where $s \in [-0.5, 1]$ is the uniaxial scalar order parameter and $\underline{I}$ is the identity tensor. A perfectly aligned and a melted nematic order correspond to $s = 1$ and $s = 0$, respectively.

The translational order is determined by the complex order parameter[9]

$$\psi = \eta e^{i\phi} \tag{18.3}$$

It describes the LC mass density spatial variations $\rho_m = \rho_0 (1 + \psi + \psi^*)$, where $\rho_m = \rho_0$ in the absence of smectic layering. The translational order parameter η quantifies the degree of smectic ordering and ϕ determines the position of smectic layers. The bulk equilibrium is a stack of smectic layers along the z-axis, where $\vec{n}$ is aligned along the smectic layer normal, ϕ is determined by $\phi = q_0 z$, where $d_0 = 2\pi / q_0$ stands for the equilibrium layer spacing. We express the resulting Landau–de Gennes–Ginzburg free energy of the system as

$$F = \int \left(f_c^{(n)} + f_e^{(n)} + f_c^{(s)} + f_e^{(s)} + f_i \delta(\vec{r} - \vec{r}_i) \right) d^d \vec{r}, \tag{18.4}$$

where δ stands for the delta function, $\vec{r}_i$ locates particle–LC interfaces and the integral runs over the sample volume. The superscripts in free energy density term determine nematic ($f^{(n)}$) and smectic ($f^{(s)}$) contributions. In the lowest possible approximation, which is needed to describe the phenomena of interest, we express these contributions as[9]

$$f_c^{(n)} = a^{(n)}(T - T_*^{(n)})Q_{ij}Q_{ij} - b^{(n)}Q_{ij}Q_{jl}Q_{li} + c^{(n)}\left(Q_{ij}Q_{ij}\right)^2, \tag{18.5a}$$

$$f_c^{(s)} = a^{(s)}(T - T_*^{(s)})|\psi|^2 + b^{(s)}|\psi|^4 + c^{(s)}|\psi|^6, \tag{18.5b}$$

$$f_e^{(n)} = L_1 Q_{ij,k} Q_{ij,k} + L_2 Q_{ij,j} Q_{ik,k} + L_3 Q_{jk,l} Q_{ik,j} + L\kappa\varepsilon_{ikl} Q_{ij} Q_{lj,k}, \tag{18.5c}$$

$$f_e^{(s)} = C_\parallel \left|(\vec{n}\cdot\nabla - iq_0)\psi\right|^2 + C_\perp \left|(\vec{n}\times\nabla)\psi\right|^2, \tag{18.5d}$$

where a summation over repeated indices is assumed, $Q_{ij,k} = \dfrac{\partial Q_{ij}}{\partial x_k}$ determine the spatial differentiation with respect to the Cartesian coordinates x_k and ε_{ikl} is the Levi–Civita alternator.

The condensation terms $f_c^{(n)}$ and $f_c^{(s)}$ determine the degree of nematic and smectic order in bulk equilibrium in the absence of elastic distortions, where $a^{(n)}$, $b^{(n)}$, $c^{(n)}$, $T_*^{(n)}$, $a^{(s)}$, $b^{(s)}$, $c^{(s)}$, and $T_*^{(s)}$ are material constants, independent of temperature T. The elastic terms $f_e^{(n)}$ and $f_e^{(s)}$ are weighted by the positive bare (i.e., independent of T) nematic (L, L_1, L_2, L_3) and smectic $(C_\parallel, C_\perp)$ elastic constant, and κ is the chirality parameter, proportional to the equilibrium pitch of the N^* helix. The nematic constants enforce a spatially homogeneous nematic ordering for $\kappa = 0$. The smectic compressibility $(C_\parallel)$ and bend $(C_\perp)$ elastic constants enforce an equilibrium layer spacing d_0 and tend to align $\vec{n}$ along the smectic layer normal given by $\vec{v} = \nabla\phi/|\nabla\phi|$. The interfacial term f_i determines the conditions at the LC–NPs interfaces.

18.3 CURVATURE-CONTROLLED PATTERNS OF TOPOLOGICAL DEFECTS

In this section, we demonstrate how different pattern of TDs in the orientational ordering could be relatively easily enforced via the Gaussian curvature field. For this purpose, we consider thin, effectively 2D nematic films. Recently, a lot of attention has been devoted to closed nematic films, referred to as nematic shells,[29–32] which are promising to play a significant role toward future applications in nanotechnology.[29] Due to the latter reason as well as for numerical convenience, we henceforth restrict our analysis to nematic shells possessing surfaces of revolution.

18.3.1 MODELING

Here we consider a nematic ordering on a smooth and closed surface. It is assumed that all LC molecules are bound to lie on the local tangent plane of a point on the surface but are otherwise unconstrained. The corresponding 2D nematic tensor order parameter can be expressed in the diagonal form as[32]

$$Q = s(\vec{n} \otimes \vec{n} - \vec{n}_\perp \otimes \vec{n}_\perp). \tag{18.6}$$

Here, s is the positive eigenvalue of Q. The unit vectors ($\vec{n}$, $\vec{n}_\perp$) are the eigenvectors corresponding to the eigenvalues ($s, -s$). The orientational field $\vec{n}$ plays a similar role as the conventional uniaxial nematic director field in 3D, and the surface normal is determined by $\vec{v} = \vec{n} \times \vec{n}_\perp$.

The free energy density of interest is[32]

$$f = a^{(n)}\left(T - T_*^{(n)}\right)Q_{ij}Q_{ij} + c^{(n)}\left(Q_{ij}Q_{ij}\right)^2 + LQ_{ij;k}Q_{ij;k}, \tag{18.7}$$

where semicolons in the latter term designate the surface differentiation.[32] Note that the cubic term in eq 18.7 is absent. Namely, in 2D it holds $Q_{ij}Q_{jk}Q_{ki} = TrQ^3 = 0$ (Tr stands for the trace operation) due to the traceless character of Q. We also neglected the surface interaction term f_i.

The condensation part of f is minimized for

$$s_b = \sqrt{\frac{a^{(n)}(T_*^{(n)} - T)}{8c^{(n)}}}. \tag{18.8}$$

In the elastic term, we use the approximation of equal Frank elastic constants[9] and consider nonchiral LCs, that is, $L = L_1$, $L_2 = L_3 = 0$, and $\kappa = 0$. An important role in our simulations is also played by the nematic order parameter correlation length, which we estimate by

$$\xi = \frac{\xi_0}{\sqrt{(T_*^{(n)} - T)/T_*^{(n)}}} = \sqrt{\frac{L}{a^{(n)}\left(T_*^{(n)} - T\right)}}. \tag{18.9}$$

We consider surfaces of revolution defined by the position vector

$$\vec{r} = \rho(v)\cos(u)\vec{e}_x + \rho(v)\sin(u)\vec{e}_y + z(v)\vec{e}_z, \tag{18.10}$$

where the Cartesian coordinate system (x,y,z) is defined by the unit vector triad ($\vec{e}_x, \vec{e}_y, \vec{e}_z$). The surface of a given geometry is parametrized by angular parameters (u,v), where $u \in [0,2\pi]$. In our study, we restrict to surfaces, which are accessible within parameterizations

$$\rho(v) = b\sin(v) + c\sin(3v), \quad z(v) = a\cos(v), \tag{18.11a}$$

$$\rho(v) = (a + b \cos(v), \ z(v) = b \sin(v). \tag{18.11b}$$

These parameterizations describe ellipsoidal or dumbbell-type geometries (eq 18.11a), $v \in [0, \pi[$) and torus (eq 18.11b), $v \in [0, 2\pi[$).

In numerical simulations,[32,33] we parameterized the nematic tensor order parameter with scalar fields $\{q_0 \ (u, \ v), \ q_m \ (u,v)\}$ as

$$\underline{Q} = q_0(\vec{e}_1 \otimes \vec{e}_1 - \vec{e}_2 \otimes \vec{e}_2) + q_m(\vec{e}_1 \otimes \vec{e}_2 + \vec{e}_2 \otimes \vec{e}_1). \tag{18.12}$$

The unit vectors $\{\vec{e}_1, \vec{e}_2\}$ determine the directions of principal curvatures of the revolution surfaces where our study is restricted to. These parameters are related with the parameterization used in eq 18.6 via equations

$$s = \sqrt{q_0^2 + q_m^2}, \tag{18.13}$$

$$\vec{n} = \vec{e}_1 \sin \vartheta + \vec{e}_2 \cos \vartheta, \ \vec{n}_\perp = \vec{e}_1 \cos \vartheta - \vec{e}_2 \sin \vartheta, \tag{18.14}$$

$$q_0 = -s \cos(2\vartheta), \ q_m = s \sin(2\vartheta). \tag{18.15}$$

Using parameterization eq 18.12 we minimized the free energy of the system. The resulting equilibrium Euler Lagrange equations for $q_m(u,v)$ and $q_m(u,v)$ read

$$\frac{\partial}{\partial v}\left(\frac{\partial(J f)}{\partial q_{0,v}}\right) + \frac{\partial}{\partial u}\left(\frac{\partial(J f)}{\partial q_{0,u}}\right) - \frac{\partial(J f)}{\partial q_0} = 0, \tag{18.16a}$$

$$\frac{\partial}{\partial v}\left(\frac{\partial(J f)}{\partial q_{m,v}}\right) + \frac{\partial}{\partial u}\left(\frac{\partial(J f)}{\partial q_{m,u}}\right) - \frac{\partial(J f)}{\partial q_m} = 0, \tag{18.16b}$$

where $J = \rho\sqrt{\rho_{,v}^2 + z_{,v}^2}$ is the Jacobian determinant of the surface, where we use the notation $A_{,v} = \dfrac{\partial A}{\partial v}$, $A_{,vv} = \dfrac{\partial^2 A}{\partial v^2}$. The equations were solved numerically using a standard over-relaxation method.

The nematic patterns were calculated deep in the nematic phase (i.e., $(T - T_*^{(n)})/ T_*^{(n)} = -0.25$, where the I–N phase transition takes place at $T = T_*^{(n)}$ in a flat nematic layer). In the center of the core of a topological defect, the nematic ordering is melted, that is, $s = 0$. Therefore, TDs are clearly visible in our simulations. The topological charge m of a defect is computed via[34]

$$m = m_1 + \frac{1}{4\pi}\oint \frac{q_0 q_m' - q_m q_0'}{q_0^2 + q_m^2} ds, \tag{18.17}$$

where the integration is performed counterclockwise along a closed loop enclosing the defect. The prime denotes a differentiation along the path and m_1 is the topological charge of a defect in the vector field $\vec{e}_1$ in case it is enclosed by the integration loop.

The Gaussian curvature of surfaces determined by eq 18.11 can be expressed as[32]

$$K = -\frac{z_{,v}(z_{,v}\rho_{,vv} - \rho_{,v}z_{,vv})}{\rho(\rho_{,v}^2 + z_{,v}^2)^2}.$$ (18.18)

Note that according to the Gauss–Bonnet theorem[35] it holds

$$m_{tot} = \frac{1}{2\pi}\oiint Kd^2\vec{r}.$$ (18.19)

Here, the integration is performed over the closed surface and m_{tot} stands for the total topological charge of TDs within the shell. By the Poincare theorem[36] it also holds

$$m_{tot} = \chi = 2(1 - g),$$ (18.20)

where χ is the Euler–Poincare characteristic of the closed surface and g is its genus (equals to the number of "handles" of the surface). For example, the spherical (toroidal) topology is characterized by $g = 0$ ($g = 1$), consequently $m_{tot} = 2$ ($m_{tot} = 0$).

For convenience in the following steps, we introduce the smeared *Gaussian curvature topological charge* Δm_{eff} and the *effective topological charge* Δm_{eff} within a surface area ΔA as

$$\Delta m_g = -\frac{1}{2\pi}\iint_{\Delta A} Kd^2\vec{r},$$ (18.21a)

$$\Delta m_{eff} = \Delta m_{tot} + \Delta m_g.$$ (18.21b)

Here, Δm_{tot} is the total charge within the integration area ΔA. If the integration is performed over the whole closed surface, one obtains $\Delta m_{eff} = 0$. If a surface possesses patches characterized by significantly different average Gaussian curvature, then, within each such patch of area ΔA, there is the tendency $\Delta m_{eff} = 0$[37]. If this is realized, the "real" topological charges totally neutralize a local smeared *Gaussian curvature topological charge*.

18.3.2   ELECTROSTATIC ANALOGY

In this subsection, we demonstrate that the electrostatic analogy[37] could be useful to predict the critical conditions, where the difference in Gaussian curvature in different areas is large enough to trigger the formation of a pair {*defect, antidefect*}.

We first introduce the elastic analogue of the electric field. For this purpose, we consider a 2D interaction potential w for a pair of nematic point defects with topological charges $\pm m$ separated at a distance ρ on a flat plane. Using a simple director field representation w is expressed by[9]

$$w = 2\pi m^2 K_f ln(\rho / \rho_c).\qquad(18.22)$$

The quantity K_f stands for the representative Frank elastic constant. The cutoff radius ρ_c estimates the characteristic size of the defect core, which is comparable to the nematic correlation length, therefore $\rho_c \approx \xi$. The magnitude of the corresponding attractive force is then

$$\sigma = |\nabla w| = 2\pi m^2 K_f / \rho := E_e m.\qquad(18.23)$$

Using this definition, the *elastic electric field* created by a topological charge m equals to

$$E_e = \frac{2\pi K_f m}{\rho}.\qquad(18.24)$$

To estimate a threshold condition to form a stable pair {*defect, antidefect*}, we consider a toroidal nematic shell in which $m_{tot} = 0$, shown in Figure 18.3. For such geometry one does not expect TDs. Intuitively, one guesses that the nematic molecules would be aligned along the azimuthal direction of this cylindrically symmetric structure (Fig. 18.3).

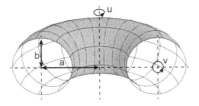

FIGURE 18.3   A torus, characterized by the outer $R_2 = a + b$ and inner $R_1 = a - b$ radius. Its surface is parameterized in terms of the azimuthal $u \in [0, 2\pi]$ and zenithal $v \in [0, 2\pi]$ angle. Meridians and parallels on the torus are described by $u =$ constant and $v =$ constant, respectively. The innermost parallel corresponds to $v = \pi$ and the outermost parallel to $v = 0$.

The Gaussian curvature of a torus parameterized by eq 18.11(b) is positive within the interval $v \in [0,\pi/2] \cup [3\pi/2,2\pi]$, it vanishes at $v = \pi/2$ and $v = 3\pi/2$, and it is negative for the remaining values of v, as it is evident from Figure 18.4a. We henceforth refer to the regions with $K > 0$ and $K < 0$ as the *positive* and *negative* region of the torus. The corresponding *Gaussian curvature topological charges* of these areas are $\Delta m_g^{(+)} = -2$ and $\Delta m_g^{(-)} = 2$, respectively. Therefore, the *positive* (*negative*) region tends to host at most four $m = 1/2$ ($m = -1/2$) TDs to compensate for the *Gaussian curvature charge*.

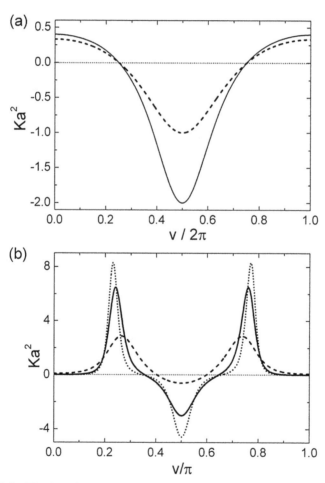

FIGURE 18.4   The Gaussian curvature K as a function of v: (a) torus: $\mu = 2$, *dashed line*; $\mu = 1.5$, *line* and (b) dumbbell structures: *dashed line* ($b/a = 0.6$, $c/a = 0.1$), *full line* ($b/a = 0.75$, $c/a = 0.25$), *dotted line* ($b/a = 0.85$, $c/a = 0.35$).

To illustrate the electrostatic analogy, we replace the torus by the circular *capacitor* as it is schematically shown in Figure 18.5. The inner (outer) plate at $\rho = R_1 = a-b$ ($\rho = R_2 = a + b$) bears the effective topological charge $\Delta m_{eff}^{(-)} = 2$ ($\Delta m_{eff}^{(+)} = -2$). Therefore, in the absence of TDs, each capacitor plate bears a charge of magnitude $\Delta m_{eff} = \left|\Delta m_{eff}^{(+)}\right| = \left|\Delta m_{eff}^{(-)}\right| = 2$, and the *elastic electric field* within it equals to

$$E_e = \frac{2\pi K_f \Delta m_{eff}}{\rho}. \qquad (18.25)$$

A possible way to reduce the energy stored within the *capacitor* is to create a pair of TDs $\{m, -m\}$, and to drag them to the opposite plates in order to screen the effective charge within them. To realize such a process, the energetic costs (to form a pair of defects and to drag them apart) must be balanced by the gain of reducing the effective charges at the plates.

(a)   (b)   (c)

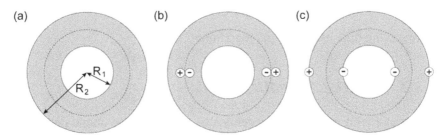

FIGURE 18.5   Schematic presentation of the electrostatic analogue of a torus nematic shell: (a) Defectless texture. The *effective topological charges* of the outer and inner part of the torus are equal to $\Delta m_{eff} = -2$ and $\Delta m_{eff} = 2$, respectively. (b) Two pairs $\{defect, antidefect\}$ bearing charges $\{1/2, -1/2\}$ are nucleated in the region where $K = 0$. (c) The *defects* and *antidefects* move toward $\rho = R_2$ and $\rho = R_1$, respectively.

Following this scenario, we set that the *elastic field* is strong enough to form a pair $\{-1/2, 1/2\}$, where $|m| = 1/2$ is the smallest possible topological charge. It is sensible to set that they are created at an arbitrary point along a line where, $K = 0$. The penalty of formation is roughly given by the condensation free energy cost of the melted ordering within the cores of TDs:

$$\Delta F_{cond} \approx a_0^{(n)} \left|T - T_*^{(n)}\right| s_b^2 \pi \xi^2, \qquad (18.26)$$

where s_b stands for the bulk degree of nematic order. The work required to pull apart a pair of these defects from the initial separation 2ξ to the final separation $2b$ is equal to

$$\Delta F_{work} \approx \frac{\pi K_f}{2} ln\left(\frac{b}{\xi}\right). \tag{18.27}$$

The energetic gain in moving defects in the *elastic electric field* within the *capacitor* is estimated by

$$\Delta F_{gain} \approx -\pi K_f \Delta m_{eff} ln\left(\frac{a+b}{a-b}\right). \tag{18.28}$$

The condition to form and separate a pair $\{-1/2, 1/2\}$ of TDs within a *capacitor* charged with Δm_{eff} is determined by $\Delta F_{cond} + \Delta F_{work} + \Delta F_{gain} = 0$. This equality yields the critical value of the ratio $\mu = a / b$:

$$\mu_c(\Delta m_{eff}) = \frac{\left(\frac{b}{\xi}\right)^{\frac{1}{2\Delta m_{eff}}} e^{\frac{1}{\Delta m_{eff}}} + 1}{\left(\frac{b}{\xi}\right)^{\frac{1}{2\Delta m_{eff}}} e^{\frac{1}{\Delta m_{eff}}} - 1}, \tag{18.29}$$

where e is the base of the natural logarithms.

18.3.3   NUMERICAL SIMULATIONS

We proceed with numerical simulations to prove that the above-described elastic analogy works well. Based on that, it is demonstrated that the Gaussian curvature could be exploited in order to form complex lattices.

We first study nematic textures within a toroidal nematic shell on decreasing the ratio $\mu = a / b$. For a relatively large ratio (i.e., $\mu \gg 1.2$) the so-called *parallel*[33] nematic structure exists, as shown in Figure 18.6a. In the (i) left, (ii) central, and (iii) right panels are plotted (i) the nematic director field direction on the (u,v) plane, (ii) degree of orientational order s on the (u,v) plane, (iii) and s on the torus. In it $\vec{n}$ is almost perfectly aligned along $\vec{e}_2$ and $s \approx s_b$ is relatively weakly distorted. Upon decreasing μ the degree of nematic ordering becomes progressively distorted within the inner region $(\rho \sim a - b)$ of the torus. At a critical value μ_c two pairs $\{-1/2, 1/2\}$ of TDs are formed. The presence and the locations of TDs are well visible in the first and second columns of Figure 18.6b. In the first column, a characteristic nematic director pattern of TDs bearing $m = 1/2$ and $m = -1/2$ are visible. Furthermore, at the centers of cores of TDs nematic ordering is melted, which is clearly visible in the second column. The *antidefects* and *defects* assemble at $\rho = R_1 \equiv a - b$ and $\rho = R_2 \equiv a - b$ as predicted by the electrostatic analogy. Furthermore, for $\{\rho = b/\xi_0 = 10, (T_*^{(n)} - T)/T_*^{(n)} = 0.25\}$ simulations yield $\mu_c \simeq 1.7$. On the other hand, the estimated eq 18.29 yields $\mu_c \simeq 1.8$ which signals that our estimate works well also quantitatively despite the drastic simplifications used in the derivation.

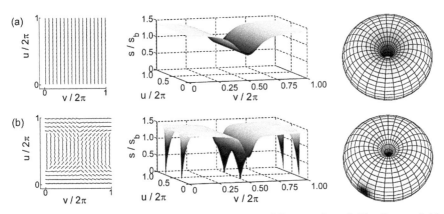

FIGURE 18.6 Nematic textures on a toroidal shell for different ratios a/b. The director field profiles and order parameter variations on the (u,v) plane are shown in the *left* and *central* panels, respectively. The spatial variations of the scalar order parameter on torus are depicted in the right panel, where darker shades correspond to lower degrees of ordering. (a) $a/b = 1/4$, (b) $a/b = 1.1$. In all cases, $b/\xi_0 = 10, (T_*^{(n)} - T)/T_*^{(n)} = 0.25$.

Next, we demonstrate on a very simple example that one can in principle form an arbitrary number of TDs where their positioning is predictable. For this purpose, we consider LC shells exhibiting a spherical topology, where we study the shapes accessible by parameterization of eq 18.11a. The spherical topology enforces $m_{tot} = 2$.

In Figure 18.4b, we plot some typical $K(v)$ spatial profiles. In a sphere, the Gaussian curvature is spatially independent, that is, $Ka^2 = 1$, where a stands for the sphere's radius. On squeezing the sphere a dumbbell structure is formed, where regions exhibiting $K < 0$ are introduced. Using our parameterization, the squeezing is realized by increasing the ratio c/a in eq 18.11a.

Changes in number and position of TDs on squeezing the shell is depicted in Figure 18.7. In a sphere, we obtain four $m = 1/2$ TDs (see the second column of Fig. 18.7a) as already predicted by Vitteli and Nelson.[30] Because K is spatially constant, the relative positions of TDs are determined by their mutual elastic repulsion (like putting four equal electric charges on a sphere). Consequently, TDs occupy vertices of a "virtual tetrahedron" touching the spherical surface in order to maximize their mutual separation.

On squeezing the sphere a spatially dependent $K(v)$ profile is established, leading to a spatially nonhomogeneous distribution of the *Gaussian curvature charge*. Consequently, the TDs move initially toward regions exhibiting maximal positive K in order to locally minimize Δm_{eff} as it is illustrated in Figure 18.7b. When a value of $K < 0$ within the neck area reaches the critical value, it triggers local nucleations of two pairs of TDs. Here each pair consists of $(m = 1/2, m = -1/2)$. A typical configuration is shown in Figure 18.7c. Two created TDs bearing $m = 1/2$ $(m = -1/2)$ are after their

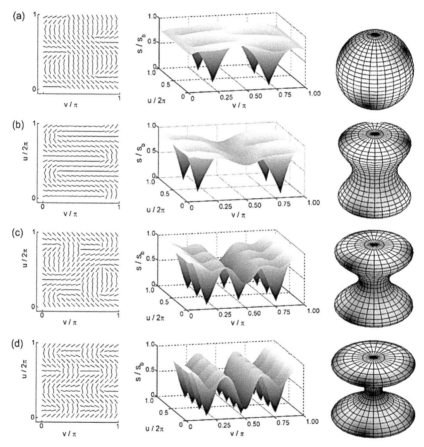

FIGURE 18.7 The nematic order parameter spatial variations within thin LC layers. The *left* column: the director field orientation within the (u,v) plane. The *middle* column: spatial variation of λ in the (u,v) plane. The *right* column: plot of λ within LC layers covering different closed geometries exhibiting spherical topology, enforcing $m_{tot} = 2$. The first row ($b/a = 1$, $c/a = 0$) and second row ($b/a = 0.6$, $c/a = 0.1$): four TDs with $m = 1/2$. The third row ($b/a = 0.75$, $c/a = 0.25$): eight TDs, six TDs with $m = 1/2$, and two with $m = -1/2$. The fourth row ($b/a = 0.85$, $c/a = 0.35$): twelve TDs, eight TDs with $m = 1/2$, and four with $m = -1/2$. $b/\xi = 6$.

creation pushed toward areas exhibiting maximal positive (maximal nega-tive) value of K in order to partially screen the effective smeared *Gaussian curvature topological charge*. On further squeezing, the geometry additional two pairs ($m = 1/2$, $m = -1/2$) of TDs appear, see Figure 18.7d. The resulting equilibrium profile consists now of four $m = 1/2$ TDs within each *positive Gaussian surface* and four $m = -1/2$ TDs within the *negative Gaussian surface*, respectively. Note that further squeezing the structures do not any more qualitatively change the configurations of TDs because $\Delta m_{eff} \sim 0$ in the neck area, where $K < 0$, and also in the surrounding areas. Namely, the inte-gration over each surface patch exhibiting positive Gaussian surface yields, $\Delta m_g \sim -2$. On the other hand, the integration over the surface part possessing negative Gaussian curvature yields, $\Delta m_g \sim 2$.

18.4 TRAPPING OF NANOPARTICLES WITHIN CORES OF TOPOLOGICAL DEFECTS

In the previous section, we demonstrated how the Gaussian curvature field could be exploited to form various templates of TDs. We showed that both the position and the number of TDs could be well controlled. In continua-tion, we show that TDs can be exploited as efficient traps for appropriate NPs. As typical examples we study BPs and TGB phases in LCs. These phases are dominated by lattices of line TDs. BP structures are shaped by TDs in the orientational order, referred to as $m = -1/2$ disclinations. In LC samples BPI, BPII, or BPIII structures exist[10] differing in the lattice struc-ture of disclinations. On the other hand, TGBs possess line TDs in transla-tional order, named screw dislocations. One typically distinguishes between crystal-like TGB structures and topologically similar structures, where screw dislocations strongly oscillate. The latter structures are referred to as chiral line liquid phases (N_L^*).[38–40] The NPs trapping efficiency of these LC struc-tures is dominantly determined by the core structure of TDs.[11] Note that the detailed LC structure has also an impact, however, it plays a secondary role. For this purpose, we henceforth focus mainly on TDs within LC structures possessing lattices of line defects, where LC distortions describing mutual interactions of TDs are considered only approximatively. The core struc-tures of these TDs have already been studied in detail using the Landau–de Gennes ($m = -1/2$ disclinations)[41,42] and the Landau–de Gennes–Ginzburg (screw dislocations) approaches.[43,44]

In the following analysis, we confine our interest to spherical NPs or radius r. It is assumed that they do not significantly disrupt the relevant gauge LC field if they are trapped within the cores of TDs. In practice, this is achieved by the appropriate surface decoration of NPs (e.g., covering them with flexible carbon chains) that can adapt to enclosing nematic director field.[27] A convenient quantity in mixtures of LCs and NPs is the mass concentration x of NPs, which is independent of T. This is defined as

$$x = \frac{m_{NP}}{m_{LC} + m_{NP}}. \qquad (18.30)$$

Here, $m_{NP} \sim \rho_{NP} V$ and $m_{LC} \sim \rho_{LC} V$ determine the total masses of NPs and LC molecules in the sample volume V, and $\{\rho_{NP}, \rho_{LC}\}$ are the corresponding mass densities. In the cases of interest, the concentrations of NPs are relatively low, typically, $x \leq 0.1$. Alternatively, the concentration of NPs is given by the volume concentration

$$\phi = \frac{N_{NP} v_{NP}}{V} \sim x \frac{\rho_{LC}}{\rho_{NP}}, \qquad (18.31)$$

where v_{NP} determines the volume of a nanoparticle.

We focus on the impact of NPs trapped to TDs on phases appearing at different temperature, encompassing isotropic, BP, N^*, TGB$_A$, and SmA type of order. We first present a theoretical analysis of the trapping efficiency of NPs to LC structures exhibiting lattices of TDs. This efficiency is manifested in the increased temperature stability range of such structures, revealing the overall stabilizing impact of NPs. Afterward we present experimental proofs of the predicted behavior. In order to emphasize the universality of these phenomena, we study the influence of the same kind of NPs on qualitatively different TD lattices. In particular, we chose BP and TGB structures that possess qualitatively different TDs. For example, $m = -1/2$ line disclinations are singular in $\vec{n}$ in the center of defects. On the other hand, in screw dislocations, $\vec{n}$ exhibits relatively weak spatial variations across the defect cores. However, the unifying feature of both kinds of line defects is the essentially melted order parameter within the TD cores.

18.4.1 THEORETICAL ANALYSIS

In the following, we estimate analytically the phase behavior within a mean field approximation, using the Landau–de Gennes–Ginzburg mesoscopic model. We represent distinct blue phases and twist grain phases by simple representative configurations, labeled as *BP* and *TGB*, respectively. In these representatives, we assume that free energy penalties are dominated by TDs-driven contributions. We henceforth label the phase transition temperatures between the competing phases (phase A–phase B) by T_{IB} (I–BP), T_{BN} (BP–N^*), T_{NT} (N^*–TGB), and T_{TA} (TGB–SmA). We label the excess free energy costs of a phase within a predefined unit cell V_u by $\Delta F^{(phase)}$ and the corresponding average free energy density by $f^{(phase)} = \Delta F^{(phase)}/V_u$. As reference energies in evaluating the excess free energies we use either (i) the isotropic phase or (ii) the bulk equilibrium nematic ordering, to study the temperature stability window of (i) BP and (ii) TGB structure, respectively.

18.4.1.1   TEMPERATURE WINDOW OF BP

We first estimate the influence of NPs on the blue phase–cholesteric (BP–N^*) phase transition. For this purpose, we consider in the free energy expression eq 18.4 only the most essential terms responsible for widening the temperature stability range of a representative BP configuration. In our estimate the structural differences among the BP structures are neglected.

The BP configuration is characterized by a lattice of $m = -1/2$ disclination lines. Within the cores of TDs, the nematic ordering is either essentially melted (i.e., $s \sim 0$) or strongly biaxial.[41,42] The core average radius is roughly given by the relevant (uniaxial or biaxial) nematic order parameter correlation length ξ_n. Our aim is to estimate the free energy cost $\Delta F^{(BP)}$ of a representative lattice cell unit of volume V_u hosting disclination lines of a total length h in the absence of NPs.

The condensation free energy penalty $\Delta F_c^{(BP)}$ of introducing the disclination line of length h is given by[11,24]

$$\Delta F_c^{(BP)} \sim a^{(n)}(T_{IB} - T)s_b^2 \pi \xi_n^2 h. \tag{18.32}$$

In general, the condensation penalty hinders the formation of BP structure with respect to the competitive cholesteric phase N^*. The BP configuration can be stabilized by the saddle-splay elastic constant contribution if the corresponding elastic constant K_{24} is positive. This elastic term can be expressed as a surface contribution at interfaces of an LC body using the Gauss theorem.[35] The corresponding free energy contribution $\Delta F_{24}^{(BP)}$ of the saddle-splay term at the interface, separating the line disclination core volume and its surrounding is estimated by[11]

$$\Delta F_{24}^{(BP)} \sim -\pi K_{24} h, \tag{18.33}$$

where $K_{24} \sim 4Ls_b^2$.[45]

The presence of disclinations introduces distortions to the nematic director field. These elastic penalties are estimated by[10,11]

$$\Delta F_e^{(BP)} \sim \frac{\pi K_f h}{4} \ln\left(\frac{r_d}{\xi_n}\right), \tag{18.34}$$

where $K_f \sim 4Ls_b^2$ [45] estimates the average Frank elastic constant and r_d approximates the distance between neighboring disclination lines running in a similar direction.

The temperature stability window $\Delta T_{BP} = T_{IB} - T_{BN}$ of the BP structures with respect to N^* below T_{IB} is determined by the condition, $\Delta F_c^{(BP)} + \Delta F_{24}^{(BP)} + \Delta F_e^{(BP)} = 0$. It follows

$$\Delta T_{BP} = \frac{4L\left(1 - \frac{1}{4}\ln(r_d / \xi_n)\right)}{a^{(n)}\xi_n^2}. \tag{18.35}$$

We next set that NPs are introduced and that they collect at the disclination lines. Consequently, the condensation free energy penalty is decreased due to the reduced volume occupied by the energetically costly essentially isotropic order, yielding

$$\Delta F_c^{(BP)} = a^{(n)}(T_{IB} - T)s_b^2(\pi\xi_n^2 h - N_{NP}v_{NP}). \tag{18.36}$$

Here, N_{NP} stands for the number of NPs within the disclination line of length h. Assuming, that the NPs affect rather weakly the remaining elastic contributions, one obtains

$$\frac{\Delta T_{BP}(\phi)}{\Delta T_{BP}} \sim 1 + \phi \frac{V_u}{\pi\xi_n^2 h}, \tag{18.37}$$

where $\phi \sim N_{NP}v_{NP} / V_u$.

18.4.1.2   TEMPERATURE WINDOW OF TGB

We continue by considering the stability of TGB-type structures. Within the mean field description we do not distinguish between the TGB_A and N_L^* phase. We henceforth use the former configuration as the representative defect structure.

Using uniaxial order parameter parameterization, we express only the most relevant elastic contributions in f_e of our interest[9,45]:

$$f_e^{(n)} = \frac{K_1}{2}(\nabla \cdot \vec{n})^2 + \frac{K_2}{2}(\vec{n} \cdot (\nabla \times \vec{n}) - q_{ch})^2 + \frac{K_3}{2}(\vec{n} \times \nabla \times \vec{n})^2$$
$$- \frac{K_{24}}{2}\nabla \cdot ((\nabla \cdot \vec{n})\vec{n} + \vec{n} \times \nabla \times \vec{n}).$$

(18.38)

Here, q_{ch} stands for the chirality-enforced wave vector, and K_1, K_2, K_3, and K_{24} determine the splay, twist, bend, and saddle-splay Frank nematic elastic constants, respectively.

For the sake of simplicity, we henceforth neglect LC elastic anisotropy and set $C = C_{\parallel} \sim C_{\perp}$ and $K_f \equiv K_1 \sim K_2 \sim K_3 \sim K_{24}$. Note that K_2 and K_3 exhibit an anomalous increasing on approaching the smectic phase ordering on lowering temperature. This effective increase is due to the fluctuation-triggered formation of clusters exhibiting a local layering that strongly resists to the imposed bend or twist nematic director field distortions. However, in the smectic phase, the smectic elasticity is dominantly dictated by the smectic elastic constants. Therefore, the values of the Frank nematic elastic constants are in the presence of smectic layers comparable to the values deep in the nematic phase.

The most important lengths are the cholesteric pitch wavelength $P = 2\pi/q_{ch}$, the equilibrium smectic layer spacing $d_0 = 2\pi/q_0$, the smectic order parameter correlation length ξ_s, and the nematic twist penetration length λ. Below $T_*^{(s)}$ values of ξ_s and λ are estimated by

$$\xi_s \sim \sqrt{\frac{C}{a^{(s)}(T_*^{(s)} - T)}} = \frac{\xi_0}{\sqrt{(T_*^{(s)} - T)/T_*^{(s)}}},$$

(18.39)

$$\lambda \sim \sqrt{\frac{K_f}{C\eta_b^2 q_0^2}} \sim \frac{\lambda_0}{\sqrt{(T_*^{(s)} - T)/T_*^{(s)}}},$$

(18.40)

where η_b determines the bulk degree of smectic ordering and $\{\lambda_0, \xi_0\}$ are "bare" characteristic length (i.e., independent of temperature).

In order to estimate the free energy costs of the competing N^*, TGB, and SmA phase we need to define the geometry of the problem. We search for the conditions for which TGB_A ordering is stable. The main geometric features of this structure are shown in Figure 18.8. The phase consists of slabs of length l_b exhibiting essentially bulk SmA ordering. The slabs are separated by GB of width $\sim \lambda$. Within GBs parallel lattice of screw dislocations reside, are separated by a distance $l_d \sim l_b$. The presence of dislocations enables the tilt between adjacent slabs giving rise to global twisting of the LC structure along the z-axis of the coordinate frame. The volume of a representative *TGB* unit cell, consisting of an SmA slab and a GB of surface h^2, is equal to $V_u = h^2 l_b$. The length of each screw dislocation is estimated by h and their number within a GB is given by $N_{scr} \sim h / l_d$.

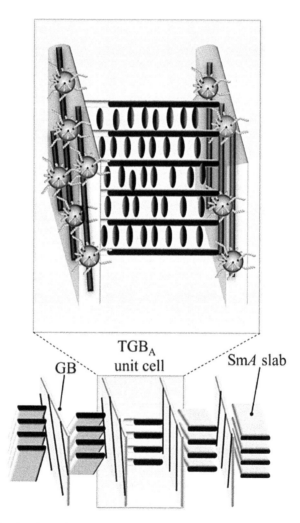

FIGURE 18.8 A schematic representation of TGB$_A$ phase, demonstrating the slabs of smectic order separated by grain boundaries. The upper part shows a blow up of the TGB unit cell. We assume that NPs are trapped within the screw dislocations, stabilizing this structure.

We proceed by estimating the free energy costs of competing N^*, TGB, and SmA structure. We first consider cases without NPs.

The LC ordering within N^* is determined by

$$\vec{n} = \vec{e}_x \sin(q_{ch}z) + \vec{e}_y \cos(q_{ch}z), \quad \eta = 0, \tag{18.41}$$

yielding

$$f^{(N^*)} = 0. \tag{18.42}$$

We describe SmA ordering within the unit cell for $T < T_*^{(s)}$ by $\vec{n} = \vec{e}_x$ and $\psi = \eta e^{iq_0 x}$. It follows

$$f^{(SmA)} = -a^{(s)}T_*^{(s)}\Delta t \, \eta_0^2 + \frac{K_f q_{ch}^2}{2}. \tag{18.43}$$

The dimensionless temperature shift

$$\Delta t = \frac{T_{NA}^{(0)} - T}{T_{NA}^{(0)} - T_*^{(s)}} \tag{18.44}$$

measures the departures from the N–SmA phase transition in a nonchiral LC and $\eta_0 = \eta(T = T_{NA}^{(0)})$.[46]

In estimating the free energy costs for the formation of TGB structure, it is assumed that within the smectic blocks ideal SmA ordering is established. Across each GB of width $\sim \lambda$ the director field is approximately rotated by an angle[10]

$$\Delta\theta \sim d_0 / l_d \sim d_0 / l_b. \tag{18.45}$$

The nematic elastic contribution reads[27]

$$\Delta F_e^{(TGB)} \sim \int d^3\vec{r}\, \frac{K_f}{2}(\vec{n}\cdot\nabla\times\vec{n} - q_{ch})^2 \sim \frac{V_u}{l_b}\frac{K_f}{2}\left(\frac{\Delta\theta^2}{\lambda} + q_{ch}^2 l_b - 2q_{ch}\Delta\theta\right). \tag{18.46}$$

The global twisting of LC configuration is enabled by GBs incorporating lattice of parallel screw line dislocations. The core-size radius of each dislocation is approximated by ξ_s and within the cores the smectic ordering is essentially melted. The resulting average smectic condensation free energy penalty within V_u is given by

$$\Delta F_c^{(TGB)} \sim -a^{(s)}T_*^{(s)}\Delta t \, \eta_0^2 \left(V_u - V_{scr}\right), \tag{18.47}$$

where $V_{scr} \sim N_{scr}\pi\xi_s^2 l_b$ is the volume occupied by the cores of screw dislocations. We neglect other elastic contributions, including the interactions among dislocations. Note that this approximation works well only in the limit $l_d > \lambda$.

It follows

$$f^{(TGB)} \sim \frac{K_f}{2}\left(q_{ch}^2 + \frac{\Delta\theta^2}{\lambda l_b} - 2q_{ch}\frac{\Delta\theta}{l_b}\right) - a^{(s)}T_*^{(s)}\Delta t\,\eta_0^2\left(1 - \frac{\pi\xi_s^2}{l_b^2}\right). \qquad (18.48)$$

For the latter purpose, we set that the triple point condition is roughly established in samples of our interest. Therefore, the LC exhibits a direct N^*–SmA phase transition at $T = T_{NA}$, where all three competing phases coexist, requiring, $f^{(N^*)}(T_{NA}) = f^{(TGB)}(T_{NA}) = f^{(SmA)}(T_{NA})$. This imposes $\lambda(T_{NA})\xi_s(T_{NA})q_0 q_{ch} = 1$ and

$$T_{NA} = T_*^{(s)}(1 - \lambda_0\xi_0 q_0 q_{ch}). \qquad (18.49)$$

The critical value of chirality wave vector enabling triple condition at $T = T_{NA}$ reads

$$q_{ch}^{(c)} = \frac{d_0}{\pi\xi_s^2}\left(1 - \sqrt{1 - \frac{\pi\xi_s^2}{\lambda l_b}}\right). \qquad (18.50)$$

Therefore, the absence of chirality corresponds to an infinite value for l_b.

If NPs are added we suppose that they are predominantly assembled within the cores of dislocations. Consequently, they decrease the condensation free energy penalty which is required to introduce TDs. Furthermore, due to their flexible tails NPs disturb relatively weakly the local LC ordering. The corresponding main free energy penalties in translational degree of order read

$$\Delta F_c^{(TGB)} = -a^{(s)}T_*^{(s)}\Delta t\,\eta_0^2(V_u - V_{scr} + N_{NP}v_{NP}), \qquad (18.51)$$

where N_{NP} now stands for the number of NPs in V_u.

The presence of NPs also introduces boundaries where the saddle-splay contributions might contribute.[11,46] Indeed, the director field spatial variation in radial direction away from the center of a screw dislocation exhibits a variation yielding finite saddle-splay contribution.[43] In the local cylindrical coordinate system attached to a screw dislocation defined by unit vector triad $\{\vec{e}_\rho, \vec{e}_\varphi, \vec{e}_z\}$, where the center of cylindrically symmetric dislocation is placed at $\rho = 0$, the director field can be well parameterized by $\vec{n} = \vec{e}_z \cos\vartheta + \vec{e}_\varphi \sin\vartheta$.

It holds $\vec{n}(\rho = 0) = \vec{e}_z$ and in the limit $\rho/d_0 > 1$ the orientational order is well described by[46,47]

$$\vartheta = ArcTan\left(\frac{1}{\rho q_0}\right). \tag{18.52}$$

The corresponding saddle-splay contribution within V_u is estimated by

$$\Delta F_{24}^{(TGB)} \sim -\frac{N_{NP} K_{24}}{2}\int(\vec{n}\times\nabla\times\vec{n}).\vec{v}\ d^2\vec{r}, \tag{18.53}$$

where the integration is carried over the surface of NPs and $\vec{v}$ stands for the outer surface normal. It follows

$$f_{24}^{(TGB)} \sim -\frac{K_{24}\phi\,a_{24}}{2rd_0}, \tag{18.54}$$

where r stands for the characteristic linear size of NP and $1\ /\ a_{24} \sim 4\pi^2$ $\left(1+1/\left(4\pi r^2\ /\ d_0^2\right)\right)^2 \left(r\ /\ d_0\right)^3$. Note that this expression is approximate. Here, the most important information is that the saddle-splay term yields a finite contribution and that its contribution depends on the characteristic size of the NP.

Therefore, in the presence of NPs it holds

$$f^{(TGB)}(\phi) \sim f^{(TGB)}(0) - \phi\left(a^{(s)} T_*^{(s)}\Delta t\,\eta_0^2 + \frac{K_{24}a_{24}}{2rd_0}\right), \tag{18.55}$$

while the impact of NPs on the other two phases is much smaller. Consequently, in the presence of NPs the stability range of *TGB* phase is increased to a finite temperature interval which is roughly estimated by

$$\frac{\Delta T_{TGB}}{T_{NA}^{(0)} - T_*^{(s)}} \sim \phi\left(1 + \frac{a_{24}}{q_{ch}^2 rd_0}\right)\frac{l_b^2}{\pi\xi_s^2}. \tag{18.56}$$

18.4.2   EXPERIMENTAL EVIDENCE

In the following, we present experimental evidence supporting the above analysis. High-resolution calorimetric results are presented in mixtures of LCs and CdSe spherical NPs.[24–27,46] LCs are considered highly chiral compounds exhibiting BPs (as pure) and having the potential to form TGB structures.

To demonstrate NP-driven widening of BP structural stability, we consider for the LC component 4-(2'-methyl butyl) phenyl 4'-n-octylbiphenyl-4-carboxylate which is referred to in the literature under the acronyms CE8 or 8SI*. Upon cooling from the I phase, it exhibits three distinct, thermodynamically stable, and well-characterized BPs. The chemical structure of CE8 and its phase transition sequence are shown in Figure 18.9a. When focusing on NP-induced TGB stability, we consider the LC compound S-(+)-4-(2-methylbutyl)phenyl-4-decyloxybenzoate (CE6). Its chemical structure together with its phase sequence is shown in Figure 18.9b. CE6 exhibits all three BPs, and a direct first-order N^*–SmA transition. Recent studies[27] suggest that at the latter transition the triple point condition is realized, where N^*, TGB_A, and SmA ordering coexist.

(a)

CE8

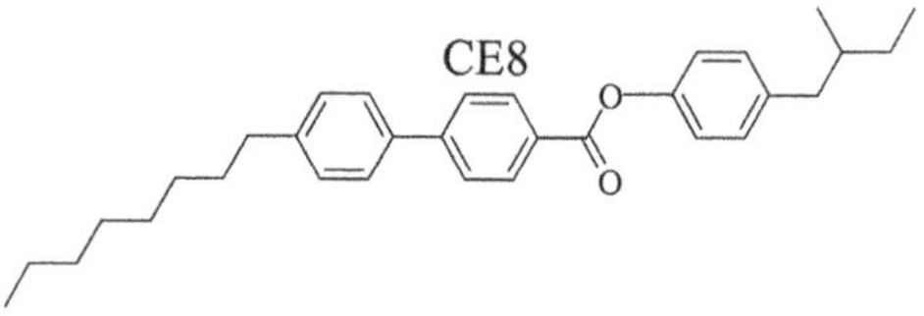

Cr - HexG*- HexJ*- HexI*- SmC*- SmA - N*- BPs - I

(b)

CE6

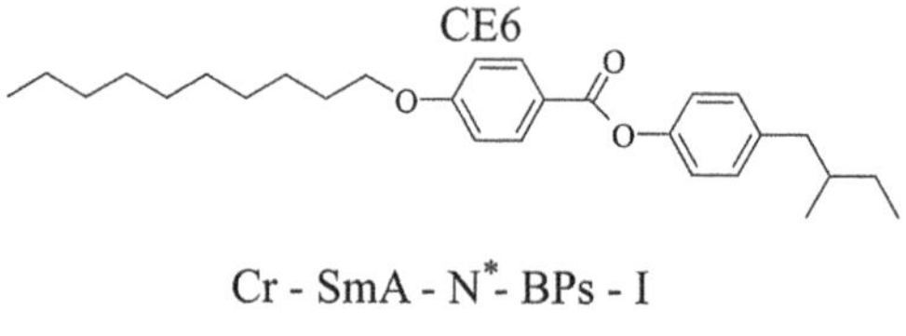

Cr - SmA - N*- BPs - I

FIGURE 18.9 The chemical formulas and the phase sequence of LC compounds CE8 (a) and CE6 (b).

We consider NPs as spherical 3.5-nm-diameter CdSe quantum dots, surface-treated with oleyl-amine (OA), and tri-octyl-phosphine (TOP) that can be produced in a monodispersed state. Such surface-decorated NPs have the capability to form stable suspensions in LCs.[10] A simple schematic representation of them is given in Figure 18.10. The preparation of mixtures with different NP concentrations is described in detail by Cordoyiannis et al.[27]

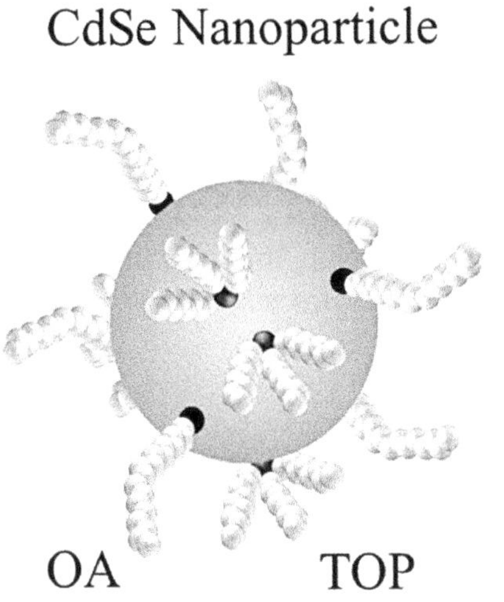

CdSe Nanoparticle

OA   TOP

FIGURE 18.10 A schematic representation of CdSe NPs. The 3.5 nm core consists of CdSe and it is coated with flexible molecules of oleylamine (OA) and trioctylphosphine (TOP).

Below, we present the phase behavior of such mixtures determined by high-resolution AC calorimetry.[48] In the presented results, in addition to the principal AC mode measurements, also the so-called relaxation or nonadiabatic scanning mode measurements were performed. The AC mode is sensitive only to the continuous enthalpy changes, but the phase of the AC temperature oscillations provides information on the nature of the transition, that is, continuous (second order) or discontinuous (first order). The relaxation mode senses both the continuous and discontinuous (i.e., latent heat) enthalpy changes. A detailed description of the technique is given by Yao et al.[49]

We first focus on the impact of NPs on the BP temperature stability range of CE8.[24] The specific heat (C_p) data obtained in bulk CE8 as well as in CE8 and CdSe mixtures upon cooling from the isotropic down to the smectic A phase are displayed in Figure 18.11. The impact of NPs is as follows. At the I-BPIII transition the C_p anomaly remains sharp and analogous to the one of pure CE8 even for the highest measured concentration, that is, $x = 0.20$. The I-BPIII transition temperature remains rather close to the one of pure CE8 for all the studied mixtures. Even a small CdSe concentration (i.e., $x \sim 0.02$) is enough to fully suppress the stability of the BPII phase. In these cases, the BPIII-BPI transition temperature is severely shifted even for the lowest studied NPs concentration and monotonously decreases with increasing x. The considerable shift of the BPIII-BPI transition to lower values combined with the mild effects upon the transition temperature of the I-BPIII leads to a substantial enhancement of the phase range of BPIII. At $x = 0.02$, the phase range is close to 10 K, while at $x = 0.20$ it reaches 20 K. The phase range is not strongly affected by thermal hysteresis effects. In addition, it is constant and reproducible for heating and cooling runs. The temperature versus $\times$ phase diagram is presented in Figure 18.12a featuring a dramatic widening of the BPIII stability regime. Note that a similar effect is observed also for CE6.[26] However, in these samples temperature windows of BP phases are narrower.

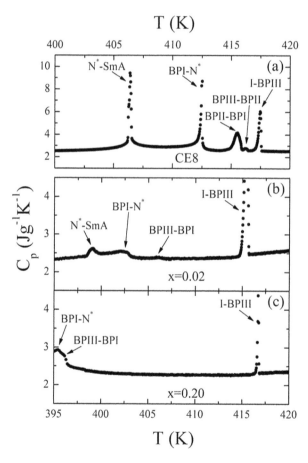

FIGURE 18.11   The temperature profiles of C_p for: CE8 (a) and its mixtures, $x = 0.02$ (b), and $x = 0.20$ (c) with CdSe NPs. These results, obtained by high-resolution AC calorimetry, nicely demonstrate the increase of BP range and, in particular, of BPIII.

The effect of NPs on the I-BPIII transition is much weaker because the penalty of introducing TDs close to this transition is relatively low. Namely, at the transition the free energies of I and BPIII phase are equal. The reason why only the BPIII phase is stabilized among all BPs is as follows. The NPs are expected to be essentially nonuniformly distributed within the disclination lines. Such a random character stabilizes the amorphous BPIII-type structure and not the more ordered BPI and BPII structures.

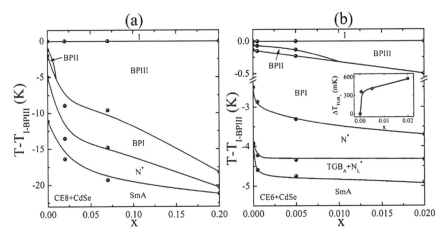

FIGURE 18.12   The phase diagrams of the systems CE8 + CdSe (a) and CE6 + CdSe (b). For CE8, a dramatic effect of CdSe NPs on BPIII is observed. In case of Ce6 + CdSe, we notice the increase of BPIII as well as the stabilization of TGB$_A$ structure. In the inset of (b), the range of TGB order is plotted as a function of NPs concentration.

When focusing on the NP-driven stabilization of TGB structure we have considered mixtures of the same CdSe NPs and CE6 LC.[27] The temperature profiles of C_p in this case were obtained using high-resolution adiabatic scanning calorimetry. This method uses very slow scanning rates (in the order of few mK/h) and has been extensively used for precise studies of phase transitions and critical phenomena in LCs.[50]

Some representative temperature dependencies of C_p, obtained upon slow cooling are shown in Figure 18.13. Bulk CE6 sample exhibits a sharp first-order N^*–SmA anomaly at $T_{NA} \sim 314.5$ K, corresponding to the direct N^*–SmA transition. The presence of any TGB-type order would result in an essential broadening of the transition and changes in the C_p pretransitional wings. Such features are absent in pure CE6. If NPs are present, a dramatically different phase transition behavior is observed. A wider double-peak anomaly is observed for all NP concentrations studied, revealing the presence of TGB-type order intervening between the $N*$ and SmA phase. The inset in Figure 18.13 shows the dual feature of C_p anomaly related to TGB ordering. This feature is characteristic for $N_L^*–TGB_A$, and $TGB_A–$SmA sequence on lowering T.[38] Note that the N_L^* phase possesses short-range positional order. This double feature, together with a higher temperature broad C_p anomaly related to the continuous conversion of $N*$ to the N_L^* phase ordering, is a typical fingerprint of these TGB structures. The corresponding phase diagram in the (x, T) plane is depicted in Figure 18.12b. The inset shows the temperature window of TGB-type of ordering, that is, the joined range of TGB_A and N_L^*.

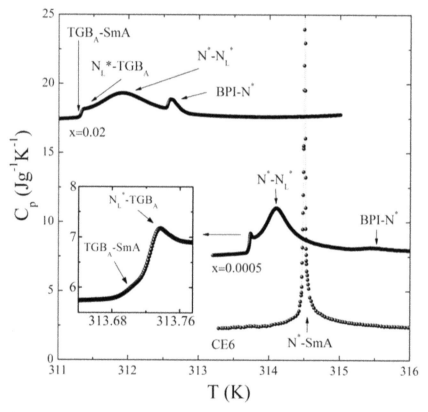

FIGURE 18.13   The temperature profiles of C_p for the CE6 and its mixtures, $x = 0.0005$ and $x = 0.02$ of CdSe NPs. The results, obtained by high-resolution adiabatic scanning calorimetry, demonstrate the stabilization of TGB_A order through the typical thermal fingerprint. The inset shows a blow up of the small anomalies related to TGB_A order, that is, the $N_L^*–TGB_A$ and $TGB_A–$SmA phase transitions for $x = 0.0005$.

Note that the stability of both the *BP* and *TGB* configuration is enabled by the combined effect of DCR and saddle-splay elasticity. The NP-driven widening of the corresponding temperature windows $\Delta T_{lattice}(x)$ of structures exhibiting lattices of TDs is given by eq 18.37 (for $\Delta T_{BP}(x)$) and eq 18.56 (for $\Delta T_{TGB}(x)$). The estimates are both in qualitative and quantitative agreement with the experimental results in the low dilution regime, where $\Delta T_{lattice}(x) \propto x$. To test the quantitative agreement, we first consider the *BP* stability. For example, for $x = 0.2$ the BPIII stability regime increases for an order of magnitude with respect to the pure sample. By using eq 18.37, $\Delta T_{BP}(x) / \Delta T_{BP}(0) \sim 10$ yields $N_{NP} v_{NP} / (\pi \xi_n^2 h) \sim 0.9$ and assuming $\xi_n \sim r$ it follows $N_{NP}/h \sim 0.2/r$. This suggests that such an increase in temperature is reached close to the saturating limit (i.e., $N_{NP}/h \sim 0.5/r$) where the disclinations are completely filled with NPs. This is in reasonable agreement with experiments showing that the $\Delta T_{BP}(x)$ dependence tends to saturate above $x \sim 0.2$. Note that above the saturation limit the DCR mechanism is not any more effective.

In deriving the expression for $\Delta T_{TGB}(x)$ we assumed that for $x = 0$ the triple point condition is realized, where N^*, TGB_A, and SmA ordering coexist as it seems to be the case in CE6. By neglecting the contribution of the K_{24} term and by setting $\rho_{LC}/\rho_{LC} \sim 1$, $l_b \sim 20$ nm, one reproduces $\Delta T_{TGB}(x) = 1$K for $x \sim 0.05$, which is roughly in line with the experimental results (Fig. 18.12b).

18.5 CONCLUSIONS

We have considered ways of forming controlled patterns of TDs in LC structures. We focused on the impact of the Gaussian curvature field, where we choose for simplicity as testing grounds 2D nematic shells. Furthermore, we have analyzed the NP trapping efficiency in lines of TDs. To stress the universality of the observed phenomena we studied line defects in (i) orientational and also (ii) translational degrees of freedom. For this purpose we studied the trapping of NPs within TDs in the representative (i) *BP* and (ii) *TGB* structures.

In studying the Gaussian curvature-driven manipulation of TDs, we used the mesoscopic Landau–de Gennes approach in terms of a 2D nematic tensor order parameter. We analyzed the impact of the Gaussian curvature field K in the case of cylindrically symmetric nematic shells. The assumption that a positive (negative) K acts effectively as a smeared negative (positive) topological charge was adopted. Such an electrostatic-like behavior has been demonstrated by Vielli and Turner[51] and Bowick et al.[37] using

XY-type structural studies on frozen curved surfaces. Based on their findings, we derived the expression for the *elastic electric field* as an analogue of the electric field. Of our interest was to estimate critical conditions to form pairs {*defect, antidefect*} in a locally defectless region. For this purpose, the toroidal geometry is considered in which TDs are not topologically required. Namely, according to the Gauss–Bonnet theorem,[35] the total topological charge within a toroidal nematic shell equals zero. Following Bowick et al.,[37] we mimic the toroidal nematic shells by concentric cylindrical capacitors. Implementing the electrostatic analogy, we derived the critical conditions to form pairs of TDs. Guided by these estimates, we stabilized numerically LC patterns exhibiting TDs, which confirms the validity of the analogy. Furthermore, we demonstrated in the case of dumbbell geometries that by varying K in structures with surface patches exhibiting both positive and negative Gaussian curvatures one can get different number of TDs at predetermined locations. Therefore, in our study, it is proven that the Gaussian curvature field could be exploited to form almost arbitrary stable or metastable patterns of TDs exhibiting diverse symmetries at least in thin LC films.

In the second part we analyzed the trapping of NPs to TDs. Theoretically we focused on typical representatives, the *BPs* and *TGBs*, that are dominated by lattices of nematic $m = -1/2$ dislocations and smectic screw dislocations, respectively. In the case of *BP* structure, we used the uniaxial Landau–de Gennes approach where the orientational order is described by the nematic director field and the uniaxial nematic order parameter. In the case of *TGB* structure we used the Landau–de Gennes–Ginzburg approach, where the translational order is described by the smectic A complex order parameter. As NPs, we considered *adaptive* spherical NPs commonly used in recent studies.[24–27] Here, the NP's *adaptive* character corresponds to a weak interaction with the relevant gauge field. More precisely, in the case of *BP* (*TGB*) *adaptive* NPs relatively weakly disturb the enclosing nematic director (smectic phase ϕ) field, as it is sketched in Figure 18.14. In practice this is commonly achieved by decorating NPs' surfaces with flexible chains.[24] We derived estimates relating the $BP^{24,26}$ and $TGB^{27,46}$ temperature stability range with the concentration x of NPs. In both structures the saddle-splay elasticity plays an important role. Note that the saddle-splay determines the Gaussian curvature of a hypothetical surface whose local normal points along the nematic director field.[28,35] In studying *TGB* structures we set that the triple point is realized for $x = 0$, where *TGB*, SmA, and N^* phases coexist. We chose this example because (i) experiments suggest that such a behavior could be encountered in CE6[24] and (ii) to demonstrate that NPs could induce the stability of structures exhibiting lattices of TDs that are otherwise unstable or metastable. In particular, we showed that apart from the DCR mechanism the saddle-splay elasticity[35,46] could be important for the TGB stabilization if $K_{24} > 0$. The reasons behind this are the LC–NP interfaces that exhibit a relatively large value of K when being close to screw dislocations. Namely, the director field enclosing dislocations exhibits a double-twist type deformation,[44] reminiscent of that in BP structures. Note that the NP–LC interfaces are absent for $x = 0$. The theoretical estimates are supported by high-resolution experimental measurements,[24,26,27,46] where the impact of surface-treated CdSe on CE8 and CE6 was studied. Experiments prove that the same kind of NPs are stabilizing qualitatively different TDs. Note that the disclinations in *BP* exhibit a singularity in the nematic director field, while the screw dislocations in *TGB* represent a singularity in the layer displacement. Therefore, the key mechanisms responsible for the stabilization appear to be universal.

(a)   (b)

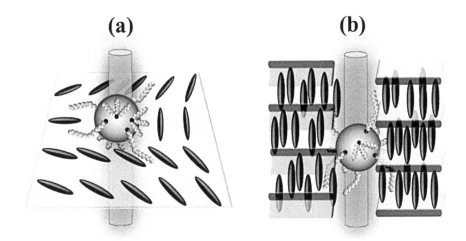

FIGURE 18.14 NPs trapped in the lines of TDs. The left part (a) of the figure shows a disclination line of BPs, whereas the right part (b) shows a screw dislocation characteristic of the TGB$_A$ structure. Due to flexible tails, NPs do not significantly disrupt surrounding relevant gauge field.

In summary, this work demonstrates that diverse controlled patterns of TDs can be formed in LC matrices. These could be exploited as templates and/or nucleation sites for NP assembling. In such a way different controlled assembly of NPs or their superstructures could be formed, like nanowires or more complex assemblies.[22,23] By changing an LC structure hosting NPs one could reconfigure the trapped NP configuration. This might be exploited in applications where flexible controlled reorganizations of NPs are needed. Vice versa, NPs could also be exploited to stabilize diverse pattern of TDs which are otherwise metastable or unstable,[27] or exist only in a narrow relevant phase space window.[10,11] Such configurations are of potential interest for applications in emerging nanotechnologies as well as biological or medical applications. In addition, specific NPs enabling structural transitions into structures hosting TDs could be exploited as sensitive detectors. TDs disrupt the relevant gauge field that typically exhibits an elastic deformation on a geometrically imposed scale, which in turn enables a relatively simple optical visualization. Furthermore, NPs could be exploited to tune a relevant LC medium close to a critical state enabling the detection of subtle changes in some other parameter of interest that can influence the state.[52] Finally, nematic shells, which we have used as typical examples to visualize the impact of curvature on TDs, could pave the way toward scaled crystals. As suggested by Nelson,[29] nematic shells immersed in an isotropic liquid could be linked by appropriate nanobinders, attached to TDs within the shells as depicted in Figure 18.15. In these structures, nematic shells correspond to atoms and TDs play similar role as valence. For example, from this perspective, spherical shells hosting four TDs exhibit sp^3 hybridization, which is characteristic for diamonds. Hence, by changing the number of TDs one could change the "valence" of such structures enabling a drastically different symmetry of the resulting structure; the symmetry being known for having a strong impact on various physical properties.

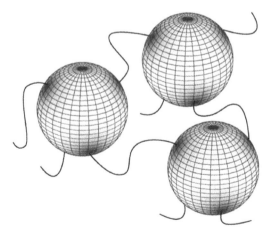

FIGURE 18.15 Schematic presentation of scaled crystals where the nematic shells play the role of atoms and the TDs determine the valency of "scaled atoms." Neighboring shells are linked via nanobinders attached to cores of topological defects.

KEYWORDS

- **liquid crystals**
- **topological defects**
- **nanoparticles**
- **blue phases**
- **twist grain boundary phases**
- **Gaussian curvature**

REFERENCES

1. Mermin, N. The Topological Theory of Defects in Ordered Media. *Rev. Mod. Phys.* **1979,** *51,* 591–648.
2. Zurek, W. H. Cosmological Experiments in Condensed Matter. *Nature* **1985,** *317,* 505–508.
3. Kibble, T. W. B. Topology of Cosmic Domains and Strings. *J. Phys. A: Math. Gen.* **1976,** *9,* 1387–1389.
4. Lavrentovich, O. D. Topological Defects in Dispersed Liquid Crystals, or Words and Worlds Around Liquid Crystal Drops. *Liq. Cryst.* **1998,** *24,* 117–125.
5. Volovik, G. E.; Lavrentovich, O. D. Topological Dynamics of Defects: Boojums in Nematic Drops. *Sov. Phys. JETP* **1983,** *58,* 1159–1166.
6. Palffy-Muhoray, P. The Diverse World of Liquid Crystals. *Phys. Today* **2007,** *60,* 54–60.
7. Bray, A. J. Theory of Phase-ordering Kinetics. *Adv. Phys.* **1994,** *43,* 357–459.
8. Bradač, Z.; Kralj, S.; Žumer, S. Molecular Dynamics Study of Isotropic-nematic Quench. *Phys. Rev. E. Stat. Nonlin. Soft Matter Phys.* **2002,** *65,* 021705-1-10.
9. De Gennes, P. G.; Prost, J. *The Physics of Liquid Crystals*; Oxford University Press: Oxford, 1993.
10. Kitzerov, H. G. Twist Grain Boundary Phases. In *Chirality in Liquid Crystals*; Kitzerov, H. G., Bahr, C., Eds.; Springer-Verlag: New York, 2001.
11. Kikuchi, H.; Yokota, M.; Hisakado, Y.; Yang, H.; Kajiyama, T. Polymer-stabilized Liquid-crystals Blue Phases. *Nat. Mater.* **2002,** *1,* 64–68.
12. Renn, S. R.; Lubensky, T. C. Abrikosov Dislocation Lattice in a Model of the Cholesteric to Smectic-A Transition. *Phys. Rev. A* **1988,** *38,* 2132–2147.
13. Dierking, I.; Lagerwall, S. T. A Review of Textures of the TGBA Phase Under Different Anchoring Geometries. *Liq. Cryst.* **1999,** *26,* 83–95.
14. Abrikosov, A. A. The Magnetic Properties of Superconducting Alloys. *J. Phys. Chem. Solids* **1957,** *2,* 199–208.
15. Poulin, P.; Stark, H.; Lubensky, T. C.; Wietz, D. A. Novel Colloidal Interactions in Anisotropic Fluids. *Science* **1997,** *275,* 1770–1773.
16. Pires, D.; Fleury, J. B.; Galerne, Y. Colloid Particles in the Interaction Field of a Disclination Line in a Nematic Phase. *Phys. Rev. Lett.* **2007,** *98,* 247801.

17. Liu, Q.; Cui, Y.; Gardner, D.; Li, X.; He, S.; Smalyukh, I. I. Self-alignment of Plasmonic Gold Nanorods in Reconfigurable Anisotropic Fluids for Tunable Bulk Metamaterial Applications. *Nano Lett.* **2010**, *10*, 1347–1353.

18. Tkalec, U.; Ravnik, M.; Čopar, S.; Žumer, S.; Muševič, I. Reconfigurable Knots and Links in Chiral Nematic Colloids. *Science* **2011**, *333*, 62–65.

19. Li, F.; Buchnev, O.; Cheon, C.; Glushchenko, A.; Reshetnyak, V.; Reznikov, Y.; Sluckin, T. J.; West, J. L. Orientational Coupling Amplification in Ferroelectric Nematic Colloids. *Phys. Rev. Lett.* **2006**, *97*, 147801.

20. Kinkead, B.; Hegmann, T. Effects of Size, Capping Agent, and Concentration of CdSe and CdTe Quantum Dots Doped into a Nematic Liquid Crystal on the Optical and Electro-optic Properties of the Final Colloidal Liquid Crystal Mixture. *J. Mater. Chem.* **2010**, *20*, 448–458.

21. Balazs, A. C.; Emrick, T.; Russel, T. P. Nanoparticle Polymer Composites: Where Two Small Worlds Meet. *Science* **2006**, *314*, 1107–1110.

22. Yoon, D. K.; Choi, M. C.; Kim, Y. H.; Kim, M. W.; Lavrentovich, O. D.; Jung, H. Internal Structure Visualization and Lithographic Use of Periodic Toroidal Holes in Liquid Crystals. *Nat. Mater.* **2007**, *6*, 866–870.

23. Coursault, D.; Grand, J.; Zappone, B.; Ayeb, H.; Levi, G.; Felidj, N.; Lacaze, E. Linear Self-Assembly of Nanoparticles Within Liquid Crystal Defect Arrays. *Adv. Mater.* **2012**, *24*, 1461–1465.

24. Karatairi, E.; Rožič, B.; Kutnjak, Z.; Tzitzios, V.; Nounesis, G.; Cordoyiannis, G.; Thoen, J.; Glorieux, C.; Kralj, S. Nanoparticle-induced Widening of the Temperature Range of Liquid-crystalline Blue Phases. *Phys. Rev. E. Stat. Nonlin. Soft Matter Phys.* **2010**, *81*, 041703.

25. Cordoyiannis, G.; Losada-Péreza, P.; Tripathi, C. S. P.; Rožič, B.; Tkalec, U.; Tzitzios, V.; Karatairi, E.; Nounesis, G.; Kutnjak, Z.; Muševič, I.; Glorieux, C.; Kralj, S.; Thoen, J. Blue Phase III Widening in CE6-dispersed Surface-functionalized CdSe Nanoparticles. *Liq. Cryst.* **2010**, *37*, 1419–1426.

26. Rožič, B.; Tzitzios, V.; Karatairi, E.; Tkalec, U.; Nounesis, G.; Kutnjak, Z.; Cordoyiannis, G.; Rosso, R.; Virga, E. G.; Muševič, I.; Kralj, S. Theoretical and Experimental Study of the Nanoparticle-driven Blue Phase Stabilization. *Eur. Phys. J. E Stat. Nonlin. Soft Matter Phys.* **2011**, *34*, 17.

27. Cordoyiannis, G.; Jampani, V. S. R.; Kralj, S.; Dhara, S.; Tzitzios, V.; Basina, G.; Nounesis, G.; Kutnjak, Z.; Tripathi, C. S. P.; Losada-Perez, P.; Jesenek, D.; Glorieux, C.; Musevic, I.; Zidansek, A.; Amenitsch, H.; Thoen, J. Different Modulated Structures of Topological Defects Stabilized by Adaptive Targeting Nanoparticles. *Soft Matter* **2013**, *9*, 3956–3964.

28. DiDonna, B. A.; Kamien, R. D. Smectic Blue Phases: Layered Systems with High Intrinsic Curvature. *Phys. Rev. E. Stat. Nonlin. Soft Matter Phys.* **2003**, *68*, 041703.

29. Nelson, D. R. Toward a Tetravalent Chemistry of Colloids. *Nano. Lett.* **2002**, *2*, 1125–1129.

30. Vitelli, V.; Nelson, D. R. Nematic Textures in Spherical Shells. *Phys. Rev. E. Stat. Nonlin. Soft Matter Phys.* **2006**, *74*, 021711.

31. Skačej, G.; Zannoni, C. Controlling Surface Defect Valence in Colloids. *Phys. Rev. Lett.* **2008**, *100*, 197802.

32. Kralj, S.; Rosso, R.; Virga, E. G. Curvature Control of Valence on Nematic Shells. *Soft Matter* **2011**, *7*, 670–683.

33. Jesenek, D.; Kralj, S.; Rosso, R.; Virga, E. G. Defect Unbinding on a Nematic Toroidal Shell. *Soft Matter* **2015**, *11*, 1–11. Submitted.

34. Rosso, R.; Virga, E. G.; Kralj, S. Parallel Transport and Defects on Nematic Shells. *Contin. Mech. Thermodyn.* **2012**, *24*, 643–664.
35. Kamien, R. D. The Topological Theory of Defects in Ordered Media. *Rev. Mod. Phys.* **2002**, *74*, 953–971.
36. Poincaré, H. Mémoire sur les courbes définies par une équation différentielle. *J. Math. Pures Appl.* **1886**, *2*, 151–217.
37. Bowick, M.; Nelson, D. R.; Travesset, A. Curvature-induced Defect Unbinding in Toroidal Geometries. *Phys. Rev. E. Stat. Nonlin. Soft Matter Phys.* **2004**, *69*, 041102.
38. Chan, T.; Garland, C. W.; Nguyen, H. T. Calorimetric Study of Chiral Liquid Crystals with a Twist-grain Boundary Phase. *Phys. Rev. E. Stat. Phys. Plasmas Fluids Relat. Interdiscip. Topics.* **1995**, *52*, 5000–5003.
39. Navailles, L.; Barois, P.; Nguyen, H. T. X-ray-measurement of the Twist Grain-boundary Angle in the Liquid-crystal Analog of the Abrikosov Phase. *Phys. Rev. Lett.* **1993**, *71*, 545.
40. Navailles, L.; Pansu, B.; Gorre-Talini, L.; Nguyen, H. T. Structural Study of a Commensurate TGB$_A$ Phase and of a Presumed Chiral Line Liquid Phase. *Phys. Rev. Lett.* **1998**, *81*, 4168–4171.
41. Schopohl, N.; Sluckin, T. J. Defect Core Structure in Nematic Liquid Crystals. *Phys. Rev. Lett.* **1987**, *59*, 2582–2584.
42. Kralj, S.; Virga, E. G.; Žumer, S. Biaxial Torus Around Nematic Point Defects. *Phys. Rev. E. Stat. Phys. Plasmas Fluids Relat. Interdiscip. Topics.* **1999**, *60*, 1858–1866.
43. Kralj, S.; Sluckin, T. J. Core Structure of a Screw Disclination in Smectic A Liquid Crystals. *Phys. Rev. E. Stat. Phys. Plasmas Fluids Relat. Interdiscip. Topics.* **1993**, *48*, R3244–R3247.
44. Kralj, S.; Sluckin, T. J. Landau de Gennes Theory of the Core Structure of a Screw Dislocation in Smectic A Liquid Crystals. *Liq. Cryst.* **1995**, *18*, 887–902.
45. Kralj, S.; Žumer, S. Freedericksz Transitions in Supramicron Nematic Droplets. *Phys. Rev. A* **1992**, *45*, 2461–2470.
46. Trček, M.; Cordoyiannis, G.; Tzitzios, V.; Kralj, S.; Nounesis, G.; Lelidis, I.; Kutnjak, Z. Nanoparticle-induced Twist Grain Boundary Phases. *Phys. Rev. E. Stat. Nonlin. Soft Matter Phys.* **2014**, *90*, 032501.
47. Loginov, E. B.; Terentjev, E. M. Smectic Dislocations. *Sov. Phys. Crystallogr.* **1987**, *30*, 4–7.
48. Haga, H.; Garland, C. W. Effect of Silica Aerosil Particles on Liquid-crystal Phase Transitions. *Phys. Rev. E. Stat. Nonlin. Soft Matter Phys.* **1997**, *56*, 3044–3054.
49. Yao, H.; Ema, K.; Garland, C. W. Nonadiabatic Scanning Calorimeter. *Rev. Sci. Instrum.* **1998**, *69*, 172–178.
50. Thoen, J.; Cordoyiannis, G.; Glorieux, C. Investigations of Phase Transitions in Liquid Crystals by Means of Adiabatic Scanning Calorimetry. *Liq. Cryst.* **2009**, *36*, 669–684.
51. Vitelli, V.; Turner, A. M. Anomalous Coupling Between Topological Defects and Curvature. *Phys. Rev. Lett.* **2004**, *93*, 215301.
52. Dubtsov, V.; Pasechnik, S. V.; Shmeliova, D. V.; Kralj, S. Light and Phospholipid Driven Structural Transitions in Nematic Microdroplets. *Appl. Phys. Lett.* **2014**, *105*, 151606.

CHAPTER 19

MORPHOLOGICAL AND DYNAMIC MECHANICAL BEHAVIOR OF GRAPHENE-OXIDE-FILLED POLYVINYLIDENE FLUORIDE

R. RAJASEKAR[1*], P. SATHISH KUMAR[2], S. MAHALAKSHMI[3], and J. H. LEE[4]

[1]*Department of Mechanical Engineering, Kongu Engineering College, Erode 638052, Tamil Nadu, India*

[2]*Department of Mining Engineering, Indian Institute of Technology, Kharagpur 721302, West Bengal, India*

[3]*Department of Mechanical Engineering, Erode Sengunthar Engineering College, Erode 638052, Tamil Nadu, India*

[4]*Department of Polymer and Nano Engineering, Chonbuk National University, Jeonju, Jeonbuk 561756, Republic of Korea*

Corresponding author. E-mail: rajasekar_cr@yahoo.com

CONTENTS

ABSTRACT

This research work analyzed the effect of graphene oxide (GO) on the morphological and dynamic mechanical properties of polyvinylidene fluoride (PVDF). The primary phase steps into preparation of graphite oxide from flake graphite by adopting modified Hummers method. Further, the dried graphite oxide had been ultrasonicated by mixing with dimethyl formamide (DMF) solvent and kept idle. The upper portion of the solvent containing particles was extracted and dried in order to extract GO. The obtained films had been observed through transmission electron microscopy (TEM), which confirmed the formation of GO from flake graphite.

PVDF–GO films were prepared by adopting solution mixing technique. PVDF was dissolved in DMF and subsequently GO with varying wt% (1.0, 1.5, 2.0, and 3.0) was dispersed in DMF. Further DMF containing GO was added to PVDF solution under continual stirring and finally casted to obtain complete dry film.

TEM images prove the homogenous distribution of GO in PVDF matrix, which may be due to polar–polar interaction between PVDF and GO. Dynamic mechanical analysis of GO-containing PVDF films shows an increase in storage modulus, subsequent decrease in peak shifting compared with pure PVDF film. Restriction in chain mobility owing to physical and chemical adsorption of PVDF chain on GO surface leads to decreased tan delta peak intensity. This corresponds to high reinforcing efficiency of GO in PVDF.

19.1 INTRODUCTION

PVDF is a polar highly nonreactive thermoplastic polymer in the fluoropolymer family that exhibits excellent chemical, mechanical, and electrical properties.[1-3] It shows different semicrystalline forms with complicated microstructures.[4] These properties combined with high elasticity and processability made PVDF one of the most studied polymers in scientific research. Since PVDF has a low density and low cost compared to other fluoropolymers, it has been used in many important commercial and technological applications such as ultrafiltration, microfiltration, microfiltration membranes, electrode binder in lithium ion batteries, nonlinear optics, microwave transducers, defense industries, biomedical applications and in potential applications such as piezoelectric film and pyroelectric materials. It is well known that nanoparticles are usually used to enhance the properties of polymer by simple blending;[5,6] the interfacial interaction between polymer and nanoparticles are crucial to the enhancement.

Among the graphene-based reinforcing materials, GO are promising nanofillers that exhibit unique structural features and physical properties and have recently generated enormous attention from industry and academia to attain high-performance polymeric nanocomposites as it has functional groups for better interaction with polymers. They have high surface-area-to-volume ratio, excellent mechanical strength, electrical conductivity, and thermal stability. GO is structurally monolayer of carbon atoms arranged in two-dimensional honeycomb lattice similar to graphene sheet. Since GO is hydrophilic in nature, it is challenging to disperse the same in weakly polar organic solvents and polymers. When incorporated approximately, GO can dramatically enhance the electrical, physical, static, and dynamic mechanical properties of polymer at extremely low filler loadings.

However, Jae-Wan Kim et al. studied the morphology, crystalline structure, and mechanical properties of PVDF composites blended with silica, the composites showed poor properties. The author considered that there was no interaction between silica and PVDF.[7] J. C. Wang et al. explained the interfacial interaction of PVDF blended with GO-coated silica hybrids using various characterization techniques. The results show limited improvement in tensile and dynamic mechanical properties of PVDF with silica hybrid-coated GO.[8] The research work deals in developing PVDF-based GO nanocomposites by adopting solution mixing technique. GO was prepared from flake graphite using modified Hummers method. Morphology, dynamic mechanical properties indifferent weight ratios of GO-loaded polymeric nanocomposites were analyzed and the obtained results were compared with control.

19.2 MATERIALS AND METHODS

PVDF with fluorine content of 59% in the form of powder was procured from Alfa Aesar. GO was synthesized from pure flake graphite by adopting modified Hummers method. First, PVDF was dissolved in DMF solvent and a desired amount of PVDF was added into as-prepared GO/DMF dispersion to prepare PVDF/GO nanocomposites using solution mixing method. The solutions were stirred at a temperature of around 60°C for 1 h to get homogenous solution. Ultrasonication was carried out for 10 min followed by continuous stirring. The content of GO in PVDF is 1.0 (F 1), 1.5 (F 2), 2.0 (F 3), and 3.0 wt% (F 4), respectively.

19.3 RESULTS AND DISCUSSION

19.3.1 TEM ANALYSIS

Figure 19.1 shows the TEM images of GO (2 and 3 wt%) dispersion in PVDF matrix by solution casting. From the images, it appears that nano-sized GO sheets are embedded in the PVDF matrix. In Figure 19.1a, the dark spots clearly indicate the nonuniform dispersion of GO in the polymer matrix. As can be seen, GO particles inhomogeneously dispersed in the polymer matrix and slight agglomeration can be noticed. In Figure 19.1b, the observed microscopic image clearly shows better GO dispersion in the PVDF polymer matrix.

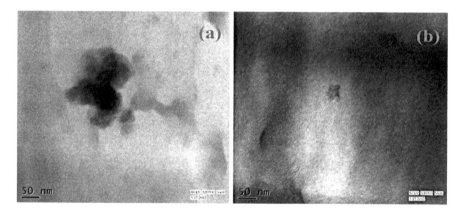

FIGURE 19.1   TEM images of (a) 2 wt% and (b) 3 wt% GO dispersed in PVDF matrix.

19.3.2 DMA

Figures 19.2a and b and 19.3 show the storage modulus (G') and tan δ plot with temperature for varying concentrations of GO-loaded PVDF nanocomposites and also show enhancement in G' and decrease in tan δ of solution casted pure PVDF compared to neat polymer. Significant enhancement in G' of F 1, F 2, F 3, and F 4 is due to homogeneous dispersion of GO in PVDF matrix. F4 shows decrease in G' at low temperature compared with F3. With an increase in temperature, the G' decreased slightly from −50°C to 90°C, and decreased rapidly above 90°C due to the melting of PVDF crystals. The increase in G' corresponds to high reinforcing efficiency and good polar–polar interaction between PVDF and GO. The peak intensity of F 1, F 2, F

3, and F 4 polymeric nanocomposites are low compared with that of pure polymer (C). With an increase in filler loading, the peak intensity of tan δ consecutively decreased to a certain extent. This may be due to good reinforcing efficiency of the filler with the PVDF matrix. The decrease in tan δ peak proves minimum heat buildup and as a result leads to better damping characteristics of the nanofiller-incorporated PVDF matrix.

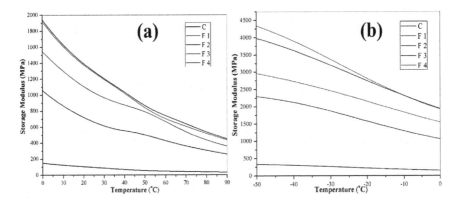

FIGURE 19.2 Storage modulus of pure and PVDF-containing GO nanocomposites.

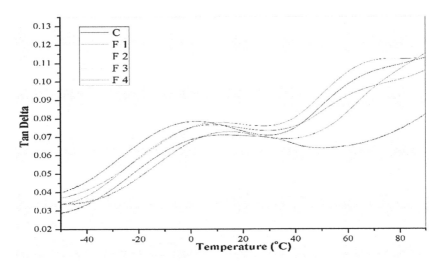

FIGURE 19.3 Tan δ of pure and PVDF-containing GO nanocomposites.

19.4   CONCLUSION

GO-incorporated PVDF nanocomposites were prepared by adopting solution mixing technique. Modified Hummers method was adopted to synthesize GO. The individually dispersed GO sheets in DMF are obtained by the solvent-exchange method. TEM analysis proves better dispersion of GO in the PVDF matrices. DMA result indicated that PVDF nanocomposites containing GO shows high G' and tan δ peak proving better damping characteristics compared with pure.

KEYWORDS

- **polyvinylidene fluoride**
- **graphene oxide**
- **storage modulus**
- **tan δ**
- **nanocomposites**

REFERENCES

1. Lovinger, A. Ferroelectric Polymers. *Science* **1983,** *220*, 1115–1121.
2. Wang, M.; Shi, J. H.; Pramoda, K. P.; Goh, S. H. Microstructure, Crystallization and Dynamic Mechanical Behaviour of Poly (Vinylidene Fluoride) Composites Containing Poly (Methyl Methacrylate)-Grafted Multi-Walled Carbon Nanotubes. *Nanotechnology* **2007,** *18*, 235701.
3. Yee, W. A.; Kotaki, M.; Liu, Y.; Lu, X. Morphology, Polymorphism Behavior and Molecular Orientation of Electrospun Poly (Vinylidene Fluoride) Fibers. *Polymer* **2007,** *48*, 512–521.
4. Chen; D.; Wang, M.; Zhang,W.-D.; Liu, T. Preparation and Characterization of Poly(Vinylidene Fluoride) Nanocomposites Containing Multiwalled Carbon Nanotubes. *J. Appl. Polym. Sci.* **2009,** *113*, 644–650.
5. Casciola, M.; Capitani, D.; Donnadio, A.; Diosono, V.; Piaggio, P.; Pica, M. Polyvinylidene Fluoride/Zirconium Phosphate Sulfo-Phenyl Phosphonate Nanocomposite Films: Microstructure and Mechanical Properties. *J. Mater. Chem.* **2008,** *18*, 4291–4296.
6. Chen, D.; Wang, M.; Zhang, W. D.; Liu, T. Preparation and Characterization of Poly (Vinylidene Fluoride) Nanocomposites Containing Multiwalled Carbon Nanotubes. *J. Appl. Polym. Sci.* **2009,** *113*, 644–650.

7. Kim, J. W.; Cho, W. J.; Ha, C. S. Morphology, Crystalline Structure and Properties of Poly(Vinylidene Fluoride)/Silica Hybrid Composites. *J. Polym. Sci. Pol. Phys.* **2002,** *40,* 19–30.
8. Wang, J. C.; Chen, P.; Chen, L.; Wang, K.; Deng, H.; Chen, F.; Zhang, Q.; Fu, Q. Preparation and Properties of Poly (Vinylidene Fluoride) Nanocomposites Blended with Graphene Oxide Coated Silica Hybrids. *Express Polym. Lett.* **2012,** *6,* 299–307.

CHAPTER 20

A THEORETICAL STUDY ON THE PHYSICO-MECHANICAL BEHAVIOR OF POLYESTER COMPOSITES USING DIFFERENT CLASSES OF NATURAL FIBER REINFORCEMENTS

A. MOHANKUMAR[1], R. PARAMESHWARAN[1],
M. MOHAN PRASATH[1], S. M. SENTHIL[1], P. SATHISH KUMAR[1,2],
C. MOGANAPRIYA[1], and R. RAJASEKAR[1*]

[1]*School of Building and Mechanical Sciences, Kongu Engineering College, Erode 638052, Tamil Nadu, India*

[2]*Department of Mining Engineering, Indian Institute of Technology, Kharagpur 721302, West Bengal, India*

*Corresponding author. E-mail: rajasekar_cr@yahoo.com

CONTENTS

ABSTRACT

Composite materials are the combination of one or two materials with different chemical or physical properties, which produce a material with characteristics different from that of the parent materials. The manufacturing of fiber-reinforced polymer composites involves incorporation of large number of fibers into the polymer matrix. The fiber-reinforced polymer composites find importance in many applications such as aircraft, military, marine, space, sporting goods, and infrastructures. This investigation evaluated the tensile, toughness, and impact properties of polyester-reinforced natural fibers with different weight percentages (10%, 20%, 30%, and 40%) by using mathematical approach and ANSYS simulation. The results obtained from mathematical modeling are compared with ANSYS simulation. Obtained results show enhancements in mechanical properties for each and every increase in fiber content. Further, for the tensile strength application of polyester, the fibers abaca, hemp, and banana exhibit better properties. For tensile modulus application, the fibers ramie, abaca, sisal are suitable. Similarly, for toughness and impact strength application, the fibers banana, sisal, and cotton are suitable.

20.1 INTRODUCTION

In present scenario, the fiber-reinforced polymers have replaced many conventional metals/materials in various applications because of its outstanding properties.[1] A fiber-reinforced polymer composite material consists of polymer matrix incorporation with high-strength fibers. The high-strength fibers are aramid, glass, and carbon. Recent researches have brought out the advantages of natural fibers (viz. low cost, low density, high specific properties, biodegradable, and nonabrasive), making it as an effective alternate for conventional fibers.[2–5]

Natural fibers are classified into three categories: seed hair, bast fibers, and leaf fibers. The various natural fibers are jute, flax, abaca, sisal, kenaf, ramie, hemp, cotton, coir, banana, henenquen, bagasse, bamboo, gomuti, and palm. Of these fibers, the most commonly used fibers are jute, ramie, flax, coir, hemp, and sisal. The specific modulus of natural fibers is found to be better than that of synthetic fibers such as aramid, glass, and carbon. The fiber-incorporated polymer composites are employed in several applications where the weight reduction is the essential need.[1]

Commonly, natural fibers consist of cellulose, hemicellulose, pectin, and lignin. Cellulose is a crystalline structure that can be assembled into microfibers with adequate steadiness. Generally, cellulose fibers contain more than 500,000 cellulose molecules. The base for high tensile strength (TS) of a natural fiber is due to H-bonds of cellulose molecules. Hemicellulose contains tiny and group of branched chain sugars containing five-carbon sugars, six-carbon sugars, and uronic acid. As hemicellulose can be easily hydrolyzed to its constituent, in contrast to cellulose, it yields more six-carbon sugars that are crystalline, strong, opposing to hydrolysis, and have amorphous structure with small strength. Lignin is of aromatic structure formed by nonreversible process of removal of water from sugar. Lignin molecules are three-dimensional structures formed by cross-linking. Lignin with hemicellulose exploits the strength of cellulose and offers benefit of flexibility.[19]

20.2 EXTRACTION OF FIBERS

20.2.1 FRUIT FIBERS

Light and hairy fruit fibers are obtained from fruits of the plant. The different fruit fibers are cotton, coir, and oil palm empty fruit bunches (EFBs). Cotton fiber, widely used in textile industry, is obtained from the fruit of cotton by labor-intensive picking process. Cotton fiber has high moisture absorption content and act as a good heat conductor. It is widely used in applications such as clothes, carpets, blankets, and mobs.[6] When large scale of extraction is accounted, the picking is carried out by machine. By adopting ginning process in picked cotton (cotton wool), the seeds, dead leaves, and other debris are washed away. Cotton fiber is weak compared to other natural fibers.

Coconut (coir) fiber is obtained from husk of the coconut fruit. Coir fibers are subtracted from the husk using retting process followed by beating and washing. Coir fibers have the ability of better reinforcement with both thermoset and thermoplastic resins. It has better adhesive property under dry conditions. Coir-fiber-reinforced polymer composites has many applications such as automotive interior, paneling and roofing, storage tank, packaging material, helmets, and postboxes.[7–11]

20.2.2   STEM FIBERS (BAST FIBERS)

The stem fibers are acquired from the stems of the plant. The length of these fibers is long because they can grow across the whole length of the stem. Few of bast fibers are jute, flax, ramie, and hemp. The jute fiber is originated in the ribbon of the stem. The jute fibers are obtained by continuously retting in water followed with beating and stripping the core. The jute fibers possess characteristics similar to wood such as high aspect ratio, excessive strength-to-weight ratio, and better insulation properties. It is used in many applications such as furniture, window, corrugated sheet, and water pipes.[6,7] The flax fibers are separated from the stem of the plant *Linum usitatissimum*. The flax fiber is obtained by the process of dressing (a process consisting of beating, shaking, and hackling). The flax fiber is strong, degrades due to sunlight, and burns when set ablaze. It has good heat-conducting properties, hard wearing, and durable. It is used in applications such as canvas, ropes, and sacks.[6] The ramie fiber is extracted from the stem of the plant *Boehmeria nivea* which belongs to nettle group. The ramie fiber is obtained by using chemicals than retting because it is a complex process due to gummy substance surrounding the fibers which may damage the fibers. Ramie fiber is very expensive and durable. It is used in applications such as curtains, wallpaper, sewing thread, and furniture curves.[6] Kenaf fibers are obtained from stem fiber of kenaf plants which contains two fibers: lengthy fibers and tiny fibers. The lengthy fiber bunches are located in cortical layer, while tiny fibers are situated in ligneous zone. The advantages of kenaf fibers are renewability, distinctive, consistent, and are bio-friendly. Kenaf fibers are used widely nowadays because they can store carbon dioxide at huge rate and assimilates phosphorous and nitrogen from the soil. Kenaf fibers are used in applications such as paper products, textiles, building materials, and absorbents.[13,15]

20.2.3   LEAF FIBERS

Leaf fibers are extracted from leaves of the plant. The leaf fibers are rough and sturdy. Different leaf fibers are sisal, abaca, and banana fibers. Sisal fiber is obtained by the process of decortications. In this process, leaves are squeezed and broken by rotating wheel with knives. Finally fibers are dried, brushed, and arranged for export. Henequen is also one of leaf fiber which is a member of agave family which is identical to sisal plant. Sisal fiber is used as a raw material in plastic industry. It can also be used as a renewable source and nonfood source of economic development for rural and farming areas. Sisal fibers have key applications such as making twines, ropes, cords, and packaging materials.[6,14]

Banana fibers are incurred from the false stem of banana plant and it is also a stem fiber. At low cost or without any investment, the banana fibers can be used for industrial purposes. Abaca fiber belongs to banana family. Abaca fiber is obtained from the leaf sheath around the trunk of the abaca plant. Abaca fiber has superior TS and flexural strength close to glass fiber. It is very sturdy and resistant to seawater. The major applications of abaca fibers are manufacturing of ropes and handicraft goods, etc.[6,16] The mechanical properties of several natural fibers are given in Table 20.1.

TABLE 20.1 Mechanical Properties of Various Natural Fibers.[17–30]

S. No	Name of the fiber	Density (g/cm³)	Tensile strength (MPa)	Tensile modulus (GPa)	Elongation at break (%)	Poisson ratio
1	Flax[19,23]	1.40–1.50	343–1035	27–80	2.7–3.2	0.21
2	Jute[19,24]	1.30–1.50	187–773	3–55	1.4–3.1	0.38
3	Abaca[19,29]	1.50	980	72	10–12	0.3
4	Sisal[19,29]	1.30–1.50	507–855	9–28	2–2.9	0.32
5	Kenaf[19,22]	1.22–1.40	295–930	22–53	3.7–6.9	0.342
6	Ramie[19,25]	1.51	400–938	44–128	3.6–3.8	0.3
7	Hemp[19,22]	1.40–1.50	580–1110	3–90	1.3–4.7	0.221
8	Cotton[19,28]	1.50–1.60	287–597	5.5–12.6	2–10	0.33
9	Coir[19,27]	1.25–1.50	106–270	3–6	15–47	0.3
10	Banana[19,29]	1.30–1.35	529–914	7.7–32	3–10	0.3
11	Henequen[19,29]	1.49	430–580	10.1–16.3	3–5	0.32
12	Bagasse[19,30]	0.55–1.25	20–290	2.7–17	0.9	0.27
13	Bamboo[17,26]	0.91	503	35.91	1.4	0.3
14	Gomuti palm[18,21]	1.29	190.29	3.69	–	0.336
15	Oilpalm EFB[20,21]	0.7–1.55	71	1.7	17	0.336
16	Palymrah palm[17,21]	1.09	180–215	7.4–604	7–15	0.336
17	Talipot palm[17,21]	0.890	143–294	9.3–13.3	2.7–5	0.336

Polymers are classified into two categories: thermoset polymers and thermoplastic polymers. In thermoplastic polymer, the molecules are not joined together but only held together by feeble minor bonds or intermolecular forces. Once the component is made, the molecules are restored and reshaped after cooling. Few of the thermoplastic polymers are PEEK, polysulfone, and polyamides. In a thermoset polymer, the molecules are connected together by cross-link and make three-dimensional structures. Once the cross-links are shaped during polymerization, there is no restoration or reshaping of molecules upon cooling. Few of thermoset polymers are epoxy, polyester, vinyl ester, cynate ester, and others.[34] Presently, thermoplastics are used as a matrix for biofibers and most generally used thermoplastics are polypropylene, polyethylene, and polyvinyl chloride, while most commonly used thermosetting matrices are phenolic, epoxy, and polyester resins.[2,3,12]

Generally, the polyester contains 50–85% of commercial polyester resins. Commercial polyester resins contain three components: polyester, monomer, and inhibitor. The most commonly used monomer for polyester is styrene. Hydroquinone is used as an inhibitor for polyester resin to increase the lifetime of the polyester. The polyesters are prepared by esterification of one or more dibasic acids with a glycol. The commonly used dibasic acid is either maleic acid or fumaric acid. The most commonly used glycols are ethylene, diethylene, and propylene.[31] Table 20.2 provides the properties of polyester resin. Unsaturated polyester resins are used in industrial and consumer application. The application of polyester is divided into two categories: reinforced and nonreinforced. In reinforced application, the resin and fibers are combined together to produce a composite with improved mechanical properties. The reinforced applications are boats, cars, shower stalls, building panels, and corrosion resistant tanks and pipes. In nonfiber-reinforced application, the filler is added into the composite for property modification. Alternatively, the nonfiber-reinforced applications are sinks, bowling balls, and coatings.[32] The properties of polyester resin are listed in Table 20.2.

The research work aims in analyzing the effect of various natural fibers with different weight fraction (10%, 20%, 30%, and 40%) on the properties of polyester resin with the aid of mathematical calculation and ANSYS simulation. The mechanical properties such as TS, modulus of elasticity, toughness and impact strength of the fiber-incorporated polyester composites are calculated using mathematical and ANSYS prediction methods and compared with the properties of pure matrix. The sample code with their compositions for various fiber-incorporated polyester composites are listed in Table 20.3.

TABLE 20.2 Properties of Polyester Resin.[2,39]

Property	Polyester resin
Density (g/cm³)	1.2–1.5
Elastic modulus (GPa)	2–4.5
Tensile strength (MPa)	40–90
Compressive strength (MPa)	90–250
Elongation (%)	2
Cure shrinkage (%)	4–8
Water absorption (24 h at 20°C)	0.1–0.3
Poisson ratio	0.4

TABLE 20.3 Sample Code with Composition of Various Fiber-Reinforced Polyester Composites.

S. No	Sample code	Composition
1	PF	Polyester + Flax
2	PJ	Polyester + Jute
3	PA	Polyester + Abaca
4	PS	Polyester + Sisal
5	PK	Polyester + Kenaf
6	PR	Polyester + Ramie
7	PH	Polyester + Hemp
8	PCO	Polyester + Cotton
9	PC	Polyester + Coir
10	PB	Polyester + Banana
11	PHQ	Polyester + Henequen
12	PBG	Polyester + Bagasse
13	PBO	Polyester + Bamboo
14	PGO	Polyester + Gomuti palm
15	POP	Polyester + Oilpalm palm
16	PPY	Polyester + Palymrah palm
17	PT	Polyester + Talipot palm

20.3   MECHANICAL PROPERTIES

20.3.1   TENSILE STRENGTH

Ultimate tensile strength (UTS) is the highest stress that a material can endure while being pulled before yielding or shattering. UTS is also named TS or ultimate strength. It is to be computed by representing stress versus strain. The highest point of the stress–strain curve is called as the UTS. It depends on several factors such as the preparation of the sample, the presence of exterior faults, and the temperature of the test environment and material.

TS of a material is given by:[35]

$$\sigma = \frac{P}{A},$$ (20.1)

where σ—Tensile strength (N/m^2)
P—Test load (N)
A—Cross-sectional area of sample (mm^2)

20.3.2   TOUGHNESS

Toughness is the ability of a material to assimilate energy and plastically distort without breaking. The Charpy and Izod impact tests are used to measure toughness. The area under stress–strain curve is termed toughness.

The toughness of a material is given by,[36]

$$U_T = \sigma \times \varepsilon$$ (20.2)

where U_T—Tensile toughness (J/m^3)
σ—Tensile strength (N/m^2)
ε—Strain of a material

20.3.3   IMPACT STRENGTH

Impact strength is the capability of the material to endure a suddenly applied load and is expressed in terms of energy. The impact strength depends on volume, modulus of elasticity, distribution of forces, and yield strength.

The impact strength of a material is given by:[37]

$$G_C = \frac{U_C}{A}$$ (20.3)

where G_C—Impact strength (J/m²)
$\quad\quad U_C$—Absorbed energy (J)
$\quad\quad A$—Cross-sectional area of sample (mm²)

20.4 MATHEMATICAL APPROACH

In this study, the direction of fiber is assumed to be unidirectional. In unidirectional composite structures, the fibers carry major load as the tensile force acts along the axial direction of the fibers. The composite specimen is selected based on the properties to be computed. The standard specification (ASTM D638-I) was used with dumb-bell section. The dimension of the composite specimen is 165 mm × 19 mm × 3 mm.

For unidirectional continuous fiber composite,[33,34]

1. Ultimate tensile strength of composite

$$\left(\sigma_c\right)^T = \left(\sigma_f\right)_{ult} V_f + \left(\sigma_m\right)_{ult} V_m \tag{20.4}$$

2. Ultimate strain of fiber

$$\left(\varepsilon_f\right)_{ult} = \frac{\left(\sigma_f\right)_{ult}}{E_f} \tag{20.5}$$

3. Ultimate strain of matrix

$$\left(\varepsilon_m\right)_{ult} = \frac{\left(\sigma_m\right)_{ult}}{E_f} \tag{20.6}$$

4. Tensile modulus of composite

$$E_c = E_f V_f + E_m V_m \tag{20.7}$$

5. Cross section of the composite

$$A_c = t_c h \tag{20.8}$$

where, $\left(\sigma_f\right)_{ult}$—Ultimate tensile strength of fiber (MPa)
$\quad\quad \left(\sigma_m\right)_{ult}$—Ultimate tensile strength of matrix (MPa)
$\quad\quad V_f$—Volume fraction of fiber
$\quad\quad V_m$—Volume fraction of matrix

E_f—Tensile modulus of fiber (GPa)

E_m—Tensile modulus of matrix (GPa)

t_c—Thickness of the specimen (mm)

h—Height of the specimen (mm)

Using above formulae, the UTS of composite, ultimate strain of fiber, matrix, and composite were calculated for fiber-reinforced polyester composites with different weight fraction of fibers (10%, 20%, 30%, and 40%). Alternatively, tensile moduli, toughness, and impact strength of the prepared composites were calculated.

20.5 ANSYS SIMULATION

In this study, ANSYS program version (14) was employed to determine the TS, toughness, impact strength of various fiber-reinforced polyester composite materials. Specific properties for both resin and fibers were imported in database of ANSYS program and then the obtained data were drawn after applying the loads. The load applied to the specimen is calculated by using mathematical model. In this study, the standard specification (ASTM D638-I) was used with dumb-bell section. Specifications used to draw the samples are given in Table 20.4.

TABLE 20.4 Specifications Used to Draw the Sample.

Test	Specimen dimension	Model	Types of element	Elements no	Nodes no
Tensile, toughness, and impact	165 mm × 19 mm × 3 mm	Linear	Solid 185 geometry, 8 nodes	58,466	13,049

Unidirectionally aligned fiber composites are a special group of orthotropic material. The unidirectional composites are also called as transversely isotropic materials because the elastic properties are identical in two to three directions. Commonly the composite specimen needs nine elastic constants but for transversely isotropic materials, the elastic properties are same in the two to three directions so that $E_{22} = E_{33}$, $G_{12} = G_{13}$, and $\gamma_{12} = \gamma_{13}$. In our study, the thickness of the specimen was little and so it is taken to be thin plate. For a thin plate, there is no out-of-plane loads and is considered to be under plane stress and isotropic.[33,38] The plane stress conditions are applied for specimen and von Mises stress and strain values are obtained for various polyester-reinforced composites with different weight fraction of fibers (10%, 20%, 30%, and 40%). The stress and strain values obtained for abaca- and oil-palm-reinforced composites at 40% weight of fiber are shown in Figures 20.1–20.4.

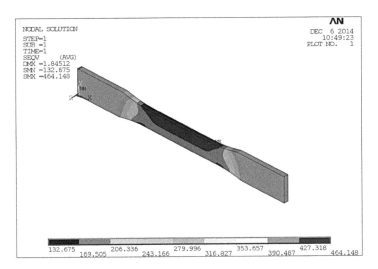

FIGURE 20.1   Stress value of abaca fiber polyester reinforced composite with 40% weight fraction.

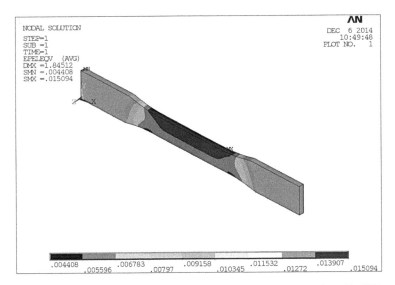

FIGURE 20.2   Strain value of abaca fiber polyester reinforced composite with 40% weight fraction.

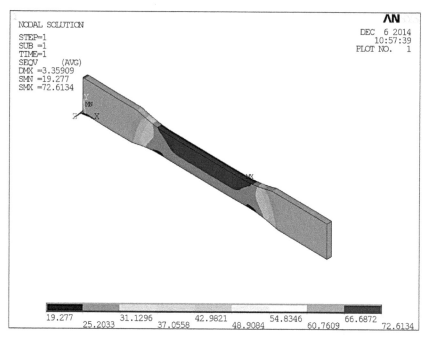

FIGURE 20.3   Stress value of oil palm fiber polyester reinforced composite with 40% weight fraction.

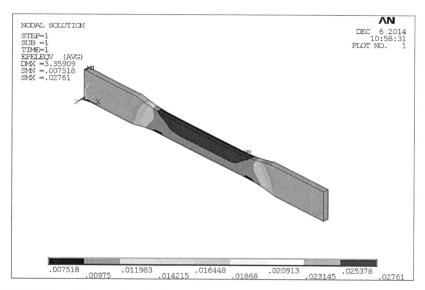

FIGURE 20.4   Strain value of oil palm fiber polyester reinforced composite with 40% weight fraction.

20.6   RESULTS AND DISCUSSION

The results attained from mathematical approach had been compared with ANSYS simulation for various polyester-reinforced composites with dissimilar weight fraction of fibers (10%, 20%, 30%, and 40%). It is found that results obtained from mathematical approach and ANSYS simulation have superior interrelation. Mechanical properties of different fibers added to polyester resin at different weight fractions attained using mathematical approach and ANSYS simulations are shown in Figures 20.5–20.12. Addition of fibers (10%, 20%, 30%, and 40%) in polyester resin results in improvement of TS, modulus, toughness, and impact strength of the polyester composites.

The TS of different weight fraction of various fibers in polyester resin using mathematical approach and ANSYS simulation is depicted in Figures 20.5 and 20.6. Abaca, hemp, and banana fiber-reinforced polyester composite shows maximum TS due to high cellulose content and H-bonding compared with all other fibers and suitable for TS application.[19] If the modeling is done mathematically, addition of abaca fiber (10%, 20%, 30%, and 40%) in polyester resin shows 140.8%, 281.54%, 422.3%, and 563.1% improvement in TS compared to pure polyester. Similarly, incorporating hemp fiber (10%, 20%, 30%, and 40%) in polyester resin results in 143, 221, 299, and 377 MPa of TS respectively. Incorporating 10%, 20%, 30%, and 40% of banana fiber in polyester resin shows 101%, 202%, 303%, and 404% increase in TS compared to pure. Instead of using mathematical approach, if simulated by ANSYS, then the addition of abaca fiber 10%, 20%, 30%, and 40% results in 159.5%, 311.23%, 461.53%, and 614.92% enhancement of TS compared to pure. Similarly, adding hemp fiber 10%, 20%, 30%, and 40% in polyester resin results in 154.1, 238, 321.8, and 405.6 MPa of TS, respectively. Incorporating 10%, 20%, 30%, and 40% of banana fiber in polyester resin results in 116.76%, 225.38%, 334.15%, and 442.76% enhancement of TS compared to pure.

The tensile modulus of diverse weight fraction of various fibers in polyester resin using mathematical approach and ANSYS simulation is depicted in Figures 20.7 and 20.8. Ramie, abaca, and flax fibers-reinforced polyester composite exhibit highest tensile modulus due to small elongation at break percentage, that is, small-strain value, over to all other fibers. Even though bagasse and bamboo fibers have small elongation at break percentage, their cellulose and hemicellulose contents are little when compared to the excelled fibers in tensile modulus values leading to lower TS, which in turn decreases the modulus.[19] Using mathematical approach, incorporation of 10%, 20%, 30%, and 40% weight fraction of ramie fiber leads to 254%, 509%, 788%,

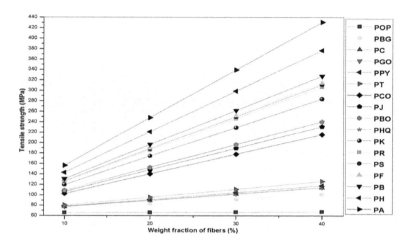

FIGURE 20.5   Tensile strength for 10%, 20%, 30%, and 40% addition of various fiber-reinforced polyester composites using mathematical approach.

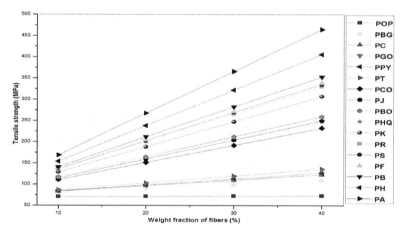

FIGURE 20.6   Tensile strength for 10%, 20%, 30%, and 40% addition of various fiber-reinforced polyester composites using ANSYS simulation.

and 1018% enhancement in modulus of elasticity over to pure polyester. With same weight fraction, the addition of abaca fiber in polyester resin shows 10.125, 17, 23.88, and 30.75 GPa of tensile modulus, respectively. Similarly, the addition of 10%, 20%, 30%, and 40% weight fraction of flax fiber displays 154%, 309%, 464%, and 617% improvement in tensile modulus compared to pure. But in ANSYS simulation, the addition of ramie fiber 10%, 20%, 30%, and 40% results in 254.7%, 509.23%, 763.69%, and 1009.23% improvement of tensile modulus compared to pure polyester. Adding abaca fiber 10%, 20%, 30%, and 40% in polyester resin results in 10.12, 17, 23.87, and 30.75 GPa of tensile modulus, respectively. Similarly, adding 10%, 20%, 30%, and 40% of flax fiber in polyester resin results in 154.46%, 309.23%, 463.69%, and 618.46% improvement of tensile modulus compared to pure.

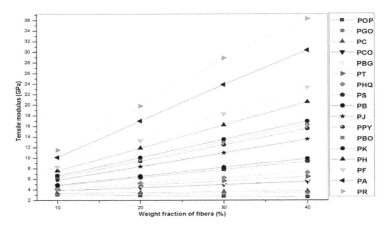

FIGURE 20.7 Tensile modulus for 10%, 20%, 30%, and 40% addition of various fiber-reinforced polyester composites using mathematical approach.

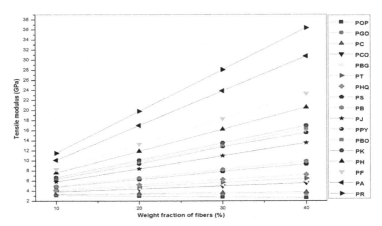

FIGURE 20.8 Tensile modulus for 10%, 20%, 30%, and 40% addition of various fiber-reinforced polyester composites using ANSYS simulation.

Alike TS and modulus, toughness and impact strength of different weight fraction of various fibers in polyester resin using mathematical approach and ANSYS simulation are depicted in Figures 20.9–20.12. As a result, banana-, sisal-, and cotton-fiber-strengthened polyester composite exhibit higher toughness and impact strength as presented in Table 20.5. This may be due to elevated cellulose and elevated elongation percentage compared to other fibers. Although, abaca and hemp have elevated cellulose and hemicellulose content, it does not have better elongation at break percentage to enhance the property with polyester resin.[19]

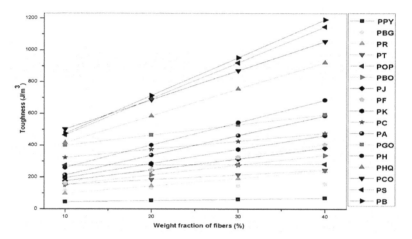

FIGURE 20.9   Toughness for 10%, 20%, 30%, and 40% addition of various fiber-reinforced polyester composites using mathematical approach.

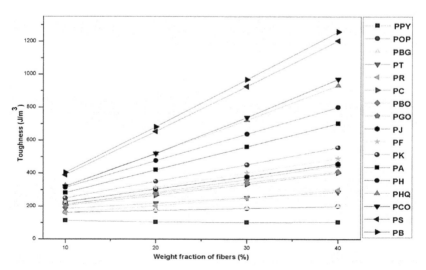

FIGURE 20.10   Toughness for 10%, 20%, 30%, and 40% addition of various fiber-reinforced polyester composites using ANSYS simulation.

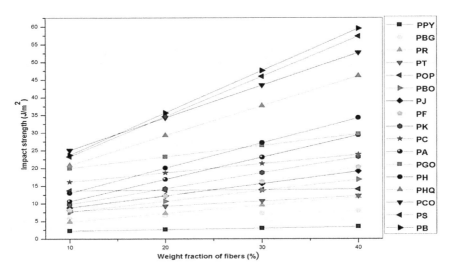

FIGURE 20.11 Impact strength for 10%, 20%, 30%, and 40% addition of various fiber-reinforced polyester composites using mathematical approach.

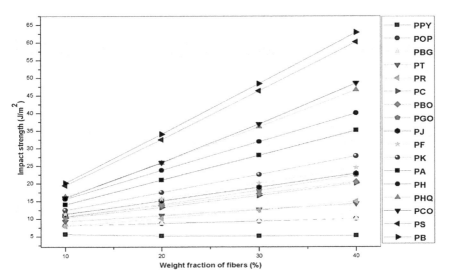

FIGURE 20.12 Impact strength for 10%, 20%, 30%, and 40% addition of various fiber-reinforced polyester composites using ANSYS simulation.

TABLE 20.5   Toughness and Impact Strength of Banana, Sisal, and Cotton Fiber-Reinforced Polyester Composites.

Fiber	Volume fraction (%)	Toughness (J/m³) (mathematical)	Toughness (J/m³) (ANSYS simulation)	Impact strength (J/m²) (mathematical)	Impact strength (J/m²) (ANSYS simulation)
Banana	10	474.881	404.214	23.74	20.21
	20	713.503	681.051	35.68	34.05
	30	952.125	967.664	47.61	48.38
	40	1190.747	1258.543	59.53	62.93
Sisal	10	466.024	390.213	23.30	19.51
	20	692.779	653.338	34.64	32.52
	30	919.534	926.021	45.98	46.30
	40	1146.288	1203.706	57.31	60.19
Cotton	10	501.584	320.067	25.08	16
	20	685.710	519.246	34.29	25.96
	30	869.836	738.124	43.49	36.91
	40	1053.962	970.501	52.70	48.53

20.7   CONCLUSION

The theoretical outcomes obtained from several fibers (jute, flax, abaca, sisal, kenaf, ramie, hemp, cotton, coir, banana, henenquen, bagasse, bamboo, gomuti, and palm fibers) evidence that the addition of increasing weight fraction of fiber content in polyester matrix results in enhancing the mechanical properties of the composite compared to pure polyester. The TS of polyester polymer composite is more prominent for abaca, hemp, and banana fibers at various loading. In mathematical approach, at higher weight fraction of fiber (40%) in polyester resin, higher TS of 431 MPa (561.3% improvement compared to pure) is attained for abaca fiber and with respect to ANSYS simulation TS of about 464 MPa (614.92% enhancement compared to pure) is obtained. Similarly, modulus of elasticity is observed to be elevated for ramie, abaca, and flax fibers because of its small deformation. Using mathematical approach, incorporating higher weight fraction (40%) of fiber content in matrix results in enhanced tensile modulus of 36.35 GPa (1018% increase compared to pure matrix) for ramie fiber and 30.75 GPa (846% increase compared to pure) for abaca fiber, whereas in ANSYS simulation the result obtained is similar to mathematical approach. Higher toughness and impact strength of various fiber-reinforced polyester composites are obtained for banana, sisal, and cotton fibers. During mathematical approach, incorporation of 40% fiber content in polyester matrix shows maximum toughness of 1190.747 J/m^3 and impact strength 58.96 J/m^2 for banana fiber which shows consistent results to that of ANSYS method where toughness of 1258.543 J/m^3 and impact strength of 59.53 J/m^2 are obtained, respectively.

KEYWORDS

- **polyester resin**
- **natural fibers**
- **mathematical approach**
- **mechanical properties**
- **abaca fiber**

REFERENCES

1. NabiSaheb, D.; Jog, J. P. Natural Fiber Polymer Composites: A Review. *Adv. Polym. Tech.* **1999,** *18*, 351–363.
2. Ku, H.; Wang, H.; Pattarachaiyakoop, N.; Trada, M. A Review on Tensile Properties of Natural Polymer Composites. *Compos. Part B: Eng.* **2011,** *42*, 856–873.
3. Groover, M. P. *Fundamental of Modern Manufacturing*, 2nd edition; John Wiley & Sons, Inc.: Hoboken (NJ), 2011.
4. Malkapuram, R.; Kumar, V.; Yuvraj, S. N. Recent Development in Natural Fibre Reinforced Polypropylene Composites. *J. Reinf. Plast. Compos.* **2008,** *28*, 1169–1189.
5. Li, X.; Tabil, L. G.; Panigrahi, S.; Crerar, W. J. The Influence of Fiber Content on Properties of Injection Molded Flax Fiber-HDPE biocomposites. *Can. Biosyst. Eng.* **2009,** *148*(8), 1–10.
6. Sfiligoj Smole, M.; Hribernik, S.; Stana Kleinschek, K.; Kreze, T. Plant Fibers for Textile and Technical Applications. *Adv. Agrophys. Res.* **2013,** *15*, 370–398.
7. Bongarde, U. S.; Shindle, V. D. Review on Natural Fiber Reinforcement Polymer Composites. *Int. J. Eng. Sci. Innov. Technol.* **2014,** *3*, 431–436.
8. Ayrilmis, N.; Jarusombuti, S.; Fueangvivat, V.; Bauchongkol, P.; White, R. H. Coir Fiber Reinforced Polypropylene Composite Panel for Automotive Interior Applications. *Fibers Polym.* **2011,** *12*, 919–926.
9. Verma, D.; Gope, P. C.; Shandilya, A.; Gupta, A.; Maheshwari, M. K. Coir Fiber Reinforcement and Application in Polymer Composites: A Review. *J. Mater. Environ. Sci.* **2013,** *4*, 263–276.
10. Yousif, B. F.; Ku, H. Suitability of Using Coir Fiber/Polymeric Composite for the Design of Liquid Storage Tanks. *Mater. Des.* **2012,** *36*, 847–853.

11. Abilash, N.; Sivapragash, M. Environmental Benefits of Eco-friendly Natural Fiber Reinforced Polymeric Composite Materials. *Int. J. Appl. Innov. Eng. Manag.* **2013,** *2*(1), 53–59.

12. Zampaloni, M.; Pourboghrat, F.; Yankovich, S. A.; Rodgers, B. N.; Moore, J.; Drazal, L. T.; Mohanty, A. K.; Misra, M. Kenaf Natural Fiber Reinforced Polypropylene Composites: A Discussion on Manufacturing Problems and Solution. *Compos. Part A: Appl. Sci. Manuf.* **2007,** *38*, 1569–1580.

13. Akil, H. M.; Omar, M. F.; Mazuki, A. A.; Safiee, S.; Ishak, Z. A. M.; Abu Bakar, A. Kenaf Fiber Reinforced Composites: A Review. *Mater. Des.* **2011,** *32*, 4107–4121.

14. Joseph, K.; Toledo Filho, R. D.; James, B.; Thomas, S.; de Carvalho, L. H. A Review on Sisal Fiber Reinforced Polymer Composites. *Rev. Bras. Eng. Agrí.* **1999,** *3*, 367–379.

15. Michell, A. Composites Containing Wood Pulp Fibres. *Appita* **1986,** *39*(3), 223–229.

16. Bledzki, A. K.; Mamum, A. A.; Faruk, O. Abaca Fibre Reinforced PP Composites and Composition with Jute and Flax Fibre PP Composites. *Express Polym. Lett.* **2007,** *11*, 755–762.

17. Sathishkumar, T. P.; Navaneethankrishnan, P.; Shankar, S. Tensile and Flexular Properties of Snake Grass Natural Fiber Reinforced Isophthalic Polyester Composites. *Compos. Sci. Technol.* **2012,** *72*, 1183–1190.

18. Ticoalu, A.; Aravinthana, T.; Cardona, F. *A Review on Current Development in Natural Fiber Composites for Structural and Infrastructure Application*, Southern Region Engineering Conference, Toowoomba, Australia, November 11-12, 2010, SREC2010-F1-5.

19. Biagiotti, J.; Pugria, D.; Kenny, J. M. A Review on Natural Fiber Based Composites-Part 1: Structure, Processing and Properties of Vegetable Fibers. *J. Nat. Fibers* **2004,** *1*(2), 37–68.

20. Yusoff, M. Z. M.; Salit, M. S.; Ismail, N.; Wirawan, R. Mechanical Properties of Short Random Oil Palm Fibre Reinforced Epoxy Composites. *Sains Malays.* **2010,** *39*(1), 87–92.

21. Intara, Y. I.; Mayulu, H.; Radite, P. A. S. Physical and Mechanical Properties of Palm Oil Frond and Stem Bunch for Developing Pruner and Harvester Machinery Design. *Int. J. Sci. Eng.* **2013,** *4*(2), 69–74.

22. Dyson Bruno, A.; Baskaran, M. Analysing the Mechanical Properties of Natural Fiber Reinforced Polymer Composites Using FEA. *Int. J. Eng. Sci. Res. Technol.* **2014,** *3*(1), 296–282.

23. Sparniņs, E.; Modniks, J.; Andersons, J. *Experimental Study of the Mechanical Properties of Unidirectional Flax Fiber Composite*, International Conference on Experimental Mechanics, Porto, Portugal, July 22–27, 2012, Ref 2743, 1–7.

24. Ahmed, K.S.; Vijayarangan, S. Elastic Property Evaluation of Jute-glass Fibre Hybrid Composite Using Experimental and CLT Approach. *Indian. J. Eng. Mater. Sci.* **2006,** *13*, 435–442.

25. Hammood, A.; Radeef, Z., Characterisation of Environmental Composites: Composite and Their Properties. Hammood and Radeef license in Tech, 2012, 12, 248–264.

26. Surya Prakash, D.; Praveen Kumar, D. Natural Fibre Sandwich Composite Panels-Analysis, Testing and Characterisation. *J. Mech. Civ. Eng.* **2013,** *9*, 58–64.

27. Sen, T.; Jagannatha Reddy, H. N. Finite Element Simulation of Retrofitting of RCC Beam Using Coir Fibre Composite (Natural Fibre). *Int. J. Eng. Sci. Res. Technol.* **2014,** *3*(1), 296–282.

28. Peel, L. D. The Response of Fiber Reinforced Elastomers Under Simple Tension. *J. Compos. Mater.* **2001,** *35*, 96–137.

29. Chandramohan, D.; Marimuthu, K. Characterization of Natural Fibers and Their Application in Bone Grafting Substitutes. *Acta. Bio. Eng. Biomech.* **2011,** *13,* 78–84.

30. Plaza, F.; Harris, H. D.; Kirby, J. H. Modelling the Compression, Shear, and Volume Behaviour of Final Bagasse. *Proc. Aust. Soc. Sugarcane Tech.* **2001,** *23,* 428–436.

31. Earl, E.; Moffett, E. W. Physical Properties of Polyester Resins. *Ind. Eng. Chem.* **1954,** *46,* 1615–1618.

32. Dudgeon, C. D.; Ashland Chemical Company. Polyester resins. In *Composites Volume 1 of Engineering and Materials Handbook;* 91–96.

33. Kaw, A. K. *Mechanics of Composite Materials,* 2nd edition; Taylor and Francis Group, Inc.: Boca Raton, 2006.

34. Malik, P. K. *Fiber Reinforced Composites Materials, Manufacturing and Design,* 3rd edition; Taylor and Francis Group, Inc., Boca Raton, USA, 2008.

35. Al-Mosawi, A. I.; Moslen, M. A.; Abhas, S. J. Using of ANSYS Program to Calculate the Mechanical Properties of Advanced Fibre Reinforced Composite. *Iraqi J. Mech. Mater. Eng.* **2012,** *12,* 673–679.

36. Balkan, O.; Demiver, H. Mechanical Properties of Glass Bead and Wollastonite-filled Isotactic-polyopropylene Composites Modified with Thermoplastic Elastomers. *Polym. Compos.* **2010,** *31,* 1285–1308.

37. Najim, A.; Hamzah, S.; Ahmed, F. *Experimental and Numerical Simulation of Impact Fracture Toughness of Polyphenylene Sulphide Basis Composite Material;* 2014.

38. Chaudhary, V.; Gohil, P. P. Stress Analysis in Cotton Polyester Composite Material. *Int. J. Metall. Mater.* **2013,** *3,* 15–22.

39. Mohammed, F. A.; Goda, G. M.; Galal, A. H. Experimental Investigation of the Dynamic Characteristics of Laminated Composites Beams. *Int. J. Mech. Mechatron. Eng.* **2010,** *10,* 41–47.

CHAPTER 21

SUPRAMOLECULAR POLYMER (UPY-PEGMA)-BASED SOLID POLYMER ELECTROLYTE FOR LITHIUM ION BATTERY APPLICATION

RAJENDRAN T. V. and JAISANKAR V.[*]

PG and Research Department of Chemistry, Presidency College (Autonomous), Chennai 600005, Tamil Nadu, India

[*]Corresponding author. E-mail: vjaisankar@gmail.com

CONTENTS

ABSTRACT

A novel supramolecular solid polymer electrolyte (UPy-PEGMA) consisting of ureidopyrimidinone-based supramolecular-polymer-doped lithium perchlorate (LiClO$_4$) as charge carrier was prepared by solution casting method. The interaction between the polymer and salt was studied using Fourier transform infrared spectroscopy (FTIR) and X-ray diffraction method (XRD). FTIR spectrum shows the interaction of carbonyl group of synthesized polymer with lithium ion. Ionic conductivity, electrochemical stability, and thermal behavior of polymer electrolyte were analyzed by AC impedance and differential scanning calorimetry (DSC), respectively. The temperature dependence of the conductivity can be correlated to the Arrhenius and the VTF equations in the temperature range from 313 K to 373 K. DSC thermogram shows the thermal stability of the polymer electrolyte. The results of the investigation reveal that the synthesized supramolecular polymer (UPy-PEGMA) blended with LiClO$_4$ can be used as a potential material for lithium ion-based electrochemical devices.

21.1 INTRODUCTION

Polymer electrolytes have received great attention due to their potential application in various electrochemical devices particularly in solid-state rechargeable lithium batteries.[1] Polymer electrolyte battery is well known for its high power density, flexibility, and plasticity of size and shape.[2,3] Polymer electrolytes are free from several disadvantages such as risk of leakage of electrolyte, internal shorting, and formation of noncombustible reaction products at the electrode surface that are observed in devices using liquid electrolytes.[4] Interest in developing solid-state batteries was motivated by the hope that the above problems would be minimized when the liquid electrolyte is replaced by an ureidopyrimidinone-based supramolecular solid polymer electrolyte. UPy-based supramolecular solid polymer system was made by hydrogen bonding between the segmental motion of different species and also Li$^+$ ion conducting pathways promoted by Lewis acid–base interactions.[5,6]

The supramolecular polymer is added with Li salt, that is, polymer salt complexes, are of technologies interest due to their possible applications as solid electrolytes in various electrochemical devices such as energy conversion units (batteries/fuel cells),[7] electrochromic display devices, and photoeletrochemical solar cells, etc.[8,9] Subsequently, the studies in this field have been reported by poly(ethylene oxide)(PEO)-based polymer electrolyte complexes using alkali salts.[10,11] The supramolecular solid polymer electrolytes have proved to be safe and potential electrolytes of the lithium batteries owing to their high ionic conductivity and good thermal stability which enables it to accommodate a high concentration of charge carriers.[12] $LiClO_4$ was used as the doping salt to furnish Li^+ ions for ionic conduction in the polymer matrix.

In the present investigation, we synthesized a new type of polymer electrolyte (UPy-PEGMA) by dispersing alkali salt in the supramolecular polymer matrix. The characterization results on the polymer electrolyte system in this study such as FTIR, XRD, DSC, and Electrochemical Impedance Spectroscopy (EIS) are analyzed and discussed in terms of ionic conductivity, morphology, and thermal stability of the polymer electrolyte.

21.2   EXPERIMENTAL METHODS

21.2.1   MATERIALS

2-amino-4-hydroxy-6-methylpyridine, 1,6-diisocyanato hexane, and dibutyl tin dilaurate were purchased from Aldrich. PEG-methaacrylate was dried at 60°C for 24 h and 2-amino-4-hydroxy-6-methylpyrimidine was dried at 70°C for 6 h before synthesis. $LiClO_4$ (Aldrich) was dried in a vacuum oven at 80°C for 24 h and stored in a desiccator prior to use. Acetonitrile was distilled at a suitable temperature under a nitrogen atmosphere prior to use. All other materials and solvents used were of analytical grade.

21.2.2   SYNTHESIS OF SUPRAMOLECULAR POLYMER ELECTROLYTE

The synthesis of uriedopyrimidinone-based supramolecular polymer method was discussed in our previous research work.[13] The supramolecular polymer electrolyte was synthesized by the addition of 1:1 ratio of supramolecular polymer and Li salt taken in a 100-mL beaker with acetonitrile (solvent) and the solution was stirred for 4 h to get uniform concentration. Li salts add into polymer matrix by dispersion method. After the reaction time, the solution was kept at room temperature for solvent evaporation. Finally, we collect the solid polymer electrolyte as a white material and is used for further characterization.

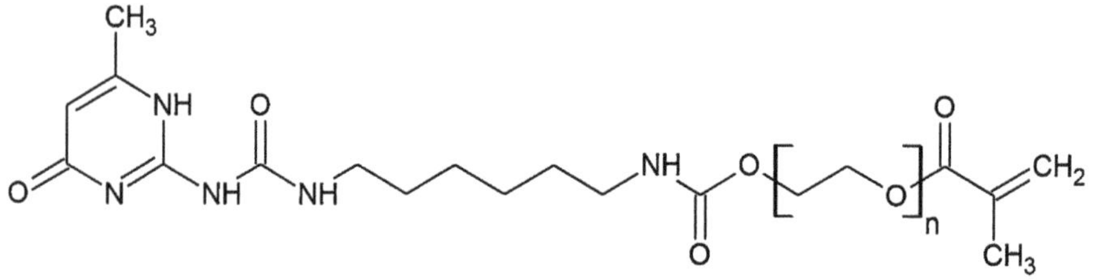

FIGURE 21.1   Ureido-pyrimidinone PEGMA-based supramolecular polymer.

21.2.3   CHARACTERIZATION

The synthesized supramolecular polymer electrolyte was characterized by some analytical methods. The formation polymer electrolyte and the frequency bands were characterized by FTIR spectroscopy. The crystalline of the polymer was analyzed by XRD. The thermal behavior of solid polymer electrolyte was discussed by DSC. The ionic conductivity of polymer electrolyte was characterized at four different temperatures (40°C, 60°C, 80°C, and 100°C) by Inductance Capacitors Resistance (LCR) meter. The E_a value calculated for polymer electrolyte from the Arrhenius equation and analysis of discharge characterization studies were used for battery application of synthesized supramolecular polymer electrolyte.

21.3   RESULT AND DISCUSSION

21.3.1   FTIR SPECTROSCOPY

The FTIR spectra of polymer electrolyte, supramolecular polymer, and Li salt are shown in Figure 21.2. The intensity of the aliphatic C–H stretching vibration band at 2870 cm^{-1} in polymer is observed to be hiding by addition of Li salt in the polymer electrolyte. The intensity peak at 1080 cm^{-1} indicates the presence of C–N stretching in polymer as well as polymer electrolyte. The peak at 1590 cm^{-1} indicates the presence of C–C stretching vibration of benzene ring in supramolecular polymer. Also the appearance of new peaks along with changes in existing peaks in the IR spectra of polymer and polymer electrolyte, respectively.

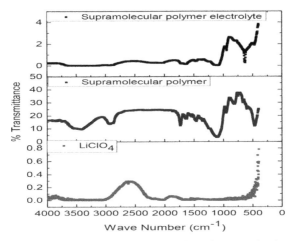

FIGURE 21.2 FTIR spectra of Li salt, supramolecular polymer, and polymer electrolyte.

21.3.2 XRD METHOD

The XRD analysis is used to determine the complexation and crystallization of the supramolecular polymer matrix. The diffraction peaks observed for 2θ values at $31.4°$ and $43.6°$ were found to be less intense in polymer electrolyte compared to pure supramolecular polymer. This indicated that the addition of $LiClO_4$ salt caused a decrease in the degree of crystallinity of the polymer electrolyte. This could be due to the disruption of the semicrystalline structure of the polymer by $LiClO_4$ salt. Hodge et al. established this type of correlation between the intensity of the peak and the degree of crystallinity.[14]

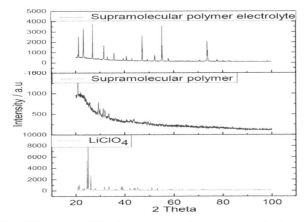

FIGURE 21.3 XRD pattern of Li salt, supramolecular polymer, and polymer electrolyte.

21.3.3 DSC ANALYSIS

DSC thermogram obtained for supramolecular polymer electrolyte in Figure 21.4 shows endothermic peaks at 99°C and 150°C. At temperature T_m, which corresponds to the melting point (97°C) of the polymer electrolyte, there is a change from the semicrystalline to amorphous phase transition. Due to phase change, the conductivity shows a sudden increase at T_m. The conductivity of polymer electrolyte is about 6.17×10^{-6} Scm^{-1} at room temperature and its value increases with increasing temperature. Further, the relative degree of crystalline nature decreases and the amorphous phase of the polymer electrolyte becomes more flexible for movement of ionic conduction in supramolecular polymer matrix.

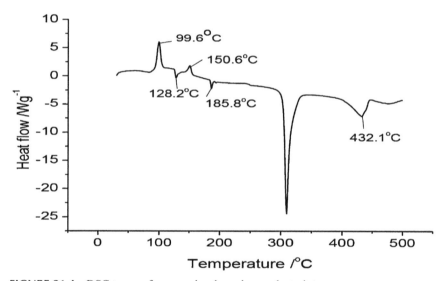

FIGURE 21.4   DSC traces of supramolecular polymer electrolyte.

21.3.4 DC CONDUCTIVITY

The complex impedance data in the case of supramolecular polymer electrolyte at four different temperatures (40–100°C) were obtained in the form of impedance diagrams and to determine the bulk resistance (R_b) of this polymer electrolyte. The R_b value is given by the intercept obtained on the real axis in the plot of real versus imaginary component of impedance. The conductivity (σ) of the polymer electrolyte was calculated using the relation:

$$\sigma = t / A \times R_b, \qquad (21.1)$$

where t is the thickness of the polymer electrolyte and A is the area of cross of the sample. The conductivity of the polymer electrolyte is about 6.17×10^{-6} S/cm at room temperature. The conductivity increased with temperature and followed Arrhenius behavior with activation energy. The activation energy was evaluated from the slope of log σ versus 1000/T plots, and the value is 0.63 eV. This may be due to a transition from a semicrystalline phase to an amorphous phase. As per the Arrhenius relation, the dependence of conductivity has the form:

$$\sigma = \sigma_0 exp (- Ea / kT), \qquad (21.2)$$

where σ_0 is the proportionality constant, E_a is activation energy, k is the Boltzmann constant, and T is the absolute temperature. In this method, the DC current is calculated as a function of time on application of a DC voltage is 0.76 V in the configuration of $LiCO_3$/(polymer electrolyte)/C. The conductivity plot is shown below.

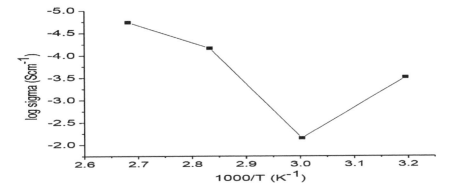

FIGURE 21.5   Temperature-dependent conductivity of supramolecular polymer electrolyte.

21.3.5   DISCHARGE PROFILES

Solid-state battery was fabricated with the configuration $LiCO_3$/polymer electrolyte system (supramolecular polymer + $LiClO_4$)/C. The surface area and the thickness of the electrolyte are 1.23 cm^2 and 135 μm. Figure 21.6 shows discharge characteristics are studied at room temperature by connecting a multimeter with polymer electrolyte system.

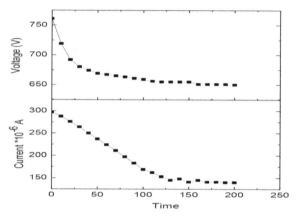

FIGURE 21.6   Discharge characteristics of electrochemical cell.

The open-circuit voltage (OCV) and SCC for the polymer battery (50:50) are found to be 0.76 V and 297 μA, respectively. The other cell parameters for this cell were evaluated and summarized below:

Cell parameters	UPy polymer + LiClO$_4$ (50:50)
Open-circuit voltage (V)	0.76 V
Short-circuit current (μA)	297
Area of the cell (cm²)	1.23
Weight of the cell (g)	1.15
Thickness of the cell	135 μm
Time for plateau region (h)	53
Power density	2.58 W/kg
Energy density	136.8/Wh/kg
Current density	241 μA

The OCV and the SCC and other cell profiles for this cell are given in table above. This suggest that UPy-based supramolecular polymer electrolyte cell exhibit improved performance and offer an interesting alternative to other reported electrolyte system for room temperature solid-state batteries.

21.4   CONCLUSION

Supramolecular polymer electrolytes based on ureidopyrimidinone group and LiClO$_4$ were prepared, and the effect of ionic conductivity and discharge characteristics was investigated. The results reported in this work stated that by combining the lithium salt with the dispersion of UPy-based supramolecular polymer electrolytes having unique properties in terms of temperature range of ionic conductivity, electrochemical stability, and thermal property can be obtained. These optimized polymer electrolytes are suitable for the development of advanced lithium polymer batteries and may be proposed for a series of relevant applications which include electric, hybrid vehicles, and energy storage materials for photovoltaics.

KEYWORDS

- **solid polymer electrolyte**
- **ionic conductivity**
- **thermal behavior**
- **electrochemical devices**
- **supramolecular polymer**

REFERENCES

1. Sun, X. G.; Liu, G.; Xie, J. B.; Han, Y. B. *Solid State Ion.* **2004,** *175*, 713.
2. Scrosati, B. *Chem. Rec.* **2001,** *1*, 173.
3. Vincent, C. A.; Scrosati, B. *Bull. Mat. Soc.* **2000,** *25*, 28.
4. Stephens, M. A.; *Euro. Polym. J.* **2006,** *42*, 21–42.
5. Croce, F.; Persi, L.; Scrosati, B. *Solid State Ion.* **2000,** *135*, 47.
6. Serraino, F.; Plichta, E.; Hendrickson, M. A. *Electrochim Acta.* **2001,** *46*, 2457.
7. Kumar, K. V.; Sundari, G. S.; Chandrasekar, M.; Ashok, M.; *J Chem. Technol. Res.* **2011,** *3*, 1203–1212.
8. Mac Callum, J. R.; Vincent, C. A. *Polym. Electr. Rev. Elsevier.* 1987.
9. Owen, J. R.; Laskar, A. L.; Chandra, S. *Super Ionic Solids and Solid Electrolytes-Recent Trends;* Academic Press: New York, 1989: pp 111.
10. Sorenson, P. R.; Jacobson, J.; *Electrochim Acta.* **1982,** *27*, 1675.
11. Cristoloveanu, S.; Li, S. S. *Eletrochem. Soc.* **1986,** *133*, 307.
12. Yun, J. H.; Kyung, J. S.; Suk, N. K.; Stephen, M. A. *Eur. Polym. J.* **2007,** *43*, 65.
13. Rajendran, T. V.; Jaisankar, V. *J. Polym. Compos.* **2014,** *3*, 1.
14. Hodge, R. M.; Edward, G. H.; Simon, G. P. *Polymer.* **1996,** *37*, 1371–1376.

INDEX